GEOLOGY
of the
MOAB AREA

by

F. A. Barnes

With mile-by-mile descriptions of
the geology and outstanding
geologic features that are visible from
the roads and a selection
of off-road vehicle trails
in the vicinity of Moab, Utah,
in the heart of southeastern Utah's
Canyon Country

----- CANYON COUNTRY -----

The heart of what should be
Colorado Plateau National Park

1993
Canyon Country Publications

This book is the THIRTIETH in a series of practical guides to travel and recreation in the scenic Colorado Plateau region of the Four Corners States

All written material, charts and photographs are by the author unless otherwise credited

Covers and interior illustrations by Rick Showalter

Canyon Country Publications
P. O. Box 963
Moab, UT 84532

ISBN 0-925685-04-6
Library of Congress Catalog Card Number 92-73225

CONTENTS

FOREWORD

Utah's Canyon Country, in the heart of the general Four Corners region, offers some of the most spectacular and readily viewable geology to be found in America. This book provides an overview of this region's geology, plus mile-by-mile guides to the general geology and geologic features that are visible from the paved highways, graveled roads and graded dirt roads in the general vicinity of Moab, Utah, in the heart of Canyon Country.

For those who have vehicles capable of traveling backcountry trails, the book also offers mile-by-mile descriptions of geologic features along a selection of off-road vehicle trails in the Moab area that travel through interesting regional geology.

The author of this book is not a professional geologist, but has been carefully observing, studying and documenting Canyon Country terrain for more than a quarter-century. That study and observation has been based on the kind of broad technical background and experience that permitted ready grasp of the basic principles of geology, and brought with it perspectives and observational techniques that enhanced their understanding.

Over these past twenty-five years, the author has worked personally with a number of professional scientists, including geologists, paleontologists, archaeologists and others, as a field contact who was able to find, identify and report for documentation and professional study many kinds of original field discoveries. These practical working relationships with professional scientists increased still further the author's interest in and understanding of these subjects.

This comprehension, plus reference to the hundreds of geologic books and reports the author has accumulated over the same quarter-century, along with his own proven observational abilities, have all been brought together in this guidebook.

Thus, while this book is not the product of a professional geologist, it is based on decades of solid, verifiable knowledge and first-hand experience, and has the further advantage of authorship by one very familiar with the region and well-practiced at reducing complex and highly technical information to a form understandable to the average reader, yet still accurate within its chosen level of abstraction.

Canyon Country Publications

Author Fran Barnes, 1992. Photo by Peggy Smith.

INTRODUCTION

The Canyon Country of southeastern Utah lies in the heart of the Colorado Plateau Geologic Province, which is a gigantic oval of ancient terrain that occupies the center of the Four Corners States -- Arizona, Colorado, New Mexico and Utah. The small but well-known Utah town of Moab lies in the heart of Canyon Country, and serves as a gathering place for tourists, recreationists and naturalists of all kinds.

To anyone journeying to Moab, or even through it on U.S. 191, the highway that travels north and south through Canyon Country, the rugged, colorful and beautiful nature of the region's spectacularly eroded terrain quickly becomes apparent. To anyone who drives the other highways, roads and off-road vehicle trails in the Moab vicinity, the unique nature of the region's landforms becomes even more apparent.

Familiarity with the eroded, colorful, vertical nature of the rock formations that dominate Canyon Country does not, however, bring understanding. The geologic strata are too complex, too fractured, offset and distorted, to make much sense without at least some form of semi-technical guidance and interpretation.

That is the purpose of this book -- to provide a general overview of Canyon Country geology, plus mile-by-mile road-logs that define and describe what travelers see around them as they explore this wondrous region of ancient geology and spectacular geologic landforms and features.

REGIONAL GEOLOGY

AN OVERVIEW

The geologic rock strata exposed in the canyon country of southeastern Utah range in age from early Pennsylvanian through the Cretaceous, with a few younger deposits in the Book Cliffs area and still younger igneous intrusions in the form of the three island mountain ranges in the region. Thus, the basic rock strata represent sedimentary deposits laid down over approximately 240 million years.

During this almost unimaginable period of time, the drifting continental land masses of Earth, with the Colorado Plateau region carried along as part of one of them, evolved from two large masses of land in the southern hemisphere and one in the north, to the single immense supercontinent known to geologists as Pangea. Pangea later broke up into much the same continents we have today, but somewhat more closely grouped. See the sidebar, COLORADO PLATEAU.

As the wide variety of sediments in the canyon country heartland of the Colorado Plateau were accumulating over this vast span of time, all of this drifting, colliding and separating of continents was affecting the global climate, the sea level and the very fabric of the continents themselves. The deep planetary forces that moved these great land masses around like pieces on a game-board strongly affected them.

Sub-sea rifting all over the globe first brought the separate drifting land masses together in mountain-forming collisions. Continent-sized slabs of oceanic crust were thrust beneath masses of land, producing earthquakes and volcanoes. Changes in the rate of sea-floor spreading brought sea-level rises that drowned immense areas of the continents for millions of years. Unimaginably slow but powerful currents of molten rock beneath the continents varied, thus changing the direction of movement and rates of speed of these land masses. This in turn brought on still more periods of faulting, rifting and volcanic activity, which often lasted for tens of millions of years.

Finally, after all of Earth's wandering continents coalesced into one immense supercontinent, titanic rifting tore this single continent asunder, splitting it apart vertically, horizontally and seemingly at random into the present North and South Americas, Eurasia, Africa, Australia and innumerable smaller fragments of continental crust.

As these continents traveled in new directions, they overrode seafloor crust and collided with other smaller continental masses. Mountain ranges and great chains of volcanoes were created where continental crust slid over seafloor crust, and where continents impacted other smaller continental masses. Where continental crust overrode oceanic crust, regional mountain-building, uplifting and subsidence were sometimes triggered, and long periods of inland volcanic activity were initiated. These tectonic activities in turn changed regional climates and erosion rates. Global and regional climates were also strongly affected as the land masses drifted across, then away from the Earth's south pole and equator. As the continents shifted, joined, then separated, ocean currents were redirected, causing additional climate changes.

All of this drifting and rending apart of formerly joined continental land masses by almost inconceivable planetary forces brought with it geologic violence on a local level. Great fault systems appeared in the basement rock of continents, entire mountain ranges formed and eroded away to low hills. Then new ranges were thrust upward along the same basement fault lines. The separated continents moved at periodically changing rates and in shifting directions, thus creating immense crustal stresses and producing still more earthquake violence, rifting and volcanic activity.

All of this geologic mayhem, this tectonic activity, took place during the last 240 million years. Most of it, however, happened so slowly that if humans had existed then, they would have noticed little except an occasional earthquake, or the sporadic eruption of a nearby volcano. Most geologic processes happen too slowly for human perception, but they did occur in the past, they are still taking place, and they will continue to transpire in the future, long after the human species has gone extinct on this small planet circling around a medium-sized sun, in a remote spiral arm of a modest-sized galaxy.

Studying a particular geologic scene, then, is like scrutinizing a snapshot, an image that represents a moment in time. That scene was vastly different in the past, it is slowly changing now, and it will be very different in the future. To understand the canyon country scene it is necessary to bear this in mind.

Over the past 240 million years everything about this region has changed, changed again and changed still further, a great many times. For millions of years, the Colorado Plateau region of what is now the North American continent was part of another immense land mass and was drifting slowly around south of the equator, in a great ocean that almost covered the entire planet.

During that prolonged period of relatively placid existence, the Colorado Plateau was either just below or just above sea level, and somewhere or another on a western coastline. Later, as this continental mass was forceably joined to others, the Colorado Plateau was alternately above and below sea level over additional long periods of time, before global changes lowered sea levels and allowed the Colorado Plateau, and what was to become canyon country, to rise above the ocean waters for a long period of time. Next, the entire region was a vast desert that endured equatorial heat and aridity for tens of millions of years.

Toward the end of the time represented by the rock strata in canyon country, rising seas again invaded the region and stayed there for more than ten million years before retreating once again. During the last several tens of millions of years, the land was again thrust far above sea level, distant volcanic activity showered the region with fallout, a nearby uplift pushed still further upward, and a great river system that originated within the latest version of the Rocky Mountains carved deeply into the ancient, uplifted rock strata. Within the past five million years, that complex drainage -- the Colorado River and its numerous tributaries, the Green, San Juan, Gunnison, Dolores and many smaller rivers -- has cut deeply into those widely varying canyon country deposits to form the magnificent canyons and unique landforms that have come to make this spectacular region famous.

For a more detailed summary of this region's geology, refer to the book, ***Canyon Country* GEOLOGY.**

THE REGION

As noted earlier, considerable geologic activity has taken place on Earth and on what is now the North American continent over the last several hundred million years, as the various geologic strata within the canyon country region were being deposited at their leisurely pace. This deposition was at times imperceptibly slow, and at other times fairly rapid, and for long periods of time even lost ground, literally. Such losses of one period's accumulated deposits before those of the next period began in earnest are called "unconformities." There are several major unconformities -- gaps in the stratigraphic record -- in canyon country.

Further, during much of the 240 million years represented by this region's geologic strata, the terrain was far from smooth and level, with each succeeding layer of sediments being laid uniformly over the older one. At times, deposition was uniform, in particular during the long eras of relatively quiescent continental drift activity, when the land in this region was largely vast coastal plains or shallow seas. During that immense span of time, which lasted for perhaps the first sixty-five million years of deposition in this region, the marine and coastal-marine land deposits were neatly layered.

Later, however, as the tectonic activities associated with more active phases of continental drift affected this region, and the emerging land was distorted, uplifted and misshapen by major violent geologic forces, the land's surface eroded in complex ways. Subsequent deposits were irregular and patchy, occurring here but not there, starting at a shifting coastline and ending at the bases of uplifts or mountain ranges, and wearing away in one area while still accumulating in another.

As a result, while it is possible to graphically depict a general sequence of rock strata in this region, readers must bear in mind that such stratigraphic charts are generalizations, that cross-sectional rock columns vary from place to place, often drastically. Most technical geology books contain numerous rock columns, each one based on actual drill cores, and each one varying in the thicknesses of the identifiable rock formations and members, if they exist there at all. With the land surface changing as sediments accumulated, deposit thicknesses also varied, and in some areas certain sediments were completely removed before the next permanent deposits began to accumulate.

Thus, the following list of the geologic strata commonly found in the Moab vicinity of canyon country is a generalization. A number of exceptions are covered in the geologic road and trail logs. First, the general stratigraphic sequence is discussed, then some of the major factors that have affected and distorted that sequence over the past 175 million years are summarized, in the main text and in special sidebars, in order to give perspective to the entire ages-long processes that have created what is now seen in this geologically unique region.

STRATIGRAPHY

Area covered. With a few exceptions, the area covered by this book is defined by Interstate 70 in the north, the Green and Colorado rivers in the west, the Abajo Mountains in the south, and Paradox Valley and the Dolores River in the east. Since this book is basically a mile-by-mile log of the geology visible from the roads and a selection of off-road vehicle trails in the Moab vicinity, it does not pretend to be comprehensive beyond the overview offered earlier, the generalized stratigraphy described next, and the major tectonic events noted later. Only geologic strata and other features visible from the roads and trails covered are described in any detail, even though these do not provide access to the entire defined region. Some special geologic features are described in more detail in sidebars that are listed alphabetically in an index in the back of this book.

Sequence. The geologic strata that remain in the area covered by this book are described in the sequence in which they were deposited, from the oldest to the youngest. Where there were breaks in deposition -- unconformities -- these are also noted. While geologists know in general where each distinct, identifiable geologic layer starts and ends, and can usually identify erosional gaps between and within such layers, they rarely know how long it took for each deposit to accumulate.

In sum, the time factor for the creation of each deposit is weak. A thin layer may easily have taken much longer to build up than a thick one. A deposit rarely accumulates at a uniform rate, but rather builds

STRATIGRAPHIC SEQUENCE

Green River Formation		Page 21
Wasatch Formation		Page 21
Mesaverde Group		Page 21
Mancos Shale	Ferron Sandstone	Page 20
Dakota Sandstone		Page 20
Burro Canyon/Cedar Mountain Formations		Page 19
Morrison Formation	Brushy Basin Salt Wash Tidwell	Page 18
Entrada Sandstone	Moab Slickrock Dewey Bridge	Page 16
Navajo Sandstone		Page 13
Kayenta Sandstone		Page 13
Wingate Sandstone		Page 12
Chinle Formation		Page 12
Moenkopi Formation		Page 12
Cutler Group	White Rim Organ Rock Cedar Mesa Lower Cutler	Page 11
Hermosa Group	Upper Hermosa Paradox Salts	Page 9

up, erodes away, gains again, then makes no net progress at all for long periods of time. Deposition is not uniform and steady with the passage of time. A deposit just a few feet thick may have taken millions of years longer to accumulate than a sandstone hundreds of feet thick. Thus, time scales can be applied only very loosely to charts of geologic strata.

Further, the thickness of an individual deposit may vary drastically from place to place, even when it is fairly continuous in general. For example, one major dune-sand deposit that tapers to nothing in the vicinity of the Utah-Colorado border is about 2,500 feet thick in the vicinity of Zion National Park. Other deposits vary in thickness because of localized uplifting and sinking, such as in the vicinity of active salt anticlines.

For the most part, geologists can identify the outer perimeters of specific deposits and formations, but there are sometimes exceptions to these generalized boundaries. Bits and pieces of a deposit may inexplicably be found far from the parent deposit. As with many scientific studies, geology is so complex that it can seem at times to be as much art as science.

This impression is often reinforced as its professional practitioners disagree with each other over both major concepts and minor details. As the literature basis for this book was being accumulated and studied, some of the identities of rock strata were changed, other new identities were added, and still others were eliminated entirely as being false and based on flawed research.

Thus, readers of this book should bear in mind that the science of geology, as applied to the canyon country region, is still evolving, changing and developing, especially in the field of stratigraphy. Understanding of this geologically complex region is not complete or cast in concrete. The author of this book has provided the best descriptions of the terrain and reconstructions of ancient events currently possible, which is about all even a professional geologist could do.

Anomalies. Within the various geologic strata there are anomalies, odd differences and variations in the deposit record. Some of these have erosional causes, others are due to tectonic activities. Still others were caused by life itself -- "bioturbation," as paleontologists and geologists label such phenomena. Some effects created by life leave no obvious traces, others do. Even so, over this planet's geologic history, life has been a major geologic factor. This continues today, as human activities perpetuate this ancient tradition. See the sidebars, HUMANITY AS A GEOLOGIC FORCE and GREAT EXTINCTIONS.

Where identifiable anomalies in the strata described are fairly common, they are noted in the following descriptions. Due to the complexity of most strata, however, only the more obvious differences and variations are noted. For additional details about each of the strata listed, serious students of canyon country geology should refer to technical literature.

Information provided. For each named unit of deposit, several items of information are noted in addition to its name. First is the "type-locality," or geographic site where the unit was initially studied and after which it was named. Next, any subdivisions in the deposit, or "members," are named in their sequence of deposition. The general composition and approximate thickness of each deposit is given, plus any anomalies or irregularities common to it. The environment which prevailed while the deposit was accumulating is described as accurately as present data permit, and any limitations to the deposit within the area of coverage are noted. Finally, any information unique to the deposit that might be of interest to the reader is offered.

HERMOSA GROUP

This group of deposits has three formations, the oldest of which is not exposed in southeastern Utah's canyon country. The next two formations of the group are described below.

PARADOX FORMATION

The Paradox Formation is essentially salts that crystallized out of sea water when it was concentrated by solar evaporation. Collectively, these white crystalline salts are called "evaporites." Some relatively thin layers of shale, or interlayered silty sediments, also occur within the marine salts. The type locality is Paradox Valley, Colorado. As the Paradox salts were accumulating, the sea bed was also gradually sinking, allowing the Paradox salts to accumulate to thicknesses as great as 5,000 feet. This basic deposit is not exposed in canyon country, but upward extrusions -- salt anticlines and domes -- are exposed a number of places, which are noted in the appropriate geologic road-logs. Others are mentioned in the sidebar, SALT ANTICLINES.

While Paradox salts were accumulating, over a vast expanse of time, this entire region was near-equatorial and beneath a shallow, intermittently land-locked sea. The climate was hot, which promoted evaporation of the sea water and the concentration and precipitation of marine salts. This immense inland sea at first extended far into what is now Colorado, but was later divided by the first rising of the Uncompahgre Uplift. The marine deposits east of that divide now bear another name. See the sidebar, UNCOMPAHGRE UPLIFT.

When the Paradox salts were eventually covered deeply by later deposits, the immense weight of this overburden caused the salts to act in a strange way. They flowed upward wherever there was a weakness in the overlying younger rock. In one sense, highly compressed crystalline salts act like a slow-motion liquid -- flowing like a very thick fluid, at a glacially slow rate. This flow in time produced many great salt anticlines, or subsurface ridges, in canyon country, and a number of smaller "salt domes."

A few of these upward extrusions of the deep Paradox salts have been exposed by erosion in certain areas covered by this book. There are a number of other such extrusions in canyon country that are not visible from any road or ORV trail.

UPPER HERMOSA DEPOSITS

Above the Paradox Formation, but still within the Hermosa Group, there are younger marine and coastal-marine deposits. To the south of the Abajo Mountains, these were once called the "Honaker Trail Formation," after an historic route down into the San Juan River gorge. In the area covered by this book, upper Hermosa Group deposits were also described as Honaker Trail Formation, but this designation was

YES -- GEOLOGISTS ARE HUMAN

Between 1933 and 1946, long before the establishment of Canyonlands National Park, geologists A. A. Baker and E. T. McKnight made detailed studies of the Colorado and Green river gorges in that vicinity, and established the identities of the rock strata there.

In 1962, geologist D. L. Baars asserted that the earlier geologic studies of that area were incomplete and in error. He claimed that much of the rock in the area was a previously undescribed formation he named "Elephant Canyon Formation," and that this was underlain by strata of the Honaker Trail Formation that was first described in the San Juan River gorge farther south.

When geologist D. B. Loope and others expressed serious doubts about the existence of the "Elephant Canyon Formation," Baars defended his position in professional papers published in 1971 and 1987. Significantly, the U. S. Geological Survey did not accept the "Elephant Canyon Formation" as a valid designation, although some geologists did to the extent that the name appeared in their published books and papers about Utah geology.

In a professional paper by D. B. Loope, G. A. Sanderson and G. J. Verville titled *"The Abandonment of the Name 'Elephant Canyon Formation' in Southeastern Utah: Physical and Temporal Implications,"* published October, 1990 in ***The Mountain Geologist,*** the journal of the Rocky Mountain Association of Geologists, this controversial matter was finally settled, once and for all --

" -- the Elephant Canyon Formation is not a valid lithostratigraphic unit. We advocate a return to pre-1962 nomenclature -- we suggest use of a temporary informal term 'lower Cutler beds' -- "

The report also confirmed earlier studies that questioned the establishment of the "Honaker Trail Formation," stating that --

" -- upper member of the Hermosa Formation' is a more appropriate name than 'Honaker Trail Formation -- "

While the suggested substitution of the term "lower Cutler beds" had not yet been formally accepted when this book was being written, it was nonetheless descriptive and hence used throughout this book pending official designation. The original term "Hermosa Formation" or Group was used for the same reason.

This summary is presented here for readers who are curious about the fascinating geology of Canyonlands National Park, and who may wonder about the discrepancies between this book and others that describe this region's geology in the Colorado River gorge. While the erroneous terms "Elephant Canyon Formation" and "Honaker Trail Formation" appear in a number of scientific papers, such papers are not in widespread distribution, nor were they written for the general public.

The following publications, however, were intended for general readership and are therefore for sale and accessible in some libraries. *They are in error to the extent that they apply the terms* **"Elephant Canyon Formation"** *and* **"Honaker Trail Formation"** *to the geology of Canyonlands National Park and the southeastern Utah region.*

Baars, D. L., **"Red Rock Country,"** Doubleday, 1972

Baars, D. L., **"The Colorado Plateau: A Geologic History,"** University of New Mexico Press, 1983

Baars, D. L., **"A River Runners' Guide to Cataract Canyon and Approaches,"** Canyon Publishers, 1987

Baars, D. L., **"Canyonlands Country,"** Canyon Publishers, 1989

Barnes, F. A., **"Canyon Country Geology,"** Wasatch Publishers, 1978

Hintze, L. T., **"Geologic History of Utah,"** Department of Geology, Brigham Young University, 1988

Huntoon, et al, **"Geologic Map of Canyonlands National Park and Vicinity, Utah,"** Canyonlands Natural History Association, 1982

Stokes, W. L., **"Geology of Utah,"** Utah Museum of Natural History, 1986

Yes -- geologists are human and, like other humans, they make errors, sometimes big ones!

recently shown to have been based on faulty research. Until they can formally be renamed, these strata are being called "Upper Hermosa" deposits. See the sidebar, YES -- GEOLOGISTS ARE HUMAN.

As the thick beds of the Upper Hermosa were accumulating, the weight of this enormous overburden slowly began to distort the horizontal uniformity of the underlying Paradox salt deposits, causing them to flow laterally and upward wherever faults in the overburden developed. This glacially-slow flowage of Paradox salts has continued sporadically ever since, and its flow-rate has even been measured in some canyon country localities, such as in the exposed salts through which Onion Creek flows, in mid-Fisher Valley.

UNCONFORMITY

There is a major unconformity between the Upper Hermosa and lower Cutler Group deposits in some areas, but the existence of this gap in the deposition record is questionable in the area covered by this book.

CUTLER GROUP

The Cutler Group has several members, each with its own geographic limitations. The type locality is Cutler Creek, Colorado. The oldest deposits, once designated as the "Elephant Canyon Formation," are now known unofficially as the "Lower Cutler" beds. Again, see the sidebar, YES -- GEOLOGISTS ARE HUMAN. Above that, depending on the area, are undivided Cutler sandstones that are generally dark red, or beds that are interlayered with the Cedar Mesa Sandstone Member in a transition zone. Some locations in Canyonlands National Park, and other areas in the Moab vicinity, have exposures of the Organ Rock Shale member, whose type locality is Organ Rock, in the part of Monument Valley that lies in Utah. In some areas, the Cedar Mesa Sandstone member is time-equivalent to the Organ Rock Shale member. The top member of the Cutler Group varies at different locations. West of the Colorado River, the youngest member is White Rim Sandstone, which has its type locality in the heart of Canyonlands National Park, where this distinct rock rims the broad terraces below Island-in-the-Sky. East of the Colorado River, and north of a certain point, the youngest Cutler deposits are dark red sandstones, although there is one remnant of White Rim Sandstone in Castle Valley, many miles east of its normal limit.

The lower, dark red Cutler deposits are largely coastal freshwater sediments and dune sands that phase into marine deposits farther west. These distinctive sandstones are hundreds of feet thick in some areas. The Cedar Mesa Sandstone that dominates the Abajo Mountains foothills is coastal dune and shallow-marine sands. It, too, is quite thick, especially south of the Abajos. It is distinctly color-banded, with reddish layers within the predominantly white sandstone, and has fractured and eroded into The Needles in the southern part of Canyonlands National Park. Cedar Mesa Sandstone readily erodes into natural arches and bridges, many of which appear in the Needles District of Canyonlands National Park, adjacent Canyon Rims Recreation Area, and in the great canyon systems eroded into this rock south and west of the Abajo Mountains. The magnificent spans in Natural Bridges National Monument are in Cedar Mesa Sandstone.

White Rim Sandstone is also coastal shallow-marine and dune sands, deposited by an arm of a sea intruding from the west. It is quite thick where exposed along the Green River gorge in the park, but thins toward the east before stopping at the Colorado River, with the one noted exception in Castle Valley. It also thins drastically to the south. It is unlikely that the present course of the Colorado and the eastern edge of the White Rim Sandstone is purely coincidental. Rather, the eastern edge of this hard sandstone probably influenced the location of the present river gorge.

All Cutler deposits were made during prolonged periods of hot equatorial climate, with the land either just below or just above sea level. The material for the deposits came from the east, largely from the Uncompahgre Uplift, which was a sub-range of the Ancestral Rocky Mountains, but not of the present relatively new range that formed in the same approximate alignment.

These last incursions of the sea that left the Cedar Mesa and White Rim members of the Cutler Group behind as they retreated, marked the beginning of a long period of dominance by dry land in the region that was to become canyon country. During that next ages-long period, the sea retreated generally toward the west, allowing freshwater sediments from the east to build up to great depths over the region.

Colorful and distinct Cutler deposits play a major part in the geologic landscape of Canyonlands National Park and the adjacent Lowlands District of Canyon Rims Recreation Area. They are also exposed in the Colorado River gorge in Richardson Amphitheater, upriver from Moab Valley.

UNCONFORMITY

There is a major unconformity between the youngest Cutler and oldest Moenkopi deposits. Evidently, for a long period of time, the last Cutler deposits, including the White Rim and Organ Rock Shale deposits, eroded away. This was probably due to further rising of the Uncompahgre Uplift, which was sporadic over an extended period of time. The higher terrain would have increased precipitation and runoff, which in turn would have carried away the most recent deposits in streams and rivers. This kind of selective cutting would explain the remnant of White Rim Sandstone left in lower Castle Valley. At about the time this major change between types of deposits occurred, a cataclysm of some kind extinguished approximately 95% of the species of life on Earth. It is highly probable the two events are related.

See the sidebar, GREAT EXTINCTIONS.

MOENKOPI FORMATION

The type locality for this fresh-water deposit is Moenkopi Wash in Arizona. While this formation has two named and three unnamed members that are technically different from each other, most geologic maps do not distinguish among these members, which to the casual observer seem similar. Moenkopi deposits are generally reddish in color, and occur either in thin layers or somewhat thicker sandstone ledges. When exposed to erosion, the thinner layers form steep slopes, while the harder sandstones develop into ledges and low bluffs. Moenkopi deposits vary in thickness, ranging from a few feet to hundreds of feet, indicating that the land at that time was not flat and uniform, partly due to the sporadic activity of salt anticlines in the area.

The climate during deposition of the Moenkopi continued to be hot and arid, with little life thriving in the region, as indicated by the red color of its sediments. See the sidebar, ROCK COLORS, for a discussion of this. The absence of life was due partly to the inhospitable climate, but mostly to the great extinction that took place prior to its period of deposition. As with all of the "great red beds" of the region, the sediments of the Moenkopi came from the iron-rich Uncompahgre highlands to the east.

UNCONFORMITY

There is a major unconformity between the youngest Moenkopi and oldest Chinle deposits. During that gap in the deposition record, either the youngest Moenkopi deposits eroded away, or the first Chinle deposits did so later. A combination of these is most likely. As with the period of erosion between the Cutler and Moenkopi, this later erosion was probably due to increased runoff from the highlands of the Uncompahgre to the east. In places, this erosional gap between the Moenkopi and Chinle confuses their present interface.

CHINLE FORMATION

The type locality for this freshwater deposit is Chinle Valley in Arizona. Chinle deposits are complex and variously consist of mudstones, silts, conglomerates, shales and sandstones, with thin deposits of coal and other organic remnants in some areas. The Chinle also contains scattered lenses of limestone and some members contain minerals from volcanic ash fallout, which is probably the source of the uranium that was mined from these deposits in many canyon country areas. The colors of Chinle deposits range from reddish to gray, with bands and lenses of other colors in some locations. In thickness it ranges from 200 feet to four times this.

As with the Moenkopi Formation, the Chinle Formation has a number of members, largely in other areas. Most of these are indistinguishable in canyon country, and geologic maps of this area show only undifferentiated deposits. Chinle sediments accumulated on a vast floodplain, left there by rivers, streams, lakes and swamps. This provided an ideal environment for various forms of plant and animal life, which left plenty of fossil remains, making the formation popular with paleontologists.

As exposed by erosion, the Chinle forms steep, ledgy gray-hued slopes or red sandstone ledges below sheer cliffs of Wingate Sandstone, and above similar slopes of red-hued Moenkopi, where this formation is exposed. In some locations, Chinle deposits are virtually identical in appearance to underlying Moenkopi sediments.

UNCONFORMITY

Geologists differ as to whether or not there is an unconformity between Chinle and Wingate deposits. This is largely because that interface differs drastically in appearance in various areas. If there is an unconformity, it is not readily apparent in the canyon country region covered by this book. Because of the quite different nature of the two deposits, the lower being heavily vegetated water deposits on a floodplain and the higher being primarily desert dune-sand, the interface would certainly not be neat and orderly. Whether by coincidence or from the same basic cause, one of the several great extinctions of life on the planet that have taken place in the past occurred at the time of this interface between geologic deposits. This author has little doubt that both the extinction and drastic change in type of deposit were due to the same cause. See the sidebar, GREAT EXTINCTIONS.

WINGATE SANDSTONE

The type locality for this red-hued rock is Fort Wingate, New Mexico, even though the rock at this first-study site later proved to be Entrada Sandstone. Everywhere in canyon country, cliffs of Wingate Sandstone soar above steep slopes of Chinle deposits. As noted above, Wingate Sandstone is almost entirely dune-sand, with the angled cross-bedding characteristic of such aeolian sandstones. Wingate can be as much as 400 feet thick, and in canyon country is rarely less than 250 feet thick. It is cross-bedded, fine-grained sandstone with occasional thin horizontal water-affected layers that tend to channel seeping water to cliff surfaces, where these seeps create ever-deepening seep-caves by chemical exfoliation. See the sidebar, ROCK EROSION. Most such seep-lines host water-loving plants and insects. Larger animals visit seeps that produce enough water to drip or run. The dominant bedding dips in Wingate indicate that this vast area of desert dunes was subjected to steady winds from the northwest, and at least some of the sand that made up the dunes probably came from that direction, too. Iron minerals in the deposit provided its reddish color.

The environment during that time was still hot, arid and equatorial. The prevailing red hue of Wingate Sandstone, and a dearth of fossils, indicate that the region was not very hospitable to life then. This was at least partly due to the extinction at the Chinle-Wingate interface, with its 60% loss of species, although not far from Moab there are several known sites with innumerable small dinosaurs tracks in water-affected layers in the top fifty feet of the Wingate. Evidently the climate was becoming somewhat wetter then, and small reptile species that survived the Triassic/Jurassic extinction were spreading into the newly-hospitable terrain.

The older Wingate deposits mark the beginning of 40 million years of aridity in the region. This prolonged period of desert climate was broken only twice for relatively brief times. See the sidebar, GREAT DESERT PERIOD.

Wingate Sandstone, like other massive desert-dune sandstones, is strong under compression but relatively weak under tension and shear. Thus, when it is undercut by erosion of the soft Chinle beneath it, Wingate Sandstone tends to collapse along near-vertical planes, whether or not there are joints there, thus forming the great lines of Wingate cliffs so characteristic of the high promontories of canyon country. The harder, water-deposited sandstone that tops the Wingate generally protects its upper surface from exposure. Where this is exposed, Wingate Sandstone erodes into rounded domes and fins much like Navajo and Entrada aeolian sandstones do. One such area is described in the geology log for the Cane Creek Canyon Rim off-road vehicle trail, elsewhere in this book. When its upper surface is exposed, Wingate tends to form potholes much like those commonly found on the upper surfaces of other aeolian sandstones such as Navajo and Slickrock-Entrada.

Although some geologists have contended that Wingate Sandstone does not readily erode to form natural openings such as arches and bridges, that is far from true. One canyon system near Moab that is walled by this rock has more than three dozen sizable natural spans in it, and numerous others exist in Wingate exposures elsewhere.

KAYENTA SANDSTONE

The type locality for Kayenta Sandstone is Kayenta, Arizona. Kayenta Sandstone consists of red, gray and lavender sandstones, with interbedded shales, limestones and siltstones, all deposited by fresh water. It is usually about 200 feet thick but varies, depending on location. This hard, ledgy sandstone covers the less durable desert-dune Wingate Sandstone everywhere in the region described, almost always protecting its upper surface from erosion. Kayenta Sandstone was deposited by streams that originated from highlands to the east, probably the Uncompahgre Uplift, which influenced all earlier strata since it first divided the great inland Paradox Basin sea.

Although Kayenta deposits were water-borne, the regional climate probably remained very arid. Fossils are scarce, although this author has found petrified wood in this sandstone in what was probably a driftwood deposit, with the trees washed down from higher terrain. Because of its stratified nature, Kayenta Sandstone readily forms natural bridges when exposed by erosion. Kayenta Sandstone marks one of the two periods of somewhat wetter climate in the 40 million years during which this region was a great, arid desert.

NAVAJO SANDSTONE

The desert reclaimed the region during the long period when Navajo Sandstone was being deposited. The type locality for this sandstone is Navajo County, Arizona. Like Wingate Sandstone, the Navajo is desert-dune deposits, but the prevailing winds, and the probable source of the sand, was from the north and northeast, although other geologists contend that the winds were from the northwest, west or southwest. Navajo is also different in several other aspects. It is largely gray-white, but exhibits hues ranging from yellow to red and even brown in some locations.

For a period of time, there were two hypotheses concerning the nature of Navajo Sandstone. One contended it was largely marine sandbars crossbedded by ocean currents, while the other held out for desert-dunes, or aeolian, with the distinct angled crossbedding caused by wind. This question was recently firmly resolved, with data acquired by this author and several other amateur geologists. A series of small petrified forests were discovered associated with dark-hued freshwater deposits in the basic sandstone, deposits that were obviously made by permanent springs, or desert oases, where trees grew and other life forms lived or gathered. Some sites also produced petrified algae. Petrified dinosaur tracks and a variety of other vertebrate and invertebrate tracks were also found in Navajo Sandstone, in "petrified playas," light-hued limestone deposits left by ephemeral desert drylakes. See the sidebar, PETRIFIED PLAYAS. These findings decisively settled the matter of marine-versus-aeolian deposits. Trees do not grow, and dinosaurs did not roam, on ocean bottoms. Even now, however, the nature of Navajo Sandstone remains mysterious in other ways, and will until these very recent findings have been fully studied.

Navajo Sandstone is as much as 2,500 feet thick in central Utah, but thins toward the east, to disappear entirely in the vicinity of the Utah-Colorado border. This thinning may have been due to a general slope in the underlying terrain, or it may have been caused by a long period of erosion following the buildup of the deposit, with the heaviest loss in the east. This loss might have been caused by the eroding away of desert sands piled up on higher slopes to the east, probably remnants of what was then the greatly reduced Uncompahgre Uplift.

While Navajo Sandstone is now considered to be

GREAT DESERT PERIOD

A little more than 200 million years ago, about 60% of the species of life on Earth were wiped out by a major catastrophe, including most of the mammal-like reptiles that were the remote predecessors of humanity. See the sidebar, GREAT EXTINCTIONS. That extinction was so widespread that it defined the end of one great geological-biological period, the Triassic, in which the ancient ancestors of humans dominated, and the beginning of another period, the Jurassic, which was the earliest phase of the well-known "Age of Dinosaurs."

Many geologists and paleontologists believe that an immense meteor that impacted with explosive violence in northern Canada about then was responsible for this mass extinction. Other scientists attribute it to different causes. The question is still not settled, although the evidence to date supports the meteor impact concept.

One major piece of evidence is the fact that at about the same time, there was a drastic change in types and rates of deposition or erosion in many regions, in addition to the extinction of many forms of life. There is little doubt that the extinction and the change in geologic deposits had a common cause. In canyon country, the change in deposition was drastic, and occurred between the Chinle Formation sediments left by lakes, rivers, streams and swamps that were rich in plant and animal life, and the arid desert-dune sands of Wingate Sandstone.

Whatever the cause or causes of these drastic and simultaneous alterations in Earth's biology and geology, the result in the region that later became canyon country was a prolonged period of almost unbroken aridity, which permitted correspondingly little plant and animal life and produced the kinds of geological deposits common to deserts of various kinds.

Whatever caused the Triassic-Jurassic extinction also turned the general region into a vast, low-lying coastal desert for almost 40 million years, with that hot, arid lowland situated on the western equatorial flanks of the supercontinent, Pangea. North America did not yet exist as a separate continent.

This GREAT DESERT PERIOD left its marks in the rocks and the erosional topography of canyon country. Its massively thick sandstones -- Wingate, Navajo and Entrada -- were almost barren of fossils, making it difficult to date them, leading geologists and paleontologists to squabble endlessly over their ages. These thick desert-dune deposits eroded into dune-like domes, and into immense, rounded rock fins when jointed by local tectonic activities. They scaled away into near-vertical cliffs when their bases were undercut. They provided the basis for the most prominent erosional features of canyon country.

The first of these desert deposits, Wingate Sandstone, stands in stark biological contrast to the life-rich Chinle deposits just beneath it. Few fossils have been found in Wingate Sandstone, but toward the end of its 7 million year deposition period a few small dinosaurs were roaming its arid dunes and its few transient mudflats, leaving hundreds of petrified tracks for paleontologists to study 200 million year later.

This prolonged GREAT DESERT PERIOD was broken briefly only twice, once between the Wingate and Navajo, once between the Navajo and Entrada. A temporary change in climate brought a time of freshwater deposits that ended the period of Wingate desert-aeolian deposition, leaving thick layers of stream sediments but not changing the general region very much biologically. These became Kayenta Sandstone. The land was revegetated to some extent, leaving a few fossilized trees, and dinosaurs roamed the region, leaving numerous petrified tracks in its freshwater sediments, but the GREAT DESERT PERIOD was far from over.

A second drastic climate change brought with it another type of desert, one somewhat like the present Sahara of northern Africa. Strong prevailing winds carried in immense quantities of sand from which most iron minerals had been removed, probably by the washing action of coastline waves. The winds also brought in marine evaporites, probably from coastal tidal flats. Navajo Sandstone was deposited, more than 2,500 feet thick in its western areas. Navajo Sandstone deposits may have been quite thick elsewhere too, but a subsequent stratigraphic unconformity destroyed the evidence.

Navajo sands may have accumulated to greater thickness than the present sandstone, then eroded away during the succeeding 16-million-year gap in the stratigraphic record. Alternately, the deposit may have reached its present thickness, then simply stayed at that level for a long, long time, its source of sand cut off by some kind of change, such as a rise in sea level that inundated the beaches that supplied its sand and evaporites.

Whatever terminated its 4-million-year period of deposition, the Navajo Sandstone desert was not entirely devoid of life. Numerous animal species, both vertebrates and invertebrates, frequented its drylakes and oases, leaving their petrified foot tracks in the mineral-rich desert muds. Trees, shrubs, algae and other forms of plantlife left their petrified remains. See the sidebars, PETRIFIED PLAYAS and PETRIFIED OASES.

Eventually, Navajo Sandstone deposits were covered by the sediments of another type, the iron-rich sands of the lowest member of Entrada Sandstone. Evidently, the world's sea level had risen due to some change in continental drift tectonics, because during the 5 million year period of Entrada deposition the canyon country region was barely above sea level. Sands and silts rich in iron and other minerals were brought into the region and worked by tidal action, leaving first the distinct deposits of the Dewey Bridge Member of

continued ----

Entrada Sandstone.

This relatively short period was followed by millions of years during which the region was again dominated by vast plains of coastal dunes, perhaps somewhat like Africa's great Namibian desert. In some areas of this deposit of the Slickrock Member of Entrada Sandstone, these dunes reached enormous proportions. Since the entire region was still near the equator and subject to strong winds and violent tropical storms, these dunes were occasionally subjected to storm-surges of seawater that temporarily inundated vast areas of dunes, leveling them and leaving the strange, harder horizontal layers that occur at intervals in this predominantly aeolian deposit.

This long period of low-lying coastal desert was eventually terminated, and the GREAT DESERT PERIOD was ended, by the invasion from the east of a tongue of an inland sea. This immense bay covered most but not all of the Moab area. The white beach sand left behind by this relatively brief marine intrusion was the Moab Member of Entrada Sandstone, which now flanks the northern slopes of the La Sal Mountains, but occurs no further south, and extends just a few miles west of U.S. 191. Its extent to the north is buried deeply under the younger deposits there, but it is prominent in the exposed Entrada Sandstone cliffs of Colorado National Monument and various other canyons on the northern nose of the Uncompahgre Uplift.

Throughout the region not covered by Moab-Entrada, the marine shoreline deposits of the red-hued Tidwell Member of the Morrison Formation buried the Slickrock-Entrada, and even left a thin layer on the Moab-Entrada where it occurs.

Thus canyon country's GREAT DESERT PERIOD of some 37 million years ended with a brief marine invasion followed by a long period in which freshwater sediments were brought into the region from the north and west, from highlands that were thrust upward by powerful continental drift tectonics. These new deposits turned the great desert once again into a land of rivers, streams, lakes and swamps that was very hospitable to life, including the dinosaurs whose petrified remains are so common in the Morrison Formation sediments then being deposited.

The GREAT DESERT PERIOD was over, but its unique deposits still play an important part in the erosional scene in the general Moab area, with great lines of soaring cliffs, countless miles of sheer-walled canyons, immense areas of weather-rounded "petrified dunes," mazes of colorful sandstone fins, and strange petrified playas and oases.

non-marine dune sands, unlike the massive aeolian strata above and below it -- Slickrock-Entrada and Wingate -- Navajo deposits are remarkably free of iron, as are most beach-sands. They also contain liberal amounts of minerals that are normally marine evaporites. These could have been carried in from marine tidelands by the same winds, or may have come entirely from the sea water that saturated the Navajo as the sea level ultimately rose and covered the Navajo with tideland deposits.

Despite the controversy over the source and nature of Navajo deposits, certain facts are inescapable. As deposited, Navajo Sandstone represents thousands of square miles of sands, averaging something like 1500 feet deep. This enormous amount of material had to come from somewhere, but exactly where remains a mystery. Only an even greater area of existing, or recent past, coastal-marine environment could have produced this quantity of relatively evaporite-rich, iron-free material.

When its upper surface is exposed, this massive aeolian sandstone erodes into rounded mounds that resemble desert dunes. Such rolling surfaces are often called "petrified dunes," even though their contours do not actually relate to those of the original dunes. This sandstone, like the older Wingate and younger Slickrock-Entrada, is weak in tension and shear. Thus, when its base is undercut, it tends to collapse along near-vertical planes, leaving cliffs much like those common to the Wingate, but with the cliff tops more rounded because the top of Navajo Sandstone is not protected by a hard, younger layer, like the Kayenta.

This sandstone also readily forms potholes in its upper surfaces when exposed to weathering. When such potholes are near cliff rims, the water that accumulates in them seeps downward and laterally, to form immense seep-caves in the cliff faces. When the potholes that form them eventually break into such deepening caves, natural arches are created. Such "pothole arches" are fairly common in Navajo Sandstone, as are other types formed from the massive rock by other erosional processes. For more details about the formation of natural spans, refer to the book, ***Canyon Country* ARCHES & BRIDGES.**

Altogether, Navajo Sandstone is one of the most fascinating, mysterious and controversial geologic strata in canyon country. Despite decades of study, it still presents unresolved questions, especially about the indigenous life and environment which prevailed during its deposition period.

UNCONFORMITY

There is a major unconformity between the last Navajo Sandstone deposits and the accumulation of the first sediments of the three Entrada Sandstone members. This unconformity may represent a period of erosion, as do most such gaps in the stratigraphic record, or it may indicate that the mysterious source for Navajo Sandstone sand was eliminated, thus beginning a long period of stability, with the existing Navajo sands shifted about by changing winds, but with no net gain or loss in total depth. This cessation in transfer of wind-borne sand into the canyon country region could have been triggered by the rise in sea level that was occurring, which would have inundated the beach-sand sources, or by a major change in direction of the winds that transported the sand into the region, or by both. The mountain-building that was going on along the entire west coast of North America could have affected the winds.

Whatever the cause, something happened to stop the transport by wind of great amounts of sand from somewhere else, and either allowed the removal of the hundreds of feet of existing deposits or held their depth to zero net gain for a long period of time. This same change eventually brought on the first of the next permanent deposits, which are now identified as the three distinctly different members of Entrada Sandstone.

ENTRADA SANDSTONE

Dewey Bridge Member. This member of Entrada Sandstone represents the second interruption of canyon country's Great Desert Period. During this time, iron-rich sediments were deposited along the shores of a sea that briefly invaded from the north, then retreated, during a period of hot, arid climate when the region was near the equator. The sediments came from the Uncompahgre Uplift to the east. The type locality for this member is the vicinity of historic Dewey Bridge, where Utah 128 crosses the Colorado River upriver from Moab Valley. Similar deposits west of the Green-Colorado rivers are called the Carmel Formation. Dewey Bridge-Entrada ranges from a few feet to more than 200 feet thick, but in the area covered by this book is generally in the lower end of this scale. It consists of stratified, iron-rich water-deposited red-brown silts and sandstones exposed along the bases of cliffs and slopes of the more prominent Slickrock Member. In some locations, such as in Arches National Park and in the bases of the Monitor & Merrimac Buttes and Big Mesa farther west, these horizontal layers have been distorted into strange waves and wrinkles. This was probably brought on by tilting of the local terrain by salt anticline activity and the resulting lateral sliding of its overburden of dune-sand, while the Dewey Bridge deposits had still not solidified.

This deposit was coastal-marine and thus highly mineralized. Eventually, some of these minerals crystallized out within the deposit as beautifully-colored agate, which eventually weathers out and accumulates on the upper surface of the underlying, more durable Navajo Sandstone. Great quantities of "Dewey Bridge Agate" generally appear wherever this Entrada member is exposed. The Dewey Bridge-Entrada exists throughout the area described in this book. Specimens of petrified wood have also been found in this deposit, but the trees probably washed down into the Dewey Bridge tidelands from higher terrain to the east, or floated in from elsewhere and were deposited by high tides or storm-surges.

Slickrock Member. This is the most prominent member of Entrada Sandstone. Its type locality is Slick Rock, Colorado. Some geologic references simplify this name into one word, others do not. Slickrock-Entrada varies in thickness to the east and west of the general Moab vicinity, and is around 300 feet thick there. It is primarily coastal dune-sand that was deposited over the tideland muds of the Dewey Bridge-Entrada during a period of hot, arid climate, as the seashore retreated. The aeolian nature of Slickrock-Entrada phases into gray-hued shallow-marine deposits west of the area described in this book. At various times during its deposition, the entire area of coverage was inundated by brief returns of the sea, probably as exceptionally high tides, or storm-surges caused by the equatorial hurricanes that were common to the region at that time. These short-term encroachments left a number of thin water-affected horizontal layers in the cross-bedded dune-sand deposits. Such tidal invasions are about the only type of water action that could leave thin, perfectly flat horizons in an area of round-topped sand dunes.

This beautiful, light-red, basically aeolian sandstone is prominent in canyon country as gracefully-curved cliffs with weather-rounded rims, and with their lower slopes just above relatively thin layers of the darker red water-deposited Dewey Bridge-Entrada. Such cliffs are often topped by remnants of the lower members of the younger Morrison Formation. Like the massive Wingate and Navajo sandstones, when undercut by erosion the Slickrock-Entrada shears away along near-vertical planes, and when its upper surface is exposed it weathers into rounded "petrified sand dunes." It also forms potholes and seep-cave arches just as readily as the other sandstones. When massively jointed by salt anticlines, as in Arches National Park, Slickrock-Entrada erodes into great parallel fins. These in turn are often eroded by water and various exfoliation processes into the immense natural arches for which this national park is famous.

One odd phenomenon that occurs in the Slickrock-Entrada, as well as in Navajo Sandstone, is what has come to be known as "Entrada Berries." These are small, slightly harder marble-sized spheres of sandstone that weather out of the aeolian rock by the thousands in some areas, for example in the western slopes of the Salt Valley anticline along the boundary of Arches National Park, an area dubbed "Klondike Ridge" in the book, ***Canyon Country* SLICKROCK Hiking & Biking.** Although geologists have yet to explain how these strange sandstone marbles form, the noted book proposes a working hypothesis.

Slickrock-Entrada sandstone is considered by some to be the most esthetically pleasing of all canyon country rock strata, with its bright-hued, sheer, color-banded cliffs, beautifully shaped slopes and intricate cross-bedding. This rock is conspicuous along several of the roads and off-road vehicle trails described in this book, and dominates the Cameo Cliffs area. For more information about this public land recreation area, refer to the guidebook and matching map titled **CAMEO CLIFFS - Biking - Hiking - Four-Wheeling.**

The long period of Slickrock-Entrada coastal dune-making represents the last phase of the Great Desert Period that began with Wingate Sandstone. During this time, the region's climate remained hot and dry, and the terrain was barely above sea level. For a while, a tongue of an inland sea to the east invaded the area, leaving a third Entrada Sandstone member in part of the canyon country region covered by this book.

Anomalous deposit. Beginning in the vicinity of U.S. 191, a deposit that is undescribed in geology books appears between the Slickrock and Moab members of the Entrada. East of that alignment, the sandstones of the Slickrock and Moab members are in intimate contact. To the west, thin-layered sediments appear between them, growing thicker toward the west, to a total thickness of more than 20 feet. The deposit varies from relatively soft and slope-forming to thin harder layering, and is generally dark brown. It is clearly visible from one ORV trail described in this book, and is high in several cliffs and buttes. Another described trail crosses a large exposure of the anomalous deposit. This deposit may represent interlayering between the Slickrock-Entrada and the extreme western extension of the Moab-Entrada and the Tidwell Member of the Morrison Formation that overlies the Moab-Entrada where it appears. A 1955 U.S.G.S. geologic map calls this anomalous deposit "Lower Summerville." In current terminology applicable to the general Moab area, this would be "Lower Tidwell-Morrison." Although this nomenclature and description may be correct, it does not adequately explain this odd stratigraphic phenomenon.

Moab Member. The type locality for the Moab Member of Entrada Sandstone is Moab Valley, Utah, because this distinct layer of sandstone dominates the rimlands of Wilson and South mesas, which loom over the valley high in the foothills of the La Sal Mountains. Even so, the rock was first studied in the Dewey Bridge area where it was more readily accessible. The Moab-Entrada is a white sandstone that is basically beach sand, with some sand cross-bedded by winds and some made flat-bedded by tidal invasions or wave action. It varies in thickness from about 100 feet in the mesa rimlands to less than that to the east and west. It was deposited by an arm of a sea that crept into the region from the east or northeast, then retreated, leaving beach-sand deposits that extend to about eleven miles west of the Grand County airport, north of Moab, to as far south as the La Sal Mountains, and to just north of Arches National Park. To the west and south of the La Sals, the Moab-Entrada phases into the Slickrock-Entrada and the first Morrison Formation deposits. Some geologists claim to have detected a small unconformity between the Moab and Slickrock members of Entrada Sandstone. This may or may not be related to the anomalous deposit noted in the preceding paragraph.

To the west and north of the Moab area, the Moab-Entrada also phases laterally into marginal-marine deposits, indicating that the Uncompahgre Uplift had finally eroded completely away, at least for a time, and that the region to the east of canyon country was largely dominated by an invading inland sea. The Moab-Entrada also occurs in and near Colorado National Monument, to the south of Grand Junction, Colorado, on the now-elevated and tilted northern slopes of the Uncompahgre.

Because of its harder nature, wherever it covers the older Slickrock-Entrada the Moab-Entrada tends to protect the softer aeolian sandstone from rain erosion, thus promoting the formation of natural arches. This has contributed to the formation of numerous natural spans in Arches National Park and has created an arch-wonderland in the vicinity of Rattlesnake Canyon, near Colorado National Monument.

The Moab-Entrada generally collapses into near-vertical cliffs when it is undercut, as it commonly is by the easily erodable anomalous deposit that occurs below it west of U.S. 191. In a few areas, this undercutting has also created small natural arches. Where the upper surface of the Moab-Entrada is exposed, it tends to erode into shallow, well-sealed potholes that are often capable of holding rain and snowmelt for weeks and even months. A few of the deeper potholes have permanent pools that host colonies of small plants and animals. These potholes, in turn, often promote seeping springs in nearby cliffs.

In some locations, the Moab-Entrada is highly mineralized, as are most shoreline deposits, and strangely colored and oddly-shaped concretions have formed within it and near its base. In other locations, green copper-bearing concretions weather out of the white sand matrix. In one area just outside of Arches National Park, the Moab-Entrada contained enough mineralized copper to encourage the start of a copper mining operation. The Klondike Bluffs off-road vehicle trail logged in this book ends near that long-abandoned mine.

The beaches of the Moab-Entrada were popular with both herbivorous and carnivorous dinosaurs, which were first venturing into the region. Evidently, at least the shoreline was vegetated and hospitable to dinosaurs. Although fossil plants are not common in the Moab-Entrada, this writer has observed at least one petrified log that seemed to be in place at the base of this member. The topmost few inches of the Moab-Entrada, however, at least in the general Moab vicinity, has numerous petrified dinosaur tracks in it. The freshly uncovered surfaces of this rock on each side of the Salt Valley Anticline have thousands of theropod tracks, quite a few sauropod tracks, and those of a few other types of dinosaurs. Several of these sites are documented in the book, **TRACKING DINOSAURS,** by Martin Lockley.

UNCONFORMITY

Some geologists show an unconformity between

the upper members of Entrada Sandstone and the next younger deposits, but others do not. Either way, the contact between the uppermost Entrada members and the next younger deposits is generally quite clean and obvious within the area covered by this book.

MORRISON FORMATION

The type locality for the Morrison Formation in general is Morrison, Colorado. Its deposits total between 700 and 800 feet thick in the area covered by this book, and represent a major change in general topography in the region, as well as a significant change in the canyon country environment while the Morrison was being deposited. Morrison deposits covered most of Montana, Wyoming and Colorado, large parts of Utah, South Dakota, Nebraska, Kansas and New Mexico, and small areas of Arizona, Texas, Oklahoma, North Dakota, Idaho and Canada. The Ancestral Rocky Mountains, including the associated Uncompahgre Uplift, had long since eroded away, allowing the Morrison deposits, or their stratigraphic equivalents with other names, to cover all or most the areas they had occupied. Of the three Morrison members that exist in the area described in this book, the oldest covered the least terrain, while each subsequent deposit covered more, indicating that they were gradually filling an immense, shallow basin.

The major climate and topography changes represented by Morrison deposits were triggered by continental drift activity. What is now the North American continent was breaking away from the supercontinent, Pangea, and drifting westward at a relatively high rate of travel. This resulted in enormous tectonic activity on the leading edge of the North American continental land mass, as lateral, eastward pressures against it grew. Rates of subduction of the overridden oceanic crust increased, bringing on earthquakes, volcanism and mountain-building. Growth of the Sierra Nevada mountain range, that began when Pangea first began to break up, accelerated. As these stupendous crustal pressures were transmitted eastward into what is now central Utah, immense expanses of land were thrust up over other land in what is called the Sevier Overthrust Belt. Other massive uplifts occurred in what is now northern Arizona. All of this tectonic activity in turn triggered long-lasting regional volcanic activity.

This elevation of the land to the west of canyon country, with the concurrent retreat of the shallow seas that had dominated the region for so many tens of millions of years, changed three major factors in canyon country. It marked the end of the deposit of the "great red beds" of the region, most of which had originated from the Uncompahgre Uplift to the northeast. It ended, for a long period of time, the domination of the region by intruding seas, and it allowed the region to be inhabited by land plants and animals, including dinosaurs, where it had formerly been inhospitable to land life of all kinds.

What had been either marine, marine tidelands, or barren coastal-marine environment became dominated by fresh water and fresh-water deposits brought into the region by a system of rivers and streams that originated largely from the new higher terrain to the west and southwest. Land that had drained west into shallow seas that intruded from the south, now became a vast landlocked basin with no outlet to the sea.

For about 15 million years the rivers carried their sands, silts and other deposits, including massive amounts of volcanic ash fallout, into this enormous valley, where the water accumulated in basins, lakes and swamps, and the deposits collected to become Morrison Formation members totaling as much as 700 to 800 feet deep. In time, Morrison deposits even covered the eroded remnants of the Uncompahgre and probably the remaining nubs of the Ancestral Rockies. Although the later rise of the present Rockies destroyed most of the evidence that the Morrison Formation covered the original Rockies, fossil-rich Morrison deposits even today lie on the eastern flanks of the Uncompahgre, though it has thrust upward still more at least four times since then. Morrison deposits are exposed over a large part of the area covered by this book, to the north and northeast, and in the higher parts of the Cameo Cliffs area to the south.

All kinds of plantlife, dinosaurs and other animals found this to be an attractive environment, and their fossils are abundant in Morrison deposits. So are the heavy minerals in some Morrison strata, such as uranium, that arrived as components of volcanic ash fallout. Where these minerals were subsequently concentrated by both geologic and other factors, they have been sought and mined in recent canyon country history.

Tidwell Member. All around the limited extent of the Moab-Entrada seashore sands, this deposit phases into the dark red tidal-flat deposits of the Tidwell Member of the Morrison Formation. In the past, this distinct deposit was called the Summerville Formation, but more recent studies have redesignated it as an early member of the Morrison Formation within the region covered by this book. The type locality for this Morrison member is Summerville Point, in the San Rafael Swell, Utah, where it was first studied. Such deposits in that area are still called Summerville.

The marginal-marine Tidwell-Morrison, and other adjacent similar deposits farther south and west, represent a brief marine invasion from the north. This produced shallow inland seas over the region in which the mineral content was concentrated by solar evaporation. The retreat of this sea marked the end of the marine and shoreline environment that had dominated the region for so long. A time of freshwater deposits began next.

Tidwell-Morrison deposits range up to 100 feet thick in canyon country, but are mostly in the lower end of this range. They consist of red-hued, thinly bedded silts, shales, sandstones and limestones. The deposit is highly mineralized and eventually formed great quantities and masses of agate within its matrix. This harder mineral is left on the eroded slopes of the Slickrock-Entrada just below it many places in canyon country, in particular the Cameo Cliffs area. It is also found on top of the Moab-Entrada, where this member exists.

Although it is not commonly known, a few dinosaurs began to appear in the canyon country region even during the time the Tidwell-Morrison was being deposited, although they doubtless found the environment inhospitable. Professor John Newberry, a member of the 1859 Macomb Expedition into southeastern Utah, discovered what was then one of the first dinosaur remains to be found in the western hemisphere, in red-hued Tidwell-Morrison deposits about 30 feet above the Entrada Sandstone cliffs that wall East Wash Canyon, east of the junction of U.S. 191 and Utah 211, the road into the Needles District of Canyonlands National Park.

Although the petrified bones Newberry collected at the site were sent to the Smithsonian Institution for later identification and study, the exact location of his historic discovery was not documented and was lost to history and science until the site was rediscovered by this author and his wife following a twelve-year search. The story of that search, and the tracing of the previously unknown route of the Macomb Expedition, is described in the book, **CANYONLANDS NATIONAL PARK - Early History & First Descriptions.** Newberry's petrified dinosaur bones were identified as belonging to a sauropod, a gigantic herbivore. Although it was, and may still be, one of the oldest specimens of sauropod to be found, these gigantic creatures doubtless existed still earlier because their petrified foot tracks have been found north of Moab in even older Moab-Entrada sandstone.

Salt Wash Member. The type locality for this member of the Morrison is Salt Wash, Utah. It consists of fairly thick layers of light-hued, stream-deposited sandstone separated by thinner layers of gravels, clays, mudstones, varicolored shales and some volcanic ash. These materials were all brought into the vast basin described earlier, where they and the water that brought them provided an environment favorable to the types of plant and animal life that proliferated in North America then. Salt Wash-Morrison sediments contain fossil remains, both plant and animal. Some of the most famous "dinosaur quarries" are in this Morrison member. The Salt Wash member covered far more of the new great basin than did the older Tidwell member.

Salt Wash deposits make up nearly half of the total Morrison Formation. When exposed by erosion, the upper surfaces of this member make broad, thick terraces of light-hued sandstone. When cut, more ledges and terraces are exposed, with each one undercut by more erodable softer strata. When exposed in depth, this member erodes into steep slopes covered with large slabs and fallen boulders of broken sandstone. In much of canyon country such slopes are cut by a profusion of uranium prospecting trails.

Brushy Basin Member. The type locality for this member is Brushy Basin, Utah. Brushy Basin deposits intertongue into the older Salt Wash deposits and consist largely of gray, red and green-hued shales, mudstones and thin layers of sandstone and conglomerate. Brushy Basin-Morrison is even more mineralized than the Salt Wash member, with most of the minerals, both heavy-metal and other compounds, coming from volcanic fallout originating from the northwest. The sands that made up the sandstones probably came from higher ground to the southwest. All Brushy Basin sediments were brought into the canyon country region, and widely distributed there, by a complex system of rivers and streams that existed for millions of years. Evidently, by this time the gradients for the rivers and streams that flowed into the region were gentle, probably because the vast basin was getting full. They brought in mostly fine-grained materials and soluble minerals, with little heavier, coarse material.

This member also contains fossil plants and animals. There are known sites rich in petrified dinosaur bones to the northeast, north and northwest of Moab. The region of vast, shallow, highly mineralized lakes, however, was not as hospitable to land plants and animals as the Salt Wash-Morrison period, nor were the chemical-rich lakes favorable to aquatic life.

When the relatively soft Brushy Basin-Morrison deposits are exposed by erosion they form smooth, color-banded slopes interrupted occasionally by thin ledges of harder rock. These gray, green, red and lavender-hued "painted desert" slopes are well-known throughout the intermountain west, as well in some areas east of the present Rocky Mountains. Rockhounds find many collectible mineral specimens in this member of the Morrison.

BURRO CANYON and CEDAR MOUNTAIN FORMATIONS

These two formations are essentially the same, and are very similar to the Brushy Basin-Morrison but somewhat lighter in color. They are generally separated from the Brushy Basin by a hard ledge of sandstone and conglomerate. They represent deposits made over the same region at the same time, but for some reason have been given different names, with the present Colorado River being the arbitrary dividing line between them. They are both the same mixture of sandstones, shales, conglomerates, colorful mudstones and minor lenses of limestone. Some geologists note that the lower units of these two formations intertongue with upper Morrison sediments in some locations, indicating that the transition between the two is sometimes not clearly defined. The type localities for these formations are Burro Canyon, Colorado, and Cedar Mountain, Utah. The deposits range up to 200 feet thick, but in the area covered by this book are generally thinner than that.

These formations represent a time of transition between the building of deep Morrison river, stream and lake sediments in a vast, land-locked basin, and another invasion of the seas that had held sway over the region for such a long period earlier. This transition was very slow, however. Burro Canyon-Cedar Mountain sediments were deposited by flowing fresh water, but were eroded away intermittently almost as fast as they accumulated, over millions of years. As noted earlier, the thickness of a deposit does not always indicate how long it took to build. Apparently, the immense land-locked basin of Morrison time developed at least some kind of outlet, and this permitted intermittent removal by erosion of Burro Canyon-Cedar Mountain sediments, with very little net gain.

In general appearance, these formations resemble Brushy Basin-Morrison deposits, with the deposits somewhat softer and lighter in color. They also erode in a similar manner, with smooth varicolored slopes and occasional sandstone ledges. Fossil plants and animals are scarce, indicating that the inhospitable environment of the Brushy Basin-Morrison continued on through the next long deposition period. The fossils that do exist are plantlife, some dinosaur bone and freshwater vertebrates and invertebrates.

Relatively little of these formations is exposed in the area covered by this book, and the exposures are not very thick. These are along U.S. 191 North and adjacent to Utah 128, near the north end of the stretch described. Other exposures occur along the Monitor & Merrimac and Hidden Canyon Rim ORV trails. Although these deposits represent a considerable length of time, they play a very small part in the present erosional scene in the general Moab area.

UNCONFORMITY

Due to the intermittent erosion taking place during Burro Canyon-Cedar Mountain times, there is an unconformity between these deposits and those of the next identifiable deposit, Dakota Sandstone. This erosional gap is fairly thick to the west, but thins out to the east, where land-surface erosion was being terminated by a slowly invading sea.

DAKOTA SANDSTONE

The type locality for this sandstone is Dakota, Nebraska, indicating how widespread it is. Dakota Sandstone is freshwater coastal-stream and shoreline deposits laid down as a sea invaded the region from the east. Some geologists consider the Dakota to be a formation, because it consists of two distinct units, the non-marine freshwater deposits left by streams, and the wave-washed shoreline sands left by the invading sea. These total more than 200 feet thick in some places, but in the Moab area, are less than half of that, where they remain at all. This is to the north and east of Arches National Park, but in that area the geologic strata are so tilted that only intermittent, slanting ridges of Dakota Sandstone are exposed.

The freshwater Dakota sediments are thin sandstones and conglomerates that are moderately mineralized, with petrified wood and thin beds of coal in some areas, indicating relatively abundant vegetation. The shoreline-marine component of the Dakota is light brown sandstone with shallow-marine fossils in some areas. When exposed, it erodes into distinctive brown sandstone ledges or ridges, undercut by its older and softer freshwater deposits.

MANCOS SHALE

The Mancos Shale type locality is Mancos Valley, Colorado. As the marine shoreline that left the upper unit of Dakota Sandstone advanced across canyon country from the east, it deposited sea-floor muds of Mancos Shale, with occasional thin beds of sandstone. Mancos Shale also has a high content of certain volcanic ash fallout minerals. When this deposit is exposed, as it is in the northern part of the area covered by this book, it has a strange physical characteristic due to these minerals. When these normally arid deposits are wetted by rain or melting snow, the minerals swell, leaving the soils with a puffed-up appearance. This swelling also causes damage to roads and other paved surfaces, and has been the source of continual problems with Interstate 70, U.S. 191, Utah 128 and the buildings and runway of the Grand County airport.

The Mancos sea eventually reached great depths, and left more than 3,000 feet of sediments over this region. One geologist has pointed out that there did not seem to have been enough higher ground around the Mancos sea at that time to provide this stupendous volume of material. The rise in sea level that brought on this final marine invasion that was to last for more than 10 million years, was triggered by a world-wide increase in sea-floor spreading, along a meandering line that wrapped around and among the continents of the world like a stupendous, fiery snake. This accelerated creation of new sea floor displaced water in the ocean basins, causing sea levels to rise drastically all over the world, flooding major areas of all the continents and inundating canyon country beneath a deep, inland sea.

When this sea was at its greatest depth, it had so little interchange of water with the open ocean that its deep, still waters lacked enough oxygen to support normal marine life, leaving only the shorelines and, to a limited extent the more oxygenated waters very near the sea's surface, hospitable to marine life. The predominantly gray and even black shales that make up this deposit are indicative of the anoxic, or reducing, deep-marine environment that dominated the canyon country region during that period.

When it is exposed by erosion, as it is along the northern stretches of U.S. 191 and Utah 128, and on both sides of Interstate 70 along the base of the Book Cliffs in southeastern Utah, Mancos Shale forms low, smooth hills of fine, gray sediments, broken occasionally by thin ledges of sandstone. It is almost uniformly gray and esthetically monotonous, yet represents a major part of the canyon country stratigraphic record.

Farther west, Mancos Shale is divided into several members, but these do not extend into this area and are not described here. A thin eastward extension of the Ferron Sandstone Member of Mancos Shale tops a low ridge west of U.S. 191 north of the Grand County airport, and appears adjacent to Utah 128 east of the Dewey Bridge area. This sandy member represents coastal-marine stream sediments deposited as the sea advanced eastward. It contains abundant shallow-marine fossils.

Except in its northernmost extent, Mancos Shale is largely gone from the Moab area of canyon country except where a few patches of the deposit have been displaced downward by faulting, such as in the southeastern end of Moab Valley. While it, and the younger deposits mentioned briefly next, did at one time cover the entire region, most of the Mancos, and all of the post-Mancos deposits and their fossil content, have long since eroded away from the region south of the present Book Cliffs that dominate the skyline to the north of Interstate 70. The younger formations are noted only because they are visible from Interstate 70 and from where U.S. 191 and Utah 128 meet this highway.

MESAVERDE GROUP

This group of deposits has several major members, but since they play a limited part in the geology of the Moab area, visible only in the distant Book Cliffs north of Interstate 70, they are described only briefly here. They represent the freshwater deposits that ultimately began to cover the marine Mancos Shale sediments as that great inland sea finally retreated. This retreat was quite slow, and was triggered by a global reduction in sea-floor spreading. This slow lowering of the sea level permitted considerable interlayering between marine and shoreline freshwater deposits, especially to the west of the canyon country region. This retreat of the sea and redomination of the land by freshwater sediments from the adjacent higher land was also aided by another coincidental continental drift event, the over-riding of a smaller plate to the west by the North American Plate. This triggered a 30-million year period of tectonic activity within and surrounding the Colorado Plateau. See the sidebar, LARAMIDE OROGENY. This distortion and uplifting of the surrounding land accelerated erosion and thus the deposit of land sediments on top of the Mancos Shale deposits. This deposition was further aided by the region once again becoming land-locked, with no significant opening to the sea.

Mesaverde deposits are primarily light brown sandstones that form the terraced cliffs visible in the Book Cliffs for several hundred feet just above the lower slopes of gray Mancos Shale.

TUSCHER FORMATION

As the Mancos sea retreated, leaving the shoreline and flood-plain Mesaverde sandstones, the still-rising higher ground continued for tens of millions of years to contribute water-borne sands, muds, shales and other types of sedimentary deposits in the resulting land-locked basin. As this was going on, the rate of deposition varied with the global climate, leaving such major geologic strata as the Tuscher, Wasatch and Green River formations. These appear in the higher levels of the Book Cliffs, to the north of Interstate 70, as a terraced, receding line of wooded cliffs that soar to mountainous heights. It should be noted that some geologists consider that Tuscher deposits are a member of the Mesaverde Group, rather than a separate formation.

Toward the end of the Tuscher deposition period, a very dramatic event occurred. A great meteor impacted the Earth in the western part of the Gulf of Mexico, triggering another major extinction of life on this planet. See the sidebar, GREAT EXTINCTIONS. A smaller companion of this meteor impacted the canyon country region in this same period of time, while the Tuscher Formation was being deposited. The erosional remnant of this impact, after some 65 million years, is near Moab, and a trip to it is a part of the Island Road geology log. See the sidebars, UPHEAVAL DOME and ROBERTS RIFT for details.

UNCONFORMITY

A major unconformity appears between the Tuscher Formation and succeeding deposits. This gap in the deposition record is probably due to erosion caused by the major climate changes brought on by the meteor impact that extinguished the dinosaurs and so many other forms of life at that time.

WASATCH and GREEN RIVER FORMATIONS

During the latter part of the long period of deposition that began with the retreat of the Mancos sea, the entire region was a series of freshwater lakes of every size, with many of them joining together during periods of wetter climate. Except for remnants high in the Book Cliffs, these deposits are completely gone from the area covered by this book, and it is not certain whether they ever did come that far south. Some geologists show the deposition of these sediments ending not far south of the Book Cliffs, while others show them going farther south. Since all of the deposits of that period are now gone from the general Moab area, there is no way to determine what is correct. Certainly, geologic deposits did cover the Moab area during the next several tens of millions of years, but we will never know exactly what kind. It is certain, however, that the Wasatch Formation existed to the west of this region, because it makes up the colorful sediments exposed in

Bryce Canyon National Park, and for many miles the Wasatch and Green River formations dominate the soaring canyon walls of the Green River's Desolation Canyon to the northwest of the Moab area.

During the period represented by these fresh-water deposits, there was considerable continental tectonic activity. While this had major effects on the Colorado Plateau and canyon country, these effects were largely indirect in the general Moab area. Great mountain ranges and uplifts formed all around the region, salt anticlines pushed upward, and immense basins sank. Volcanic activity of all kinds flourished around the perimeter of the Colorado Plateau, but had little effect on its interior beyond fallout and increased sedimentation.

LATER TECTONIC ACTIVITY

There were other great tectonic activities in the area covered by this book, but they did not take place while any of the remaining geologic deposits there were being made. The La Sal, Abajo and Henry mountain ranges thrust upward through ancient sedimentary layers to lofty heights. See the sidebar, LACCOLITHIC MOUNTAINS. The Uncompahgre Uplift, which grew again during the Laramide Orogeny, grew further at least twice, once about 10 million years ago and again as recently as 1 million years ago. The first of these uplifts contributed greatly to the present erosional phase in the canyon country region. The second either triggered a change in the course of the Gunnison River, which once flowed through Unaweap Canyon, near Gateway, Colorado, or rose since that diversion. See the sidebar, UNCOMPAHGRE UPLIFT for more details.

SUMMARY

The foregoing descriptions of the geologic deposits and activities that affected the Moab area of canyon country are, of course, just a brief conceptual summary. The full story is much more complex, and professional geologists spend lifetimes trying to fully understand even short chapters of that story. Further, as studies of the relatively new subjects of plate tectonics and continental drift make progress, the startling findings in these fields must be integrated into the older bodies of geologic and paleontologic knowledge in order to bring even better understanding. Geology is not a static science. Its complexity has grown at an enormous rate over the last three decades, as the concepts of continental drift have been accepted, sometimes very reluctantly by more conservative geologists.

As noted earlier in this book, to the outsider, the study of geology can at times resemble an art rather than a science, but that, perhaps, is more a reflection on human nature than on the science itself which, in the long run, holds the key to understanding what this ancient planet and its thin patina of fragile life is all about.

TECTONIC FEATURES

The geologic term "tectonic" refers to any physical deformation of the earth's crust. In canyon country, such deformations take the form of:

monoclines (tilted or sloping geologic strata);

anticlines (upthrust ridges of strata);

synclines (sagging strata);

basins (broad areas of sagging strata);

faults (cracks where the strata are offset);

joints (cracks where the strata are still aligned);

rifts (cracks where the sides are separated and contain foreign fill-material);

grabens (sunken areas between roughly parallel fault lines);

uplifts and swells (gigantic anticlines); and

the tilted terrain produced by **intrusive igneous activities** (volcanoes that did not quite break through the surface of the land as it was at that time).

Several of these terms are defined further in sidebars elsewhere in this book. Some examples of these tectonic features are given below, but no attempt has been made to be comprehensive. Canyon country is too geologically complex for that. Most of the features described are mentioned in the geology road and trail logs in this book.

Monoclines:

The geologic strata north of The Knoll, on Island-in-the-Sky, generally slope downward toward the north. The strata east and west of U.S. 191 between La Sal Junction and Church Rock tilt down toward the east. The entire Colorado Plateau geologic province has an approximate 6 degree tilt toward the northeast. There are innumerable smaller monoclines throughout the region covered by this book.

Anticlines:

There are numerous anticlines in canyon country, and several major ones in the area covered by this book. Some of those large enough to have been given names are Moab Valley Anticline, Castle Valley Anticline, Fisher Valley Anticline, Hatch Point Anticline, Gibson Dome Anticline, Cane Creek Anticline, Lockhart Anticline, and Salt Valley Anticline. Some of these anticlines were once continuous with similar ancient salt anticlines south of the La Sal Mountains, before the mountains thrust upward, separating them. See the sidebar, SALT ANTICLINES.

Synclines:

There are about as many synclines as anticlines in canyon country, with a significant number large enough to bear names. Some of these in the area covered by this book are Kings Bottom Syncline, Courthouse Syncline, Grays Pasture Syncline, Hatch Rock Syncline, and Indian Creek Syncline. Sometimes upthrusting anticlines produce corresponding synclines along their flanks. A roughly circular syncline surrounds the remaining evidence of the meteorite impact crater that formed Upheaval Dome in Canyonlands National Park. This particular syncline formed as a part of the reflex reaction to the extreme, short-time compression created by the meteor impact. See the sidebar, UPHEAVAL DOME.

Basins:

Although little now shows on the surface, virtually the entire Colorado Plateau, including the Moab area, lies over an ancient marine embayment now known as the Paradox Basin, in which marine salts were precipitated over a long period of time by solar evaporation. The older horizontal deposits beneath this ancient basin sagged slowly downward as the salts accumulated, ultimately allowing them to build to thousands of feet deep. In its earlier phases, this basin was much larger. It was then cut into two segments by the earliest rises of the Uncompahgre Uplift, with the southwestern part now called Paradox Basin and the northeastern part now called Maroon Basin. Some of the sagging in these basins was related to the simultaneous rise of the uplift. See the sidebar, UNCOMPAHGRE UPLIFT.

Faults:

There are innumerable faults in the general Moab vicinity. Most are in the vicinity of the La Sal Mountains, the Abajo Mountains, various salt anticlines and other anticlines and synclines. Surface deposits, soils, sands and landslide debris and vegetation leave few faults visible on the surface of the ground, except in the walls of cliffs. One fault in the Moab area major enough to bear a name is the Moab Fault, which is mentioned in the appropriate road logs.

Joints:

As with faults, there are countless joints in the Moab area. Most were formed by the same kinds of tectonic activities that formed the faults, but without horizontal or vertical displacement of the adjoining geologic strata, and few are visible on the surface for the same reasons. None have been given names, although there is a "Joint Trail" hiking route in the Needles District of Canyonlands National Park that goes through a series of barely-separated joints in Cedar Mesa Sandstone.

Rifts:

While there are doubtless other rifts in the Moab vicinity, the only one that has been named is Roberts Rift, which lies to the northwest of Moab and cuts across the several major canyons that lie between the Windows Section in Arches National Park and the Island-in-the-Sky just north of Dead Horse Point. For details about this fascinating tectonic feature see the sidebar, ROBERTS RIFT. This miles-long rift is a separated joint that was filled by material forced upward from much deeper, as well as water-soluble minerals that subsequently seeped in from the surrounding rock.

Grabens:

There are a number of grabens in the southern parts of Canyonlands National Park and Canyon Rims Recreation Area, but most of them in the recreation area are not visible on the surface, as with most faults and joints, nor are the park's grabens apparent from any of this book's road logs. Some of the off-road vehicle trails described in the book, ***Canyon Country*** **OFF-ROAD VEHICLE TRAILS - Canyon Rims & Needles Areas,** and depicted on the map, ***Canyon Country*** **OFF-ROAD VEHICLE TRAIL MAP - Canyon Rims & Needles Areas,** travel the length of immense grabens in the Needles District of the park. Utah 211, described in this book, goes through part of a large graben as it descends into Upper Indian Creek Canyon.

Uplifts and swells:

Most of canyon country has been affected by the Uncompahgre Uplift, whose present exposure angles northwest-southeast just east of the Utah-Colorado border. Most of the iron-rich red-bed deposits in canyon country came from this uplift. The spine of immense Monument Uplift runs generally southward from the Needles District of Canyonlands National Park and is mentioned in this book's road logs. Monument Uplift extends on south through Utah and into northern Arizona. The flexing of the earth's crust by this monstrous uplift created most of the joints, faults and grabens in the recreation and park areas noted, and this in turn allowed the formation of the numerous sandstone buttes, domes, mesas, ridges, fins, needles, towers and other erosional features that dominate the Needles District of the park. The Elk Ridge highlands, visible from several of the roads and trails described in this book, are the lofty spine of this uplift. On clear days, the eastern reef of the San Rafael Swell is visible to the west

continued on page 25

LARAMIDE OROGENY

Barely two decades ago, when geologists attempted to explain such major tectonic activities as earthquakes, volcanoes, uplifts and mountain-building, they were limited to discussing WHAT happened and WHEN, plus making a few groundless conjectures about WHY such geologic events occurred. Before the 1970s, knowledge of planetary geology was so limited that the causes of such events were not encompassed by the contemporary state of knowledge. Many geologists tried to construct logical causes for the readily observable results, but most failed in at least one critical area -- their concepts did not account for the stupendous amounts of energy required for tectonic events such as mountain-building to take place.

In recent years, however, despite much resistance from more conservative geologists, the concepts of continental drift and plate tectonics have come to be accepted as valid, as part of the science of geology, because these concepts can be used to create a coherent picture from a wide spectrum of facts that otherwise defy explanation -- like the various tectonic events that have shaped canyon country over the past 320 megayears -- and the forces involved in continental drift can easily provide the energy necessary for these types of geologic events.

One of the greatest of these is the Laramide Orogeny, the prolonged period of intermittent tectonic activity that produced the present Rocky Mountains, the Uinta Mountains, the San Rafael Swell, the Kaibab, Echo Cliffs, Monument, and Defiance-Zuni uplifts and at least two of the many upthrusts of the ancient Uncompahgre Uplift.

To put the Laramide Orogeny into perspective, it was the last of a series of three major mountain-building periods that produced the outstanding tectonic features that make the western part of the North American continent so geologically spectacular.

This series of orogenies consisted of the Nevadan, the Sevier and the Laramide, with these overlapping on the time scale and working from the west coast eastward over a very long period of geologic time. All of this powerful tectonic activity can now be related to various stages in the separation of the North American continent from Pangea, the supercontinent that broke up into the present continents, or to phases and changes in North America's subsequent drift westward.

During the Nevadan, the westward-drifting continent swept up a whole archipelago of islands that lay off-shore to the west of North America when the continent first broke loose from Pangea, adding these land masses to make up the present west coast. The Sierra Nevada range also began to thrust upward during this period, although that long process of mountain-building overlapped the Sevier Orogeny.

During the Sevier, which lasted from about 100 to 75 million years ago, a wide band of terrain cutting north and south across what is now central Utah was forced by crustal pressures from the west to overlap, or fold, creating the "Overthrust Belt," but this had little effect on eastern Utah. Then, from about 75 to 45 million years ago, the Laramide Orogeny took over.

It should be noted that while geologists assign different time scales, definitions and degrees of overlap to the three western orogenies, they now generally agree on the relationship between these tectonic activities and continental drift events.

The several tectonic phases during the Laramide were triggered by periods during which the North American plate changed directions and rates of travel, with these changes caused by planetary forces still not thoroughly understood. This was further complicated by a time when the Colorado Plateau resisted a southerly change of direction by the continent that surrounded it, causing the Plateau to rotate in a relatively clockwise direction and creating still more tectonic activity within and adjacent to the Plateau. See the sidebar, COLORADO PLATEAU, for a description of this distinct and unique region of ancient planetary crust.

The results of these shifts in continental travel were: the creation of a new Rocky Mountain range, in approximately the same alignment as the long-gone Ancestral Rockies; the formation of the Uinta Mountains range and the various regional uplifts noted earlier; the sagging of basins adjacent to several uplifts; additional rises in existing uplifts; the Moab Fault; and long periods of volcanic activity to the west of the Colorado Plateau.

Since the greatest tectonic effects occurred on or beyond the perimeter of the ancient and highly durable Plateau, geologists have conjectured that this durability held tectonic activities within the Plateau to a minimum, producing only the smaller uplifts found there, although these are small only when compared to the Rockies and Uintas. It is probable that the unique east-west orientation of the Uinta mountain range, which is on the northern perimeter of the Colorado Plateau, was caused by temporary northward pressures on the Plateau during one change of direction in continental drift, probably the one during which the Plateau resisted rotation.

It is interesting to note that at present the Green River, which originates in the Wind River range in Wyoming, cuts through the Precambrian red-granite spine of the lofty Uinta range, just as the Gunnison River once cut through the Uncompahgre Uplift, although the creation of the Green River gorge through the Uintas was more complex.

Of course, all of the upthrust activity that took place during the Laramide, within and all around canyon country, brought increased sedimentation into the region, depositing the Tuscher, Wasatch, and Green River formations, but these deposits are now gone in the area described in this

continued ----

book, except in the Book Cliffs.

While the major activities of the Laramide Orogeny were in the distant past, recent tectonic movements in the same region, including two new uplifts of the Uncompahgre, have led some geologists to conclude that this orogeny is still continuing, although such conjecturing is more a matter of definition than precise science.

The continent is, of course, still drifting generally westward, but that drift is no longer a relatively simple matter, so perhaps it is time to name a new phase of orogeny, a phase that humanity will never see end. Presently, North American continental drift is complicated by impact with the immense Pacific plate, and that interface is a gigantic transform, or sliding, fault, rather than the earlier, relatively simple subduction, or over-riding, of smaller plates. California's infamous San Andreas Fault is part of that tectonically active interface.

Today, although the only Laramide-age tectonic feature clearly visible from the road and trail logs in this book is the Uncompahgre Uplift, the canyon country region is surrounded by other tectonic structures of similar age and origin.

The San Rafael Swell, glimpsed in the distance from several logs in this book, lies to the west. It is crossed by Interstate 70, Utah 24 provides an excellent view of its eastern "reef," and U.S. 6 & 50 skirts its northern flanks.

Monument Uplift lies to the south, and is crossed by Utah 95 and by U.S. 163 west of Bluff, Utah. Its jointed northern nose is visible in the Needles District of Canyonlands National Park from Utah 211 and the Lavender Canyon ORV trail, both described in this book.

Other highways penetrate the Circle Cliffs in south-central Utah, the Kaibab and Echo Cliffs uplifts in northern Arizona, and the Defiance-Zuni uplifts in adjacent corners of Arizona and New Mexico. And, of course, numerous highways and roads traverse the various ranges of the spectacular Rocky Mountains, from New Mexico to Canada, all courtesy of the Laramide Orogeny and human developments made to take advantage of the boundless beauty and wealth it created.

as a jagged row of sharply tilted sandstone slabs resembling shark's teeth. This swell, or immense uplift, is about 50 miles long as exposed, and even longer beneath the surface. The many other uplifts and swells in the Colorado Plateau are not visible from the area described in this book.

Intrusive igneous activities:

There are three major mountain ranges in southeastern Utah that were formed by intrusive igneous activities, as well as others in the general Four Corners region. The La Sal, Abajo and more distant Henry mountain ranges are visible from various roads and off-road vehicle trails described in this book.

The formation of the Abajos, long after the birth of the Monument Uplift, further affected the present topography within Canyonlands National Park and Canyon Rims Recreation Area. The added effects were more tilting, additional jointing and faulting, and probably the formation of more grabens.

After megayears of erosion, parts of the igneous cores of the La Sal and Abajo mountains are exposed to view, usually as great slopes of gray rocky rubble that looks like granite but is actually "quartz diorite porphyry" or, in non-technical terms, molten rock of a certain composition that has cooled very slowly under great pressure, thus allowing its mineral content to crystallize.

EROSIONAL FEATURES

The geography and magnificent topography of canyon country are the products of its ancient geologic strata, regional and continental tectonic activity and approximately ten million years of erosion, with most of that erosion taking place during the last two million years.

This long period of erosion was not smoothly continuous, but slowed, halted, reversed and accelerated many times over the ten megayears since it started, generally because of regional climate changes and attendant plant community changes, but to some extent because of very slow but ongoing geologic processes such as general regional uplifting, major global climate changes and other associated factors. See the sidebar, GLACIERS OF THE LA SAL MOUNTAINS.

The net result of this intricate erosional process, however, is a region that is incomparable in topographic form, color and natural beauty, a region that has inspired many writers to invent new combinations of superlatives, beginning with the first white men to see its heartland, on the 1859 Macomb Expedition, continuing with members of the Powell Expeditions a decade later, and carrying on into today and, hopefully, into endless tomorrows.

The foregoing parts of this chapter have described in mundane terms the basic geologic deposits and shaping tectonic forces that are the foundation of this incomparable and unique beauty. The next section describes the shapes and forms that time and the elements have carved from these geologic fundamentals, but words can only go so far in describing terrain like this. They can at best only explain, and perhaps inspire readers to go see for themselves. For this reason, the erosional features are also described in mundane terms, but with the urging of this author that readers visit the general Moab vicinity and see for themselves what this magnificent area is all about.

Access to representative erosional features is

continued on page 29

GLACIERS OF THE LA SAL MOUNTAINS

A little more than 2 million years ago, Earth entered a prolonged ice age from which it is still emerging. Ice masses began building in both polar regions and slowly increased toward the equator. On land, this ice buildup took the form of glaciers -- great masses of ice that flowed slowly but inexorably down the slopes of mountains and across open terrain.

This ice age continued until appreciable regions of the world's continents were covered by deep layers of slowly-flowing ice, in some places thousands of feet thick. In North America, this ice sheet covered most of what is now Alaska and all of Canada, and in the Midwest extended south of the present Great Lakes to the Missouri and Ohio rivers. Even as recently as 20 thousand years ago, the Great Lakes region was still under several thousand feet of ice.

This major ice age was divided into four periods of continental glaciation, and these were reflected in glacial activities in isolated western mountain ranges, including the Rocky Mountains and the La Sal and Abajo mountains of Utah.

This ice age is now slowly ending, and the great masses of ice that covered such immense regions are retreating, but geologists and climatologists recognize that Earth is still within what is merely the latest of a series of great ice ages that have been identified. Two continental land masses -- Antarctica and Greenland -- totaling 10% of the world's land, are still buried deeply under ice. Alaska still has a number of active glaciers and more southerly mountain ranges have only very recently emerged from the latest episode of glaciation.

The great ice age from which Earth is currently emerging was not the result of a smoothly continuous climate change, but rather a series of surges or pulses of colder climate, with each followed by a warming trend. Earth is presently well into one such shift toward warmer and drier. This is evidenced by the desertification of large regions of Africa, a process which has been accelerated by human activities, and other continental regions are suffering prolonged drought or significant reductions in annual precipitation.

All of this global and continental glaciation and the triggering cyclic changes in climate produced corresponding effects in the general Moab vicinity. They strongly affected the region's past climate and have been essential to the creation of its present outstanding topography.

During the last 2 million years, the La Sal Mountains have had nine distinct episodes of glaciation, with each being followed by a period of warmer climate. Studies have shown that these pulses of local glaciation were directly related to similar increases and decreases in glacier-forming elsewhere in North America, including the nearby Rocky Mountains.

In Utah, all mountain ranges with elevations above 10,000 feet were glaciated at least some during this period, although the Abajo Mountains, being lower, had far fewer and smaller glaciers than the La Sals.

Geologists have divided the nine episodes of glaciation in the La Sals into four distinct groups, with these groups separated by exceptionally long interglacial intervals and related to the four major periods of continental glaciation noted earlier.

The four glacial "formations" in the La Sals are called -- from oldest to youngest -- Harpole Mesa, Placer Creek, Beaver Basin and Gold Basin, and were named after the "type locations" in the La Sals where they were first studied. Each complete cycle of the nine glacial episodes was related to a cycle of climate changes, with attendant changes in regional vegetation, alluviation and erosional cutting.

Periods of cooler and wetter climate initiated each glacial pulse. The glaciers then built during the following colder and wetter periods, and began receding as the climate phased from cold and wet to cold and dry. As the climate entered the following warmer periods, the glaciers disappeared and various stages of alluviation began, followed by increasing erosion. Major cutting took place next, with the severest erosion taking place as the climate ultimately became warm and dry. Then, greatly reduced vegetation communities permitted flash-flooding. In each case, the climate then initiated the next glacial cycle by again turning cooler and wetter.

While these climate fluctuations produced cyclic glacial periods, they were not extreme changes in general temperature and humidity, but variations in the annual averages of only a few degrees in temperature and a few percentages in relative humidity, over periods of centuries or millennia.

As each glacial cycle ran its course, regional vegetation followed a related pattern. As the glaciers built, the elevation zones of major plant communities dropped and the timberline in the mountains lowered. In the La Sals, the timberline was as low as 9,000 feet during the first three of the nine glacial episodes. Conversely, as each set of glaciers retreated, plant communities advanced into higher elevations. During the interglacial periods between the third and fourth and the fifth and sixth episodes, the timberline exceeded 12,700 feet. Trees completely covered the La Sals.

Before the encroaching ice age reached the mountain ranges of Colorado and Utah, the terrain in the vicinity of the Abajo and La Sal mountains was nothing like at present. The mountains were there, but were not as exposed and did not have the rugged, steep contours they do now. The deep canyons that now radiate from the Abajos had yet to form. The series of salt anticline valleys that now surround the La Sals did not yet exist. See the sidebar, SALT ANTICLINES. The ancient salt anticlines and associated synclines had long since ceased to be active and had been

continued ----

buried deeply beneath Morrison, Burro Canyon - Cedar Mountain, Dakota, Mancos Shale, Mesaverde, Tuscher, Wasatch, Green River and even younger deposits that have eroded completely away in this region.

As noted earlier in this book, the Wasatch and Green River formations may not have extended as far south as the La Sals, but certainly now-missing deposits of equivalent age did.

The Abajo Mountains thrust upward through this deep overburden about 28 million years ago, and the La Sals appeared some 4 million years later. Both appeared during the time when the younger, unnamed deposits were accumulating. See the sidebar, LACCOLITHIC MOUNTAINS. Water runoff from the young ranges starting the first slow localized erosion, but this barely had any effect on the surrounding arid plains. Significant regional erosion did not begin for another 14 million years, when the entire western part of North America was raised thousands of feet by a major plate tectonic event.

When that uplift occurred, the major rivers of the region -- the Colorado, Dolores, San Juan and Gunnison -- were created and began to cut into the sediments accumulated over some 300 million years. The additional Uncompahgre uplifting that occurred about then added still more water to these rivers, all of which originated in the Rocky Mountains. See the sidebar, UNCOMPAHGRE UPLIFT.

For the next 8 million years, the rivers cut deeply into the younger deposits that had buried the salt anticlines, carrying away sediments from the slowly eroding surface of the land. By the time the global ice age started in the polar regions, this slow surface erosion had carried away all the younger deposits down to about the Burro Canyon - Cedar Mountain formations, and the Morrison Formation in some areas, leaving some of the stocks and laccoliths of the mountains exposed.

The Colorado and Dolores rivers had established their present courses, with the Gunnison still flowing into the Dolores just northeast of the La Sals. The Colorado River northwest of the La Sals was flowing through its narrow gorge about 800 feet below the general surface of the land at that time, and its major tributaries were equally entrenched. See the sidebar, ENTRENCHED MEANDERS.

The La Sals were not nearly so sharply defined, and were surrounded by gently sloped plains and several broad, shallow valleys, one of them high above the present Castle Valley. A large but placid stream flowed through this valley, then plunged abruptly into the Colorado River gorge.

With the glaciers of the La Sals on the way, the stage was set for the final act in the creation of the present topographic scene, and the main players were the climate changes brought on by the global ice age and the glaciers these changes produced.

As the first of the nine glacial episodes began, ice accumulated in the higher V-shaped valleys of the La Sals, then spread down the lower slopes and across the adjacent open lowlands. As these first glaciers melted during the following interglacial period, drainages developed that cut directly across the present locations of streams and canyons, and glacial till -- rock and sediments moved downward from the mountains by the glaciers -- was deposited broadly across the lower terrain. Some of this first glacial deposit still tops the higher ridges adjacent to the present Mill Creek Canyon, near Moab Valley, such as Johnsons Up On Top, an unnamed extension of this ridge, and the highest levels of the Mill Creek Triangle, the immense wedge of eroded sandstone that lies between the two main branches of this stream.

This first glaciation was large, but its full extent is not known because its original deposits were almost completely obscured by the two even larger glaciations that followed. It is known, however, that one arm of ice reached the vicinity of the present Black Ridge, although at that time the ridge had not yet been defined by the present surrounding canyons and lowlands.

The warming trend that followed this first glacial pulse brought renewed life to the mountains. Supported by the moist and warming climate, vegetation climbed the mountain slopes, although the timberline did not attain very high elevation. As the melting accelerated, the flowing waters first deposited rich alluvium on the lowlands, then later began cutting into these same deposits, but the cutting streams were not yet in the same configuration as present water courses.

The second and third glacial episodes were severe, but followed much the same pattern as the first. An estimated 85 square miles of glaciers developed during the second episode, with ice from 150 to 250 feet deep. One glacier was 9 miles long. During this time, volcanic activity to the west left ash fallout in the glaciers of this region. These eruptions may have triggered this episode, or at least intensified and prolonged it. This second and most severe glacial episode shaped the valleys and canyons of the La Sal Mountains for all the succeeding glacial pulses. None of the later glaciers managed to completely cover the original deposits of this one.

Other regional changes also took place about then. There is some evidence that the entire region was uplifted further, but it is certain that the Uncompahgre uplifted another time, diverting the Gunnison River northward to join the Colorado River about where it still does, but leaving the Dolores River in its present gorge.

In the following interglacial time, the vast amounts of water produced by melting regional glaciers deepened the Colorado River gorge another 100 feet. Tributary drainages cut canyons from 200 to 400 feet deep, and the nearby salt anticline valleys began to take shape, with their surfaces subsiding as seeping ground water carried salts dissolved from the underground saltridge summits into the nearby rivers. Again, however, the timberline remained at a low elevation. The global ice age was in full sway.

The third glacial episode was not as severe as

continued ----

the second, but its glaciers covered some 50 square miles of land to depths of 400 feet. The following interglacial time was longer than the two previous ones, marking the division between the Harpole Mesa and Placer Creek sequences. Vegetation again ascended the mountain slopes, and trees ultimately conquered the highest peaks.

The first of the two following Placer Creek glacial episodes covered only about 30 square miles, even though one glacier reached 8 miles in length. The timberline receded to about 9,500 feet. The second Placer Creek episode was about the same size as the first, and was followed by a second long interglacial time, during which trees once again covered the highest peaks of the La Sals. Again, the same cycle of melting, alluviation, then erosion followed each period of glaciation.

The two glaciation episodes of the following Beaver Basin sequence were still milder, indicating that the global climate that had brought on the continental glaciation was slowly warming, forcing the equally slow retreat of the continental ice masses. The first Beaver Basin episode failed to fill all of the cirques and valleys created by earlier glaciers. Ice covered only about 8.7 square miles and glaciers descended only to about 10,270 feet elevation. The second episode produced about half as much ice as the first, and that descended only to about 10,630 feet.

During the following interglacial time, the timberline climbed to about 12,000 feet, then retreated again to about 11,000 feet as the first of the two Gold Basin glacial episodes began. Between these final two mild episodes of glaciation, the timberline gained only about 500 feet, but it lost only 200 feet during the last one, affirming the dwindling nature of the series. The last Gold Basin glacial episode lasted only about 1,000 years, covered only 0.2 square miles with ice, and ended in the mid-1800s, just prior to the first settlement of Moab Valley.

The warming period following this ninth glacial episode of the La Sal Mountains -- the present -- may or may not be "interglacial." Only time will tell. The climate is still following the same cycle of cool-and-wet, cool-and-dry, warm-and-dry, but with the global climate warming and most continental glaciers in full retreat, there may not be a tenth glacial episode in the La Sals, at least for a long time.

Although the last six glacial pulses were each milder than the first three, they still had a profound overall effect on the lower terrain. The melting ice deepened the river gorges and their tributaries. Surface erosion removed more older deposits. The present system of canyons developed and, most spectacularly, the present salt anticline valleys surrounding the La Sal Mountains became what they are now, due largely to the action of ground water from the melting glaciers. As more and more subterranean salt was dissolved and carried into the rivers, the valley surfaces sank, and undercut canyon walls slumped, creating the magnificent cliffs of Moab, Castle, Fisher, Paradox and the other anticline valleys that surround the La Sal Mountains.

Humanity first appeared in the general region about 2,900 years ago, during the last sequence of La Sals glaciation, during the first Gold Basin episode. Evidence of the earliest known human habitation of the region was charcoal found at hearth sites that was carbon 14-dated at about 900 BC. The site had evidence of lithics, but was pre-pottery. Other pre-pottery sites have been found in the area dating as recently as AD 200. The first white settlers reached Moab Valley in 1855, shortly after the last ice had disappeared from the latest glacial episode.

Evidence of the La Sal glacial episodes can be seen everywhere in the general Moab area covered by this book. Most of the area's outstanding topography was created by the glaciers and their runoff. Glacial moraines and numerous glacial cirques -- glacier-rounded valleys -- were left in the higher elevations of the mountains. To geologists who are able to identify the alluvial deposits associated with various glacial episodes, the entire region surrounding the La Sals is a chaotic patchwork of rocks, soils and coarser alluvium from the nine glacial pulses. Alluvial gravels from the various glaciers were deposited over the entire La Sals vicinity and are visible from several of the described roads, including U.S. 191 South, Utah 46, Utah 128, Castle Valley Road, Black Ridge Road and the La Sal Mountain Loop Road.

In Hatch Ranch Canyon, near Hatch Wash Road, recent cuts in alluvium deposits have revealed two unusual strata. The older is some 8 vertical feet of a peat-like deposit, doubtless accumulated during one of the last wet interglacial times. A younger layer contains countless freshwater snail shells, indicating a still more recent wet interval.

Present vegetation communities also reveal the latest climate trend. Small, scattered groves of aspen, Douglas fir and Ponderosa pine still survive in sheltered locations far below their present elevation zones. A pioneer cabin in upper Harts Spring Draw, in the Dry Valley District of Canyon Rims Recreation Area, was built from aspen logs, with the trees cut nearby, 1,000 feet below the present elevation zone for aspen. Another nearby canyon on Harts Point still has a few remnant aspen at the same low elevation, in a spring-watered alcove.

Fluctuations in climate and corresponding changes in the elevation zones of plant communities are normal events. Even aside from ice ages and glacial cycles, global and continental climates vary almost constantly, but over such long periods of time that the changes are barely discernible to humanity. Thus, certain federal and state land administration agencies have falsely assumed that the plant communities that dominated the canyon country region in the historic past were "normal," and that any changes must be fought vigorously.

Such fights have all too often taken the form of ill-advised "vegetation manipulation" programs whose goals were to combat the observed "invasion" of the pinyon-juniper plant community into

continued ----

what historically were rich grasslands. The agencies either did not recognize that this change was normal for the region's slowly-changing climate, or refused to take this into account.

The resulting "chained and reseeded" pinyon-juniper forests have, as a result, not reverted back to the rich grasslands that dominated the region during pioneer times, the grasslands whose destruction was hastened by excessive livestock grazing. Rather, the devastated pinyon-juniper forests simply came back, albeit slowly, with little or no grass. Such futile and ecologically unsound land management practices continue even today.

Blind indifference to natural geological and climatological processes still dominates both federal and state land management practices, and the present natural erosion cycle that is following a glacial episode is being exacerbated by other human activities that accelerate the natural process. See the sidebar, HUMANITY AS A GEOLOGIC FORCE. As noted in one of the geologic reports on which this sidebar on glaciation was based, and whose title is listed below:

> *"(In present time) .. deposition of fine alluvium has practically ceased and an interval of arroyo-cutting has produced narrow, deep gullies. The unusual depth of arroyo-cutting may locally be due to the influence of man."*

What effect will human activities have on future global climate? Some scientists predict a trend toward atmospheric warming, brought on by the human release into the atmosphere of great quantities of "greenhouse gases" such as carbon dioxide. Others fear global cooling, perhaps brought on by atmospheric haze and both natural and human-generated dusts. And, despite the current reduction of international political tension, there is still the possibility of a "nuclear winter" brought on by the atmospheric dusts produced by an exchange of nuclear weapons.

One thing is certain, however. Human activities are affecting the global climate and will continue to do so indefinitely. If this produces significant warming, the glacial ice that still remains from the current ice age will melt and raise the sea level to disastrous heights. If human activities produce significant atmospheric cooling, this will bring still another cycle of glaciation to canyon country, and to the entire planet. A war-induced nuclear winter will probably bring on a new planet-wide extinction of life.

In any event, climate change is inevitable. At best, humanity can affect or initiate natural trends, but not eliminate them. At worst, humanity can start a major climate cycle that will eliminate itself and countless other species of life.

Reference: "Quaternary Stratigraphy of the La Sal Mountains, Utah," by Gerald M. Richmond, U.S. Geological Survey Paper 324, 1962, U.S. Government Printing Office.

described in the geology road log part of this book.

PRESENT EROSIONAL PHASE

The long siege of tectonic activity in this region that occurred during the Laramide Orogeny was the beginning of a major change in the canyon country region. The Colorado Plateau was uplifted, distorted, folded, buckled and warped, and was surrounded on all sides by even more violent volcanic activity and mountain building. This set the stage for a major erosional phase that continues today, a major "unconformity" in the making. This phase began slowly about 10 million years ago. It was about then that geologists have found and dated the first traces of what was to become the mighty Colorado River, high in the western slopes of the Rocky Mountains, and it was about then that the Uncompahgre Uplift between the Rockies and canyon country rose once again, providing more regional precipitation and more eroding runoff water.

For long periods of time, the infant river did not penetrate very far into the broad, arid plains of canyon country, but with each passing century the waters that joined it from countless creeks and tributary rivers traveled farther. Eventually, they flowed all the way to what is now Arizona, into an enormous inland basin, but not yet through the Grand Canyon. That came later, when the river was diverted from its original course by a process called "stream piracy." This change in the river's course was aided by the slow uplifting of the entire Arizona region, which raised the Colorado's inland basin. Indeed, the Colorado might not have been diverted and thus created the Grand Canyon had it not been for this general rise in terrain.

Once the river had a course all the way to sea level, it cut more rapidly than ever, and began to shape the region as it now appears. The Colorado and its many tributaries carved deeply into the ancient geologic strata described earlier, cutting down eventually into the upwelling Paradox salts in many locations, and removing all of the younger sediments south of the Book Cliffs and most of the deposits of the last great marine invasion of the region. This shaping process was accelerated still further by an additional rise of the Uncompahgre Uplift about one million years ago.

Since this geologically recent erosional phase began, approximately 5,000 feet of rock has been removed from the Colorado Plateau in general, including the Moab vicinity. This stupendous carving away of rock that took almost a quarter-billion years to accumulate has produced the stone fantasyland that has attracted visitors to the region from all over the world.

MAJOR EROSIONAL LANDFORMS

The erosional features of canyon country take on many and diverse forms, some with names that are in the general vocabulary, some that are not. Among the former are **mountains, plains, valleys, cliffs, plateaus, basins, islands, reefs, oases, playas, peninsulas and slopes,** but even some of these take on special meaning in the high-desert terrain, ancient rock formations and unusual erosional modes of the Colorado Plateau.

For example, the mountains there are **laccolithic mountains,** which formed in a unique way. See the sidebar, LACCOLITHIC MOUNTAINS. The **valleys** are generally long and narrow, and may not slope all the way to a lower end, like conventional valleys, because they are the erosional surface indications of subsurface salt ridges. See the sidebars, PARADOXICAL VALLEYS and SALT ANTICLINES. The **cliffs** of canyon country are like few elsewhere. They are usually sheerly vertical, hundreds of feet high and may meander around complex drainage lines for dozens, even hundreds of miles, virtually unbroken. In canyon country, such cliffs literally define its various distinct regions, districts, geographic areas and life communities.

The terms **island** and **peninsula** are used much as they are with land surrounded more or less by water, but in canyon country the terms mean high terrain surrounded by much lower terrain, with the two separated by sheer cliffs and steep slopes. The term **reef** can mean a sharp ridge of rock that is prominent but also not associated with water. For example, the sharply tilted rock strata of the eastern monocline of the San Rafael Swell are called the San Rafael Reef. The word **oasis** commonly means a desert spring. Read the sidebar, PETRIFIED OASES, to see how the term is used to define a unique canyon country land form. The Spanish word **playa** means beach, but geologists use the term to describe desert drylakes. See the sidebar, PETRIFIED PLAYAS.

The word **plateau** is used much like in other regions, but in canyon country it also has additional special meaning. The "Colorado Plateau" is an immense, unique geologic area that is elevated but, paradoxically, is surrounded by still higher mountains. It is also a **basin** in that it is essentially the lower terrain drained by the Colorado River and many of its tributaries. See the sidebar, COLORADO PLATEAU.

There are a few open areas that slope gently, and are generally set with massive outcrops of rock, but these are not **plains** as the term is used in the American Midwest. They are just small, relatively level areas to which the regional term **flat** is applied, as in the Sand Flats area near Moab. There are also **slopes** in canyon country, but few resemble the smooth, vegetated terrain of elsewhere. Rather, in canyon country slopes are great masses of rock fallen from still higher sheer cliffs, and are all too often barely negotiable on foot. Even the smoother slopes that occur in a few geologic strata are not just dirt covered with low-growing grasses and other vegetation, but embankments of barren shales or arid-land minerals broken by sandstone terraces that are too steep for safe walking.

There are also quite a few erosional landforms that have few counterparts elsewhere, at least in the shapes they take in canyon country, such as **mesas, buttes, canyons, gorges, paradoxical valleys, spires, monuments, arches, bridges, windows, tunnels, fins, towers, walls, terraces, balanced rocks, potholes, seep-caves, entrenched meanders, abandoned meanders, confluences, hanging valleys, pouroffs, jumps, grottoes, plunge-pools, wash bottoms, petrified dunes, portals, rincons, ripplerock, slickrock, spring-seeps, tunnels** and other less-used special terms.

As applied to canyon country erosional forms, **mesas, buttes, monuments, spires and towers** are generally defined by smooth, near-vertical sandstone walls. **Peaks** may mean the tops of mountains, but can also mean spires of rock, such as the Six-shooter Peaks. **Valleys** are not very wide and may also have steep or sheer walls. **Canyons** have steep or sheer walls too, but are narrower than valleys. **Gorges** are deeper and narrower yet, with largely sheer walls. **Paradoxical valleys** are salt anticline valleys that are crossed by rivers that emerge from one steep wall, cross the valley, then leave it by another gap in the cliff, which is known as a **portal.** **Hanging valleys** are valleys or canyons that end abruptly at a still lower cliffline, sometimes with their drainages going away from that cliff, rather than pouring over it into the lower terrain.

Terraces are generally stacked, receding ledges of thick sandstone that bear immense loads of boulders and slabs fallen from higher ledges, with the sum total being terribly steep slopes of broken rock. **Balanced rocks** are simply that, but are often so gigantic and precariously perched that they seem to defy gravity.

Potholes are deep depressions in sandstone that often hold water. These can range in size from a few inches deep and a foot or so wide to great cylindrical holes more than 50 feet wide and 30 feet or more deep. If such potholes are near a cliff-rim, water seeping from the potholes may form **seep-caves** in the cliff, either at its base or higher, depending on the nature of the rock. When such potholes eventually break through into their seep-caves, they form pothole arches.

Fins and **spires** in canyon country are generally smoothly eroded columns or slabs of sandstone that is mostly very uniform in color and texture. These may range from a few yards to hundreds of feet tall, and may be only a few feet thick. Fins sometimes erode into spires, and spires occasionally erode further into balanced rocks.

Arches, bridges, windows and tunnels are variously shaped natural holes in native rock. The sandstones that dominate canyon country readily form such natural rock openings, making the Colorado Plateau a never-ending land of discovery for those fascinated by such strange erosional phenomena. For more on this subject, refer to the book, ***Canyon Country* ARCHES & BRIDGES.**

Entrenched meanders, abandoned meanders, rincons and confluences are terms used to describe special features in river gorges. When a river or large stream crosses gently sloped terrain, it loops around, or meanders, traveling many miles to gain fewer miles of net progress. In time, such meanders cut deeper and deeper, that is, they become entrenched. When an entrenched river or stream eventually shortcuts one of its loops, that loop is called an abandoned meander. The word applied to such features by early Spanish explorers was rincon. There are many such features in canyon country, including a large one in the Colorado River gorge that can be viewed from two geology logs elsewhere in this book. A confluence is where two major rivers meet, or where smaller drainages meet a river.

Pouroffs and jumps are places in rock-bottomed canyons where their drainages descend step-fashion to a lower level. Such sudden drops in sandstone canyon floors may be steep, sheer or even undercut, and can range from a few feet to hundreds of feet high. They often have strangely carved rock **grottoes** in their lips, and immense **plunge-pools** at their bases. The floors of canyons or more open drainages that are normally dry are sometimes called **washbottoms,** especially if they are sandy.

Slickrock is another name for an appreciable exposure of bare rock. In canyon country, that is usually sandstone. Elsewhere it may be granite or some other durable rock. **Ripplerock** is rock that was once sand in a streambed that was rippled by flowing water, then buried under still more sand, which preserved the ripple patterns in the rock that formed over time. Such rock sometimes delaminates millions of years later, when exposed by erosion, revealing the original water-ripple pattern. In some geologic strata, there are occasional petrified animal tracks in such ripplerock, tracks left by animals wading in the shallow water.

These and other special terms are used throughout this book in the road and trail logs. In some cases, the above basic definitions of terms are amplified in the logs, or readers can gain understanding by studying the feature being described. Others terms are further defined in the sidebars that are referenced where appropriate throughout the book. The alphabetical index of sidebars at the end of the book tells where to find them.

COLORADO PLATEAU

The Colorado Plateau geologic province is a rough oval of continental terrain that is about 300,000 square miles in area. This oval is centered on a spot some 50 miles west of the Four Corners, where the states of Arizona, Colorado, New Mexico and Utah meet. The Colorado Plateau is the geologic and scenic heartland of these four states. Utah's spectacular canyon country, and the general Moab area, lie at the center of the Colorado Plateau.

The geology of the Plateau is so unique that the region contains nine national parks, sixteen national monuments, all or parts of three national recreation areas, numerous state parks, several public recreation lands (BLM), many prehistoric, historic and scenic highlights, and innumerable places of special interest to the general public, including wilderness areas and hundreds of miles of wild, scenic and recreational rivers such as the Colorado, Green, San Juan and Dolores.

There is disagreement among historians as to who originally gave the Colorado Plateau its name. An early U.S. Army Corps of Engineers explorer named Lt. George M. Wheeler claimed, in one of his many reports written in the 1870s and '80s, to have originated the term, but it appears in even earlier reports written by Dr. John S. Newberry, who was the physician and general scientist for expeditions that deeply penetrated the unexplored region in 1858 and 1859, and who wrote the official government reports for these expeditions. Newberry also wrote the first scientific descriptions of the Colorado Plateau.

Geologic provinces are defined as being areas of land with unique geologic features and erosional landforms, distinct geographic boundaries, and related internal geologic history. The Colorado Plateau has all of these characteristics.

It is also very old. As a distinct piece of relatively stable continental crust, the Colorado Plateau is more than 500 million years old. It has also accumulated several vertical miles of additional overburden in the form of land and marine deposits over the last half billion years, and it has survived at least two major phases of continental collision and separation, the first being when Pangea coalesced more than 200 million years ago, and the last being when the present North American continent finally parted from Europe and headed west.

The Colorado Plateau has also endured numerous distortions by continental drift tectonic activities, including extreme lateral compression, twisting and stretching, periods of internal and peripheral mountain-building and volcanic activity, and general uplifting from over-riding a detached piece of subducted oceanic plate.

These plate tectonic activities produced within the Plateau numerous enormous uplifts and anticlines, a number of large synclines, eight intrusive mountain ranges, regional tilting, and countless major faults and joints, yet left the general geologic integrity of the province barely affected. These same violent tectonic activities left the terrain surrounding the Colorado Plateau far more severely affected, with intense volcanic activity, mountain-range building and geologic strata highly distorted and folded, demonstrating the Plateau's exceptional durability.

At one point, the Plateau was subjected to so much lateral pressure from the southwest that it was pushed some 70 miles northward, shoving up an entire mountain range ahead of it -- the Uinta Mountains of northern Utah. Such lateral crustal pressures also affected several of the region's major uplifts, sharply tilting the strata in their eastern monoclines, but leaving their western monoclines with relatively gentle slopes. This difference in tilt is obvious from Interstate 70, as it cuts across the entire San Rafael Swell, and from Utah 95, as it crosses the Monument Uplift in southeastern Utah.

There is also evidence to support the idea that at one point the Colorado Plateau temporarily resisted a change in the direction and rate of travel of the surrounding North American continent, and that this resistance produced much of the violent tectonic activity in the borderlands between the two, and may have given the Plateau its roughly 6-degree stratigraphic tilt, downward to the northeast. This general slope in geologic strata is readily apparent many places within Utah, and is visible in the Moab area in the general vicinity of the Island-in-the-Sky peninsula and the adjacent terrain to the north. The slanted strata are also apparent to boaters going through Desolation and Gray canyons on the Green River. Despite this general stratigraphic tilt, however, the present erosional surface of the Plateau is approximately level, except for its major tectonic and erosional landforms.

Scientists have speculated as to the reason why the Colorado Plateau has remained stable over such a long period of time. They know that, as a block of continental crust, it is somewhat thicker and was a little cooler in the past than surrounding crust, but are not certain these factors fully explain its exceptional durability. The matter is still an unresolved mystery.

Over the life of the Plateau as a distinct block of continental crust, its magnetic orientation and its relationship with the Earth's equator changed constantly, as it drifted around the globe, captive within larger masses of land. Over the last half-billion years, the Plateau has crossed the equator at least twice, and it has been near-equatorial for all but the last 100 million years. This left the region exposed to intense heat and aridity and to violent equatorial storms for much of its stratigraphic past. Most of this time, the Plateau was also coastal, or even under shallow seas, with no land to the west to protect it from hurricanes and other types of tropical weather. See the sidebar, GREAT DESERT PERIOD.

continued ----

Geographically, the Colorado Plateau is not at present so much a plateau as an immense elevated basin -- the drainage basin for the Colorado River and most of its tributaries. For the most part, it is surrounded by still higher terrain. In the geologically recent past, as the entire western half of the North American continental plate was elevated as it rode over a detached block of oceanic plate, the region changed drastically. It evolved from a low-lying region accumulating geologic deposits, to a higher and more arid one, thus introducing the erosional phase that has prevailed for some 10 million years, but with the most severe erosion occurring in the last 5 million years. During this time of "unconformity in the making," about 5,000 feet of Plateau deposits have eroded away.

All of the tectonic activity that the Colorado Plateau has endured over the last half billion years has affected the way in which it erodes. Its mountainous borderlands and interior ranges squeeze water from passing clouds, and the resulting streams and rivers carry the silts toward the sea or, most recently, to reservoirs behind the many dams that impede the Colorado River's natural flow. Surface erosion from rain, melting snow and running water has cut deeply into the region's mountain flanks, uplifts, plateaus and anticlines, creating the spectacular system of canyons that give the region its name and exposing great cross-sections of its geologic strata. Glaciers have helped shape the slopes of the higher mountain ranges and plateaus.

As this erosion continues, the igneous cores of laccolithic mountains, the tilted sandstone strata of uplifts, anticlines and monoclines, and the shapes of massive overthrusts, synclines, rifts, joints, faults and meteor impact craters are exposed to view, leaving the region an open geologic "book" like no other in the world. The background chapters, the road and trail geology logs, and the special sidebars contained herein were all written to help the general public read that "book."

COLORADO PLATEAU GEOLOGIC PROVINCE

HOW TO USE THE ROAD LOGS

INTRODUCTION

This book was written for use by people exploring southeastern Utah, from a base in or near Moab, who are curious about the spectacular geologic panoramas, stratigraphic sequences, tectonic features and erosional landforms that are so prominent from every road and off-road vehicle trail in the region. The book's road logs were designed to be mile-by-mile guides to these obvious geologic phenomena, and to describe other related but less conspicuous aspects of the area's geologic strata and processes.

Southeastern Utah's landscapes are geology in the raw. Megayears of erosion have left the ancient rock strata of the region exposed to view, and this view remains largely unobscured by the dense vegetation that so often hides basic geologic strata and structures elsewhere. This is because the arid high-desert climate that prevails throughout canyon country up to about 8,000 feet elevation can be tolerated by only a few hardy trees, shrubs, grasses and annuals. In the higher mountainous areas, dense vegetation prevails because of increased precipitation.

The exposed nature of southeastern Utah's geology makes it relatively easy to study, even during leisurely drives along the region's highways and secondary roads. If enough such drives are taken using this book's road logs, after reading the previous chapter on regional geology, and while reading the referenced sidebars of related information, readers can obtain a good general grasp of the intricate geologic history of this unique region. It is possible, of course, to use the road logs without reading the chapter on regional geology or the referenced sidebars, but this will limit understanding of the road log descriptions.

Readers who find that the use of this book leads to a deeper interest in the intricacy and beauty of Colorado Plateau geology will find other books about regional geology in Moab-area book outlets. Descriptive guidebooks on other aspects of the canyon country scene are listed in the back of this book.

The following paragraphs describe the format of the road and off-road vehicle trail geology logs in this book, together with related information and practical suggestions for the use of the book.

ROAD LOG FORMAT

The logs used to describe the primary and secondary roads and a selection of off-road vehicle trails in this book are standardized in format to present information to readers in the most useful sequence. The various parts of the road log format are listed and described in the following paragraphs.

ROAD OR TRAIL NAME The official name or number of the road is given. With state and federal highways, the numbers given on official highway maps are used. In the case of secondary roads and off-road vehicle trails that have no official designations, the name assigned and standardized in the ***Canyon Country*** series of guidebooks and maps is given. In most cases, these unofficial names are in common use in the Moab vicinity, and in many cases have been placed on backcountry directional signs. In every case, the ORV trail names correspond to the names used in the guidebooks and maps that are recommended for use in conjunction with this book.

SUMMARY This format section provides a summary of the geologic scene along the indicated road or ORV trail, in order to help readers choose which to explore within the time available. In some cases, suggestions are made for ways to combine two or more roads or trails into one continuous trip.

ROAD SURFACE The general surface of the road or ORV trail is given in this section of the descriptive format. For more detailed information about conditions along specific ORV trails, readers should see the referenced guidebooks, where the same trails are described in greater detail for general exploration and recreational purposes.

VEHICLE Here, the types of vehicles that can safely travel the described road or ORV trail under good weather conditions are noted. In a few cases, parts of certain ORV trails can be traveled by carefully driven low-geared, high-clearance highway vehicles such as pickup trucks and vans, but such trails can be traveled for their full lengths only by such off-road vehicles as mountain bikes, dune buggies, ATVs, dirt bikes and four-wheel drive cars designed for backcountry travel. Most of the described ORV trails should not be attempted by low-clearance highway vehicles that have four-wheel drive for use under adverse highway conditions.

MAPS This section lists maps that are useful or essential for traveling the described road or ORV trail, although in the case of numbered federal or state highways, standard highway maps are suitable. On the secondary roads and ORV trails, however, the listed maps are essential for navigation, even though a navigational summary is given in each log. Readers who plan to use the logs in this book that describe either secondary two-wheel drive roads or ORV trails are advised to acquire and use the map or maps listed, in order to assure good navigation. With ORV trails and secondary roads, the listed maps and guidebooks should be considered necessary supplements to this book.

GUIDEBOOKS This section of the geology log lists the books that match the referenced maps, as well as other books that contain useful supplementary information. As with the listed maps, the small companion guidebooks noted should be considered as essential for the safe navigation of some secondary roads and all ORV trails. With one minor exception, the road and ORV trail names used in the referenced book-map pairs are identical to those used in this book. The referenced maps and guidebooks can be obtained from many commercial outlets in southeastern Utah, ordered through most book stores anywhere, or purchased by mail from the publisher of this book.

MILEAGES

To Start Here, the distance from the center of Moab to the beginning of the road or ORV trail is given in miles.

Road or Trail This gives the actual one-way length of the road or ORV trail as it is described.

Return This provides the total return distance, assuming a return via the same route, with a few exceptions. In some cases other return routes are suggested in later sections of the log format.

Total The total round-trip mileage to Moab is given, assuming a return trip via the indicated route, with no exploration of other roads or ORV trails, although such supplementary travel is sometimes suggested in the summary or later in the log.

TIME Here, an estimate of the time required for round trip travel on the described road or ORV trail is given as an aid to trip planning, although in some cases suggestions are given elsewhere in the format for loop routes or supplementary travel that will take longer. Since the amount of time travelers will spend along each road or ORV trail will vary considerably, the times given are only general estimates.

ROUTE A brief summary of the general route to be traveled for the minimal exploration of the described road or ORV trail is given here, although this will not be adequate if a day's journey also includes other roads, trails or side trips.

NAVIGATION More detailed navigational instructions are given here, beginning at Main and Center streets in Moab, but these instructions are all that is needed only with primary highways and a few of the secondary roads. With other canyon country secondary roads and all ORV trails, the referenced maps and their companion guidebooks are essential for trouble-free navigation.

ROAD AND ORV TRAIL LOGS Each geology road and ORV trail log begins at mile 0.0 of the road or trail, and ends either where the road or trail dead-ends or where by definition it ends by joining another road or trail. Each log begins with a description of the geologic scene viewed from where the road or trail starts. Each log also has numerous stops along the way designated by mileages from the start. Some of these stops are at developed pullouts, some are at wide shoulders where it is safe to stop for a short time, with descriptions of the geologic scene visible from such stops. Some logs also indicate approximate mileages where particular geologic phenomena are visible along a stretch of road or trail without stopping. Almost all logs refer to sidebars of information in this book that provide more details on specific subjects that are not fully discussed in the log. Such sidebars can be found by looking up their titles in the alphabetical sidebar index in the back of this book.

Tukuhnikivista Arch

PRACTICAL USE OF THE ROAD LOGS

The mileages given in the road logs were taken from the author's trip meter, which was set to zero at the beginning of each described road or trail. While the trip meter in the author's vehicle was not officially calibrated, it was checked against the mileage posts along U.S. 191 north of Moab and found to be quite close to these. There is no guarantee, of course, that the mileage posts were placed by a crew driving a vehicle with a carefully calibrated odometer.

Most highway vehicles have odometers that are not highly accurate, and that are rarely if ever checked for accuracy. Further, even a precision odometer or trip meter will vary as a vehicle's tires wear, with different tire pressures, or when a different size or brand of tire is installed. Thus, those using this book can expect their trip meters or odometers to vary somewhat from the one used by the author while doing the field research for this book.

To help compensate for these anticipated mismatches in mileages, landmarks at or near many stops are also given in the road and trail logs. Once a traveler determines how his or her vehicle's meter compares with that used by the author, it should be fairly easy to compensate by adding or subtracting mileage increments along the way, and recalibrating at stops made at unmistakable landmarks such as road junctions.

The author has made every reasonable effort to make the logs in this book as accurate as practically feasible, given the differences in the mileage meters in highway vehicles. In addition to the many times they had been traveled earlier while researching other books and maps in the ***Canyon Country*** series, each road and ORV trail described in this book was also driven at least twice by the author while writing this book. The geology log was recorded with a laptop computer on a first trip. That log was then edited, augmented, printed out and later verified and refined during a second trip.

To use the geology road logs, drive to the beginning of a road or ORV trail, using the appropriate map and the navigational guidance given in the log. There, either set the vehicle's trip meter to zero and follow the log's mileages, or note the vehicle's starting odometer mileage and add the log's mileages to that along the way. The landmarks mentioned in many descriptions will aid in finding the stops where the descriptions should be read. With the few roads that are exceptionally long, mileage differences will accumulate, but the landmarks given should still help relate the various scenes to the log descriptions.

Wherever it is feasible, it is suggested that a vehicle's driver keep track of mileages and find the stops described, while someone else reads aloud the text that goes with each stop or stretch of road. Such a cooperative approach will allow drivers to give the necessary attention to safe driving while still relating the log's descriptions to the terrain in view. Most such descriptions were made from pullouts or wide shoulders in order for the driver to safely stop and study the scene being described.

GEOLOGIC STRATA VISIBLE FROM THIS ROAD

Formation	Member	Page
Mancos Shale	Ferron Sandstone	Page 20
Dakota Sandstone		Page 20
Burro Canyon/Cedar Mountain Formations		Page 19
Morrison Formation	Brushy Basin	Page 18
	Salt Wash	
	Tidwell	
Entrada Sandstone	Moab	Page 16
	Slickrock	
	Dewey Bridge	
Navajo Sandstone		Page 13
Kayenta Sandstone		Page 13
Wingate Sandstone		Page 12
Chinle Formation		Page 12
Moenkopi Formation		Page 12
Cutler Group	White Rim	Page 11
	Organ Rock	
	Cedar Mesa	
	Lower Cutler	

ROAD NAME: **ARCHES ROAD**

SUMMARY: This road is the main route into Arches National Park. It offers outstanding views of a broad spectrum of geologic strata as they have been tilted, distorted and stratigraphically displaced by the uplifting and faulting associated with a salt anticline. In addition, several natural arches are visible from the road, and even more are accessible by short hikes. The park has a wide variety of exceptionally beautiful erosional landforms, especially in the dominant red-hued Entrada Sandstone. The road first climbs to higher terrain, then descends to travel through a broad valley formed by a large syncline. It then climbs again to high terrain that affords outstanding panoramic views in all directions before descending again, this time to cross the lower end of the Salt Valley anticline. From there, it travels near the spectacular sandstone fins of the Fiery Furnace. The road and its various spurs provide access to a number of excellent short hikes that go to still other unique geologic features.

ROAD SURFACE: paved.

VEHICLE: any vehicle.

MAP: ***Canyon Country* OFF-ROAD VEHICLE TRAIL MAP - Arches & La Sals Areas.**

GUIDEBOOK: ***Canyon Country* OFF-ROAD VEHICLE TRAILS - Arches & La Sals Areas.**

MILEAGES:

TO START: 5.2 miles, from Main and Center, Moab.
ROAD: 30.8 miles, one-way to end, as described.
RETURN: 30.5 miles, from end of described road.
TOTAL: 66.5 miles, from Main and Center, round trip.

TIME: one-half day, plus time taken for hikes along the way. Allow a full day for a good look at the geology along this road and its described spurs, plus a sampling of short hikes to various viewpoints and arches.

ROUTE: U.S. 191 north, into Arches National Park on the Arches Road to its end, driving the major scenic loops and spur roads along the way.

NAVIGATION: drive north from Main and Center in Moab for the indicated distance, then turn onto the road into the park. Stop at the visitor center, then continue into the park, pulling into the various described spurs and viewpoint loops and taking the side roads into the Windows Section and Delicate Arch hiking trailhead and viewpoint. From the end of the main road in Devils Garden, drive directly back to Moab via the shortest route or, optionally, explore the park further on foot from various pullouts.

GEOLOGY ROAD LOG:

0.0 The Arches Road begins here, at the entrance to Arches National Park. The white rock ahead at road level is Navajo Sandstone. The road switchbacks above the park visitor center at the interface between the Navajo Sandstone and the Dewey Bridge Member of Entrada Sandstone. The sheer cliffs above that are the Slickrock Member of the Entrada. The gray rubble to the west of the highway is Upper Hermosa Group fossiliferous shallow-marine deposits.

0.2 Entrance Station - pay the required fee and turn right to the visitor center. After stopping there and studying the displays, return to the Arches Road.

0.3 Back at the main road, turn right.

0.9 Stop here at a paved pullout that offers a view south into Moab Valley. To the west of the main highway, the red sand dunes are erosional products left there by the prevailing winds that funnel through the narrow gorge. The high cliffs to the west of the highway are Wingate Sandstone topped by remnants of layered Kayenta Sandstone. The Navajo Sandstone on which the Arches Road is traveling is out of sight above the Wingate cliffs because of the Moab Fault offset. Estimates of that offset range from 1700 to 2600 feet.

The spires, columns, and vertical faces in the Entrada Sandstone above the Arches Road are typical erosional phenomena of this sandstone. The smooth-walled cliffs and spires are Slickrock-Entrada and the darker layered rock is the Dewey Bridge-Entrada. Note also the arch-topped alcoves in the vertical walls. Both members of the Entrada visible here are prone to form natural rock openings -- arches, bridges, windows and tunnels -- because of this erosional mode, especially at the interface between them.

1.3 From a pullout on the right, note the large arc of rock in the cliff to the east, and the loose boulders at the base of the cliff. Note also, at the right end of the arc in the cliff, an angular fault which offsets the strata a few feet. The vertical scoring in the white rock beside the road was caused by drilling the Navajo Sandstone in preparation for blasting operations during road construction.

1.6 From a paved pullout on the right, step out of the car and study the Park Service display. Note where the main highway below curves to the north, closely paralleling the Moab Fault. Note also the great amount of dune sand on each side of the railroad track across the canyon. The horizontal strata immediately above the upper Wingate cliffs to the west is Kayenta Sandstone. The slopes below the cliffs are Chinle Formation. Below that, mostly obscured by rubble fallen from above, are Moenkopi and Cutler deposits.

1.7 The road tops out here, with Navajo Sandstone on both sides and red Entrada sandstones above that on the left. For the next mile or so, note the massive amounts of rubble along the cliff bases that has accumulated over thousands of years as the cliffs have slowly eroded, collapsed and receded.

2.6 Turn left to the Park Avenue viewpoint and hiking trailhead. From there, take the short walk to the viewpoint. An optional hike descends through the sheer-walled valley to a pick-up point beside the main road at the lower end. In Park Avenue, the two Entrada Sandstone members have eroded into majestic walls, spires, alcoves and balanced rocks. After returning from the viewpoint, drive back to the main road and turn left. The brushy terrain to the right of the highway for the next mile or so is dune sand held in place by native vegetation.

3.0 Turn right on a scenic loop to the La Sal Mountains viewpoint. From the end of the loop, the panorama begins with Entrada Sandstone cliffs on the left, then detached buttes of Entrada Sandstone. The pale rock in the near-distance of the complex panorama is Navajo Sandstone. In the foreground, the reddish brushy soils are thin layers of decomposed Entrada Sandstone and dune sand held in place by vegetation. On the horizon, from left to right, the Balanced Rock and the other red sandstone masses in the Windows Section of the park are also Entrada Sandstone, with South Window visible. The tree-covered slopes to the left of the mountains are largely Navajo and Kayenta sandstones. Below that, the upper rim of the Colorado River gorge is visible.

The La Sal Mountains, which are intrusive igneous in nature, loom above the clifflines of Wilson Mesa and South Mesa. The rims of these broad benchlands are the Moab Member of Entrada Sandstone. The steep cliffs and slopes below that are Slickrock-Entrada, partially covered with rocky debris. The Dewey Bridge-Entrada is exposed beneath the rubble only a few places. The lighter rock below that is Navajo Sandstone. In the middle distance, the crenelated red sandstone wall is Dewey Bridge-Entrada. After studying the described scene and the Park Service interpretive display, return to the main road and turn right.

3.3 For the next several miles, the main road travels more or less at the interface between the Navajo and Entrada, with the sheer red cliffs being Slickrock-Entrada standing on bases of layered Dewey Bridge-Entrada.

4.3 Pull into the parking area and viewpoint to the right and study the Park Service display. This pullout is for picking up hikers who have descended through Park Avenue. Note Sheep Rock ahead, and an arch to the left. The tall monoliths in this area are all Entrada Sandstone. Again, in the open country beyond Sheep Rock, sparse vegetation is holding dune sand and decomposed rock.

5.3 Note that the road has been descending into a broad valley. This depression is the surface evidence of Courthouse Syncline, a gigantic sag in the rock strata that created the valley and much of the miles-long Courthouse Wash drainage that cuts through the southwest corner of the park. The syncline also continues southeast, roughly parallel to Moab Valley, creating another sag as it crosses the Sand Flats area east of Moab.

5.5 Note the arch on the skyline to the left.

6.8 From a pullout on the right, note that the cliffline to the left is entirely Slickrock-Entrada, with the deposits between the road and the cliff being decomposed, brush-covered Dewey Bridge-Entrada. Slickrock-Entrada erodes into rounded slopes when not undercut, but sheers away to form vertical cliffs when its base is undercut. Note also the sand dune on both sides of the road, and at the base of the cliffline to the left where the wind has piled it. The rounded tops of the Navajo Sandstone to the right are sometimes called "petrified dunes." Navajo Sandstone is desert aeolian dune-sand rock, but the "petrified dunes" do not correspond to the shapes of original sand dunes.

7.8 From a narrow pullout on the right, note the arch in the cliffline to the left, and the extreme distortion of the Dewey Bridge-Entrada to the left-front. Geologists believe that this "wrinkling" was caused by the general terrain tilting after the Dewey Bridge sediments had been covered deeply by younger deposits, but before the deposits had turned to rock, thus causing the Dewey Bridge to wrinkle. This tilting was probably associated with one or more of the nearby salt anticlines.

8.1 From a paved pullout on the right, note the several free-standing towers to the left and ahead. The upper tips of these towers are Slickrock-Entrada, with the main towers being Dewey Bridge-Entrada. Note that, although the road has been climbing steeply, the Dewey Bridge-Entrada is now exposed below the Slickrock Entrada, where it was hidden earlier, indicating that the strata are climbing more steeply than the road on this side of the Courthouse Syncline.

8.6 Note the vegetated seep-caves in the left cliffline.

8.8 From the paved pullout on the right, note that the Slickrock-Entrada to the left of the road has eroded almost completely away, with only small amounts of the Dewey Bridge-Entrada below it still showing.

9.4 From a paved pullout on the right, note that the skyline to the right is dominated by Dewey Bridge and Slickrock-Entrada, with the Balanced Rock to the left and several arches in the Windows Section visible to the right. Toward the mountains, several picturesque detached towers, some with remnants of Slickrock-Entrada on top and Dewey Bridge in the middle, all stand on bases of white Navajo Sandstone.

9.5 Elevation 5000 feet. Note that the Arches Road started at the interface of the Navajo and Entrada sandstones at elevation 4080 and is still in that interface, indicating that these strata and the road are 920 feet higher here than at the park entrance.

9.8 Park at the Balanced Rock pullout. The Balanced Rock is Slickrock-Entrada, standing on a base of Dewey Bridge-Entrada. This also applies to the massive towers on either side of it. Across the road, the rounded white rock in the picnic area is the top of the Navajo Sandstone. The graded dirt road here is the beginning of the Eye-of-the-Whale off-road vehicle trail. The branching Willow Spring Trail was once the main route into the park.

10.0 Turn right on the Windows Section spur road. The bluff straight ahead is Entrada Sandstone on a base of Navajo, with the slopes to the right being dune-sand held in place by vegetation. The lower slopes to the left that drain toward Salt Wash Canyon are decomposing Navajo Sandstone.

11.2 Turn left on the Garden of Eden viewpoint loop. From the first pullout on this loop, the fins and towers of red sandstone to the south are Entrada Sandstone standing on bases of lighter-hued Navajo Sandstone. The lower parts of the fins and towers are Dewey Bridge-Entrada while the smoother, upper parts are Slickrock-Entrada.

11.3 Park where the scenic loop ends. The view looks down into lower Cache Valley, with the Fiery Furnace to the left. Note the tilted strata in the distance, to the northeast. The Book Cliffs are on the skyline to the north. Start here for an interesting hike to the east of the large slickrock mass that extends south from here. This is the route to delicate and lovely Ribbon Arch.

11.5 Back at the Windows spur road, turn left. A few yards beyond this junction, watch for large arch openings that appear ahead in the Entrada Sandstone of the Windows Section. Turret Arch is silhouetted against the distant La Sals. Columns and spires of red-hued Entrada Sandstone perched on white Navajo Sandstone bases appear near the road as it descends toward the Windows Section parking area.

12.8 Park at the Windows Section parking area for short hikes to North and South Windows and Turret Arch. All of the arches in this vicinity are in Entrada Sandstone, with most of them developed at the interface between the Slickrock-Entrada and Dewey Bridge-Entrada.

13.0 Park for the hike to Double Arch and others in the vicinity. Note the deep caves in this area that were formed by seeping water. This process helps form arches in the Entrada. Contrary to a popular myth, wind has very little to do with forming natural arches and bridges. Water is the architect. For more information about how natural arches, bridges, windows and tunnels form, refer to the book, ***Canyon Country* ARCHES & BRIDGES.**

14.8 Stop at the Pothole Arch Viewpoint. The arch is in Slickrock Entrada. It was so named because it was formed by a pothole that eventually broke through into its seep-cave.

15.3 Turn right on the main Arches Road.

16.4 Turn right toward the Panorama Point overlook. Park at the end of the loop and study the panorama. Salt Valley to the north is the highly eroded surface remnant of the Salt Valley anticline. See the sidebar, SALT ANTICLINES. Salt Valley is bounded on each side by the tilted strata of the anticline rim. The eastern rim, with its countless spires of red Entrada Sandstone, is called the Fiery Furnace. It was given this name because of the bright reddish reflectivity of sunlight in the rock maze, rather than in reference to excessive temperatures. The white rock that tops some of the Fiery Furnace fins is the Moab Member of Entrada Sandstone. The similar but fewer fins in the distance, in the left rim of Salt Valley, are called Klondike Bluffs. The tall, slender spire opposite the bluffs is called the Dark Angel because its surface is darkened with desert varnish. See the sidebar, DESERT VARNISH.

Beyond the Fiery Furnace, the distant cliffline is the Book Cliffs, which lie north of Interstate 70. To the east beyond the gray slopes of rubble, sloping Entrada Sandstone forms the cliffs of Salt Wash. Farther to the right, the small peaks of sandstone on top of sloping Slickrock-Entrada are in the Delicate Arch area. One of them is the arch, viewed edgewise. Beyond there, to the right in the distance, the upper slopes of the Uncompahgre Uplift are silhouetted on the skyline. The broken cliffs to the right of that are Dome Plateau and the cliffs above Fisher, Professor and Castle valleys, beyond the Colorado River. The La Sal Mountains loom against the sky above these.

16.8 Turn right on the Arches Road. From this junction, the road descends steeply into Salt Valley. As it descends, the eroding rock to the right of the road is Wingate Sandstone that has been severely fractured and tilted by the faulting associated with the Salt Valley anticline.

17.3 A paved pullout on the right offers a good photo viewpoint for Cache Valley straight ahead, and the Fiery Furnace to the north. Cache Valley was allegedly named by early settlers who hauled supplies from the railhead at Thompson Springs down Salt Valley, then to the end of this valley. There, they cached the wagons and packed the supplies over the high terrain via a steep, narrow trail, then forded the Colorado River to reach their ranches in that vicinity. Note how the geologic strata in the lower end of Salt Valley have been tilted, distorted, offset and fractured by the anticline. The gray

continued on page 44

SALT ANTICLINES

The unique sequence of geologic events that created the salt anticlines in the heart of the Colorado Plateau began long ago, some 320 million years in the past.

There was no North America then, and the supercontinent, Pangea, had not yet even formed. The large masses of continental crust that were later to become North America and Europe were slowly drifting northward across the ancestral Pacific Ocean, with the captive Colorado Plateau geologic province positioned on the equator. The entire region was under shallow seas, and had been for millions of years, with hundreds of feet of marine deposits already accumulated since dry land had last been exposed.

This very long dominance of the region by tropical seas began to change slowly when a state-sized block of continental crust began to sag downward beneath part of the Colorado Plateau, centered about where the La Sal Mountains are now. This sagging was probably caused by some obscure continental drift process that is still a mystery to geologists, but the long-lasting results of that sinking of the ocean floor have been studied intently over the last six decades because understanding them could produce clues to petroleum resources.

As the ocean floor sagged downward, there was a corresponding rise around the perimeter of the sea-floor depression. Uplifts now known as the Front Range, Uinta, Emery, Kaibab and Zuni-Defiance slowly rose to just above sea level, isolating an immense basin from the surrounding sea, a basin that covered what is now most of eastern Utah, western Colorado and the adjacent corners of Arizona and New Mexico. The hot, tropical sun then began to evaporate the trapped sea water.

As this process continued, and the water level in the basin dropped, various chemicals that were dissolved in the sea water began to precipitate in layers on the basin floor. In sequence, these were carbonates, salt, anhydrite, dolomite and limestone, with salt comprising from 74% to 90% of the total deposit. In some areas, potash was deposited along with the salt. Both eventually became useful products that are presently being mined in canyon country. See the sidebar, POTASH. The carbonates tended to be deposited in the shoreline shallows.

The last deposit in the sequence was a black shale that remained after the water level in the basin evaporated to the point where the entire low-lying area became a vast saltwater marsh fed by small amounts of sediments and fresh water from the surrounding land. The decay of all this swamp vegetation and its associated ecosystem produced the black shale, which eventually transformed into the petroleum products that are occasionally found in these ancient marine deposits.

Over subsequent millennia, the basin continued to sink, and the isolating barrier of ridges continued to rise, but at intervals sea water again flooded the basin, probably because the global sea level was rising and falling. Each time the basin was again flooded, a new cycle of evaporation began. Studies have shown that the flooding probably came through two wide but shallow inlets, plus intermittent flow over low reefs, with some of this probably due to tidal surges caused by the periodic tropical storms that the region endured.

Altogether, 29 complete cycles of flooding-and-evaporation in this immense land-locked basin have been measured, although the total amount of flooding varied with each cycle, and some cycles were incomplete. There are indications of 5 more cycles in what is now the deepest part of the basin, thousands of feet below ground level and difficult to study. The evaporation-cycle deposits originally accumulated to a depth of from 5 to 6 thousand feet, but concurrent and subsequent events obscured that otherwise straightforward deposition record.

The same massive faulting in the deep, underlying rock strata that formed the immense basin also produced another effect about two-thirds of the way through the series of evaporation cycles. Another uplift ridge divided the basin into two parts. This ridge was the first appearance of the Uncompahgre Uplift, and the two land-locked dead seas it created are now known as Paradox Basin and Maroon Basin. The Paradox Basin measures about 80 by 200 miles, but is actually a large, irregular oval in shape.

The first appearance of the Uncompahgre Uplift barely divided the two basins. There may have been some remaining connection between them over subsequent millennia. Even so, the nature of the deposits in the two halves of the original basin changed from then on. Since Maroon Basin is not within the region covered by this book, and was not involved in the creation of the salt anticlines in the Moab area, the rest of this description applies to the Paradox Basin.

This first rise of the Uncompahgre barely lifted it above sea level, and thus the new terrain contributed very little sedimentary material to the deposits of the continuing evaporation cycles. Eventually the uplift was dominated by the cyclic sea-water flooding, temporarily reuniting the Paradox and Maroon basins. After several more such evaporation cycles, however, the Uncompahgre rose again, this time to a considerable height, permanently dividing the original basin.

This second rise of the Uncompahgre triggered a new series of geologic events, although three or four more evaporation cycles did occur before the Paradox Basin was filled by freshwater sediments from the Uncompahgre highlands. As the Uncompahgre slowly rose, the adjacent part of the Paradox Basin sagged to even greater depths. The transition period brought a mixture of land sediments and marine deposits. These are now known as the upper Hermosa, earlier called the Honaker Trail Formation.

During this and subsequent periods, the tectonic forces that formed the original landlocked sea and thrust up the Uncompahgre were also having other effects on the cyclic marine deposits. The growth of the uplift, plus other forces that are still not completely understood, were creating lateral pressures that caused the deep evaporite deposits to buckle in long ridges, with the ridges parallel to the axis of the Uncompahgre Uplift. These ridges were probably related to parallel faulting in the deep, underlying "basement rock," and were the first manifestations of this region's salt anticlines.

Eventually, the sinking of the Paradox Basin ceased, the uplifts around the basin's margin stopped rising, and deposits from the Uncompahgre filled and covered the basin, leaving its margins exposed to occasional invasions from the seas beyond the basin. These predominantly land deposits, shoreline minerals and sediments, and tidal plain dunes, mixed with intermittent, thin layers of marine deposits, are now called the Cutler Group.

As the upper Hermosa and Cutler deposits accumulated, the weight of this immense overburden caused the Paradox Basin to sink further, and the Paradox salts to begin the slow "plastic flow" that is

continued ----

typical of such thick salt deposits when placed under pressure. The flow was upward along the same series of parallel ridges formed earlier, and was probably augmented by still more shifting of the underlying basement rock faults and the rising Uncompahgre. The present rows of parallel salt anticlines in eastern Utah and western Colorado were being created.

The Paradox Basin continued to sink and the Uncompahgre Uplift continued to rise fairly steadily through the periods of Hermosa and Cutler deposition. The Uncompahgre, however, continued to rise intermittently even into the geologic present. See the sidebar, UNCOMPAHGRE UPLIFT.

The parallel ridges of the various salt anticlines also continued to rise, slowly for the most part, but sometimes so rapidly that sediments accumulating on their steep slopes slid downward. As these ridges of ancient marine salts rose, newer deposits ended on their flanks, or covered them only thinly. Geologists can determine when the salt anticlines were active by the ages of the younger deposits that thinned or were interrupted in their vicinity. When the anticlines rose more rapidly, pushing the overlying deposits above the level of the surrounding terrain, the crests of the anticlines eroded, leaving "angular unconformities" below subsequent deposits that geologists could also use as time indicators.

As the salt ridges rose, adjacent areas sank, as synclines. This allowed these parallel depressions to accumulate thicker deposits, which in turn promoted flow of the salts beneath them into the adjacent anticlines. Eventually, the Paradox salts near the rising anticlines were largely squeezed into these ridges by the weight of thousands of feet of accumulated overburden. With this, the salt anticlines phased into another major stage, during which they remained stable and the deposits that accumulated above them were relatively uniform. This stage was reached about the time that Morrison Formation sediments began to accumulate, and continued through the Mesaverde Group deposition period.

Thus, the salt anticlines remained fairly stable for a time, with only occasional periods of activity. Continental drift events triggered a period of violent tectonic action along the western boundary of the Colorado Plateau about 100 million years ago. Known as the "Sevier Orogeny" it had little direct effect on the Plateau's interior. A subsequent period of tectonic activity called the "Laramide Orogeny" did affect canyon country, however, although not severely at first. See the sidebar, LARAMIDE OROGENY.

In mid-Laramide, at least two meteors hit the planet. The larger one impacted in the Gulf of Mexico and caused major changes in global climate. See the sidebar, GREAT EXTINCTIONS. The smaller one hit the Colorado Plateau and may have triggered additional tectonic activity in the canyon country region. See the sidebar, UPHEAVAL DOME.

Whatever the cause, the basement rock faults that had first created ridges in the ancient Paradox salt again grew active, forcing corresponding activity in the overlying salt anticlines. The Uncompahgre Uplift thrust upward again, after having been covered by Morrison and later deposits. Other old regional uplifts were reactivated and new ones began. Still later, the rise of the La Sal Mountains directly through the center of the salt anticlines triggered considerable localized flow of salts within the anticlines.

At the peak of their rising, the Moab area salt anticlines were underground ridges of Paradox salts that were as much as 17,000 feet above their original marine limestone base. Drill holes in the Paradox Valley salt anticline, near the deepest part of the Paradox Basin, have penetrated 14,000 feet of salts, and geologists believe an additional 3,000 feet have already been removed from that anticline by erosion. Farther away from the Uncompahgre, the salt anticline ridges are not as deep because the bottom of the Paradox Basin slopes steeply upward.

As an historical footnote, Paradox Valley was named in 1875 by a member of the Hayden Survey. The name was chosen because the Dolores River "paradoxically" entered the valley through a gap in the cliffline, then left through another such gap in the opposite cliff. Most rivers flow lengthwise through their valleys. See the sidebar, PARADOXICAL VALLEYS.

Then, about 24 million years ago, the three intrusive-igneous stocks of the La Sal Mountains thrust upward. See the sidebar, LACCOLITHIC MOUNTAINS. Before then, several of the present salt anticline valleys were linked by their fault systems. Castle and Paradox valleys were linked in a straight line. Salt, Cache, Fisher and Sinbad valleys were connected by curving fault lines. Moab and Gypsum valleys were aligned, and the Cane Creek and Lisbon anticlines were associated. Pine Ridge, on the south side of the La Sal Mountains, is an uncollapsed portion of the crest of the continuous Moab-Gypsum anticline. Utah 46, described in this book, crosses Pine Ridge.

One stock of the La Sals separated the Moab and Gypsum anticlines, while another wedged in between the Castle Valley and Paradox Valley anticlines. In each case, the hot, igneous intruding material penetrated the salt anticline ridges and forced the salts there to flow again, creating new zones of faulting, uplifting and sagging. Some of this localized flow continues even today, as the plastic underlying salts adjust to the changes in overburden triggered by the mountain-building, and erosion selectively removes the disturbed geologic deposits.

During the long period of time in which the salt anticlines were rising, dormant, being thrust upward again, and then distorted by intruding mountains, some erosion was taking place, especially during the periods marked by erosional unconformities between major geologic deposits. Each such period brought some erosion to the crests of the various salt anticlines, as the higher, fractured rock was carried away by surface water. Subsurface water also played a part, as it dissolved the readily soluble salt in the uppermost layers of the anticlines and carried it back toward the sea in local streams and rivers.

As the salt was dissolved and carried away, the remaining less-soluble "caprock" minerals -- primarily gypsum (created by hydration from the anhydrite), shale and limestone -- slowly sank, carrying with them the fractured overlying strata in the anticline summits to form "grabens," or sunken areas between rock walls. These grabens eventually became the present salt anticline valleys, but only after another fairly recent tectonic event -- the raising of the entire region between the Sierra and Rocky mountains by another major continental drift event -- and a major period of erosion brought on by a prolonged ice age.

Some geologists believe this continental elevation of from 7 to 10 thousand feet was caused by the smaller Farallon Plate to the west being forced beneath the North American Plate by the Pacific Plate. Others think the rise was due to the following impact of the two big plates, or to a change in direction of travel of the North American plate that occurred about then.

Regardless of its cause, the rise in elevation of the Colorado Plateau, together with another rise of the Uncompahgre Uplift, triggered the current phase of salt anticline valley formation. The increased regional elevation brought with it a change of climate, with increased precipitation in the adjacent higher terrain. That in turn caused rivers and streams to cut across the general area above the ancient Paradox Basin, beginning the creation

continued ----

of the present extensive canyon system. The Colorado and Dolores rivers acted as knives, cutting downward through the Moab, Paradox and Gypsum valley anticlines. See the sidebars, ENTRENCHED MEANDERS and TIME TRAVEL TOPOGRAPHY.

Other nearby anticlines eroded more slowly, with the biggest variable being the amount of water flowing through them and in what manner. The region's water flow varied drastically with the coming and going of a series of glacial episodes. See the sidebar, GLACIERS OF THE LA SAL MOUNTAINS.

As increased amounts of water seeped or cut downward across the crests of the various salt anticline ridges, the complex process of salt valley formation continued at an accelerated rate. As salt was carried away, the anticline caprocks sank, adjacent rock strata jointed and bent downward, valley walls slumped, subterranean "canyons" developed in the salt strata, and broken rock and sediments sagged to fill them. Lateral drainages that flowed in both directions from the crests of some anticlines, and tributary drainages that developed along some valley walls, further complicated the valley-forming processes.

The parallel jointing in some salt valley walls in turn weathered into mazes of sandstone fins, such as the Navajo Sandstone fins that wall Moab Valley, and the Entrada Sandstone fins that make up both the Fiery Furnace and Klondike Bluffs areas adjacent to Salt Valley in Arches National Park.

As a result of this complex set of factors, plus many others too numerous to mention in this summary, each salt anticline valley developed in its own manner and produced its own unique configuration, but the major factor in all cases was water -- how much, when and where -- because it was water that dissolved and carried away the salt component of the salt anticline summits. Without this, the several spectacular salt anticline valleys that are such prominent features of canyon country would not exist.

Today, the less soluble caprock minerals and sediments of several salt anticlines are exposed in the bottoms of their valleys. Such exposures are noted in three road logs and one trail log in this book. Paradox salts are visible from U.S. 191 North, Cane Creek Road, Utah 46 - Colorado 90, and the Onion Creek - Thompson Canyon - Polar Mesa ORV loop route. This route provides an outstanding view of exposed Paradox salts that have been highly distorted by more than 300 megayears of salt-flow and tectonic activity.

1

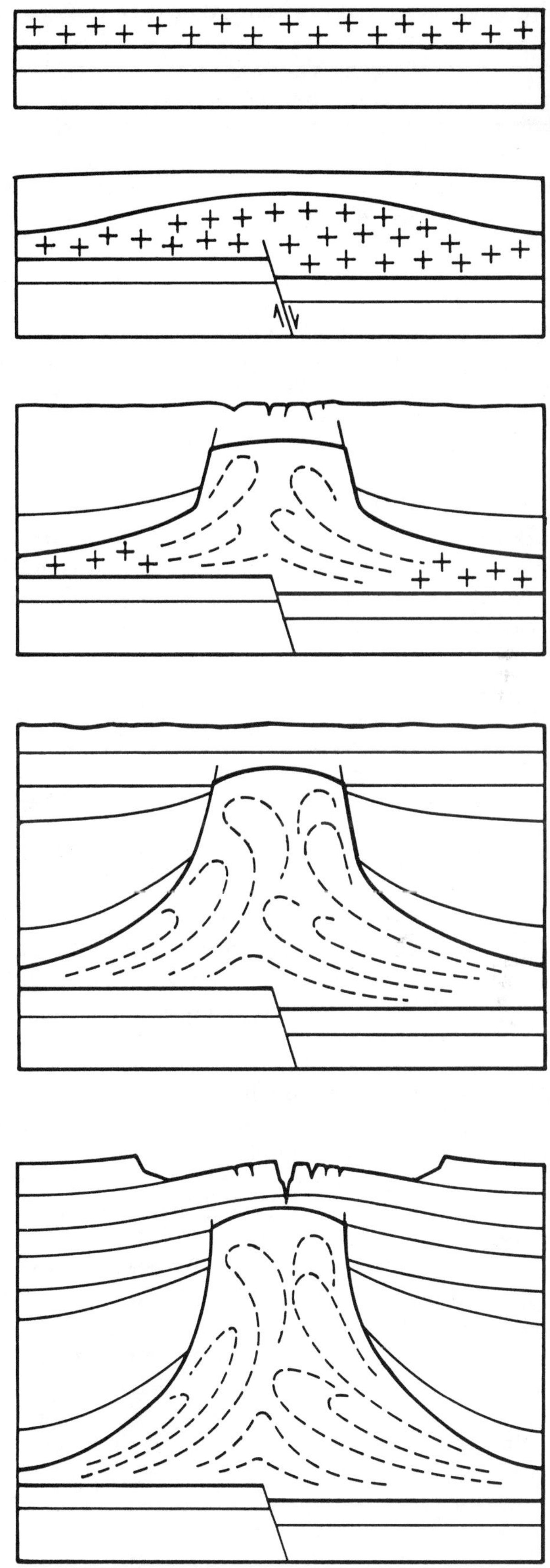

2

3

4

6

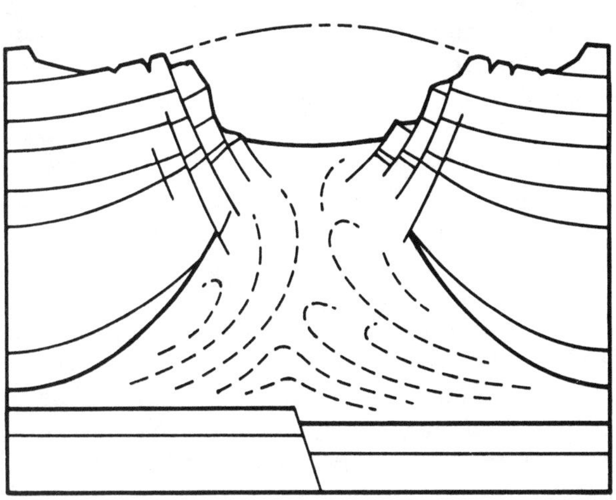

5

Sequence showing typical salt anticline and salt valley formation

deposits are actually much younger than the red and white sandstones that are above them many places.

18.1 Stop at Cache Valley Overlook. From here, the red sandstone peaks that were mentioned earlier are silhouetted on the skyline to the right. The nearest one is Delicate Arch edgewise, with its opening barely discernible. Note the large quantities of dune sand in this lower end of Salt Valley.

18.2 Turn right on the spur road to Delicate Arch. The gray rocky hills to the left-front are a mixture of Dakota Sandstone and Cedar Mountain Formation, which resembles the older Brushy Basin Member of the Morrison Formation. The road descends into Cache Valley through rounded gray deposits of Mancos Shale, on both sides of the road.

19.2 Along this stretch of the road and just ahead, the ridges of tilted rock are Cedar Mountain Formation. The red sandstone on the skyline is Slickrock-Entrada. The valley floor is largely decomposed rock, sand and alluvium held in place by native vegetation, with the tall exotic shrub Tamarisk along the drainages.

19.5 Pull into the parking lot to the left and park for the hike to Delicate Arch and to study historic Wolfe Ranch. Note the large seep-cave and black-banded cliffs to the north, which are Slickrock-Entrada. The black streaks on the cliff-face are desert varnish that was created by water runoff.

19.6 From the parking lot, turn left on the road to a viewpoint of Delicate Arch from below. **Note:** This spur is included in the road-log. If the road is closed by flooding in Salt Wash, or the spur is not driven, subsequent mileages will need to be adjusted.

20.0 The dark ridge of tilted rock to the left of the road is Dakota Sandstone. As the road curves around the end of the ridge, the equally tilted Cedar Mountain Formation stratigraphically below the Dakota comes into view. The low, rounded mounds in the valley to the right and right-front are remnants of Mancos Shale. The cliffs ahead and to the right of the valley are Wingate Sandstone above slopes of red-hued Chinle Formation deposits. A major fault line runs along the base of the red deposits. This offset has aligned the Mancos Shale with the much older Chinle. Such mismatching of geologic strata is prevalent in this end of the Salt Valley anticline, where its fracture zone angles southeastward to connect with those of the Castle and Fisher valley anticlines south of the Colorado River.

20.9 The spur ends at a parking area. The clifflines and Delicate Arch to the north are all Slickrock-Entrada. Notice the line of vegetated seep-caves in the cliffs, and how the cliff strata are dropping lower toward the west. To the northwest and northeast below the red cliffs, the white slickrock slopes are Moab-Entrada. This reversal of the normal stratigraphic sequence is due to a fault line at the base of the red cliffs. The red sediments lying on the white sandstone are the Tidwell Member of the Morrison Formation. Note the decomposing masses of agate. This hard mineral commonly forms in the highly mineralized Tidwell. The tilted rocky deposits around the parking lot are the very similar Brushy Basin-Morrison and Cedar Mountain Formation. Hike up the marked foot trail for an interesting view of Delicate Arch beyond a deep canyon that is not visible from the parking area. The canyon drains a large area of tilted Navajo Sandstone.

23.6 Back at the main Arches Road, turn right.

24.6 The road crosses the Salt Wash drainage.

25.2 From a narrow paved pullout on the right, notice the view to the right. The red sandstone on the skyline is Slickrock-Entrada. Several fault lines between the road and those cliffs have offset the strata severely, making detailed strata identification difficult. All of the lower strata, however, are younger than the high Entrada cliffs, and include Mancos Shale, Cedar Mountain Sandstone, and Morrison Formation deposits.

25.9 Turn right on the Salt Valley Overlook loop. From the end of the loop, lower Salt Valley and Cache Valley are visible to the south. A jagged ridge of red Slickrock-Entrada sandstone dominates the eastern horizon. Note how that ridgeline dips down to the south because of distortion caused by the Salt Valley anticline. To the northeast of the pullout, the red sandstone at road level is also Slickrock-Entrada.

26.3 Turn right on the main road.

26.5 Turn right toward the Fiery Furnace parking and view area. From the end of the adjacent ridge of rock, the view to the south is spectacular. Beginning on the left, the eroded red sandstone cliffs and fins are Slickrock-Entrada. Beyond the lower end of the ridge is lower Salt Valley, with its jumble of tilted and faulted strata. Beyond the gray Mancos Shale hills in the valley, the red cliffs and slopes are Wingate Sandstone set above Chinle Formation slopes that are covered with rock debris. The gap in the cliffline is the canyon through which Salt Wash, Cache Valley and Salt Valley all drain toward the Colorado River gorge. The more distant red cliffs are also Wingate Sandstone, with the La Sal Mountains on the southern skyline. The Slickrock-Entrada domes and spires of the Windows Section of the park, which was visited earlier, are on the southwestern skyline.

The western skyline is the eroded southwestern rim of Salt Valley, which is largely Wingate and Kayenta sandstones that have been tilted and fractured by uplifting and subsidence, then obscured by decomposed rock, sand and desert vegetation. The white rock in the immediate vicinity is Moab-Entrada that has been displaced downward by a fault.

27.1 Turn right on the main road. The beautifully eroded cliffs, fins and spires to the east of the road, on to where it ends, are Slickrock-Entrada, with bases of darker red Dewey Bridge-Entrada exposed in some places.

29.0 From the parking pullout here, hiking trails go to Sand Dune Arch and Broken Arch. Broken Arch is visible in the distance, at the far end of the lefthand cliffline. Sand Dune Arch is out of sight in the slickrock maze to the right of this gap in the cliffline.

29.5 The road to the left drops down into Salt Valley, providing access to a hiking trail into Klondike Bluffs, as well as several off-road vehicle trails. In this area, the lower basin of Salt Valley is largely decomposed rock, sand and alluvium, with a few protruding outcrops of Paradox salts and other ancient strata. Just ahead, watch for Skyline Arch on the right.

29.8 The hiking trail to Skyline Arch begins at a pullout on the right here. The arch is in Slickrock-Entrada. It is also accessible from the campground on the other side of the ridge.

30.3 The road enters the Devils Garden one-way loop. Devils Garden is a large picturesque meadows area set among gigantic weathered fins of Slickrock-Entrada. There are two picnic-ground pullouts just around the bend.

30.5 The spur to the right enters the Arches National Park campground. Continue ahead on the loop road.

30.8 Park at the end of the Devils Garden loop for the hike to Landscape Arch and many others. See the Park Service trail guide for details about this hike. Landscape Arch is one of the largest natural arches in the world and also one of the most fragile. A large piece fell from its underside in 1991, leaving it even more in danger of complete collapse. Some arch researchers feel that seismographic blasting in the general vicinity of the park in support of mineral-search activities may have triggered the recent collapse, and could eventually cause destruction of the arch itself.

GEOLOGIC STRATA VISIBLE FROM THIS ROAD

Formation	Member	Page
Morrison Formation	Brushy Basin	Page 18
	Salt Wash	
	Tidwell	
Entrada Sandstone	Moab	Page 16
	Slickrock	
	Dewey Bridge	
Navajo Sandstone		Page 13
Kayenta Sandstone		Page 13
Wingate Sandstone		Page 12
Chinle Formation		Page 12
Moenkopi Formation		Page 12
Cutler Group	White Rim	Page 11
	Organ Rock	
	Cedar Mesa	
	Lower Cutler	

ROAD NAME: **CASTLE VALLEY ROAD**

SUMMARY: This scenic road travels the length of Castle Valley, one of several sheer-walled valleys in the area created by erosion of massive salt anticlines. Several fascinating geologic phenomena make this short drive exceptionally interesting. There is a colorful remnant of White Rim Sandstone in one wall of the valley. Another wall has eroded to leave rows of tall sandstone towers standing like sentinels between Castle Valley and adjacent Professor Valley. The geologic strata in the upper valley were severely distorted by the upthrusting La Sal Mountains. A gigantic cone of igneous rock stands in mid-valley. The upper valley is buried under countless tons of glacial deposits, and its lower end is obstructed by strange-looking barren red hills.

ROAD SURFACE: paved.

VEHICLE: any vehicle.

MAP: ***Canyon Country* OFF-ROAD VEHICLE TRAIL MAP - Arches & La Sals Areas.**

GUIDEBOOK: ***Canyon Country* OFF-ROAD VEHICLE TRAILS - Arches & La Sals Areas.**

MILEAGES:

TO START: 18.5 miles, from Main and Center, Moab.
ROAD: 11.0 miles, one-way to end as described.
RETURN: 29.6 miles, from end of described road.
TOTAL: 59.1 miles, from Main and Center, round trip.

TIME: one-half day, plus time taken for stops along the way. Allow a full day if this road is taken as a part of a loop drive from Moab via Utah 128, Castle Valley Road, the La Sal Mountain Loop Road and back to Moab via either Spanish

Valley Road or U.S. 191 South, or for a combination trip taking Utah 128 to where its description ends, then returning to visit the Fisher Towers and explore this road.

ROUTE: Main Street (U.S. 191) to Utah 128, then upriver on this state highway to its junction with Castle Valley Road.

NAVIGATION: Drive north on U.S. 191, up the Colorado River gorge on Utah 128, then up Castle Valley on Castle Valley Road. For the full day trips noted above, follow the geology logs for the indicated roads, using the indicated map and book for guidance.

GEOLOGY ROAD LOG:

0.0 As this road branches from Utah 128, lofty Parriot Mesa dominates the skyline to the left, with hills of Moenkopi Formation deposits on the right. The vertical cliffs of Parriot Mesa are Wingate Sandstone, topped with thin remnants of layered Kayenta Sandstone. The slopes below the sheer cliffs are Chinle Formation deposits, then Moenkopi below that. In this vicinity, Chinle and Moenkopi deposits are very similar in appearance, making it difficult to tell them apart. The low bluffs on both sides of the road are Moenkopi. Note the characteristic way that Moenkopi deposits erode when exposed in open areas, with the softer sediments eroding into steep slopes that are terraced by layers of harder sandstone, with occasional low bluffs of the harder deposits.

The present road climbs through the Moenkopi hills toward the main part of Castle Valley. This grade was named Pace Hill after an early settler. Near the summit of the grade, remnants of older road alignments are visible as abandoned switchbacks, rock retaining walls and safety cables. To the right in the distance, segments of the lofty Castle Valley western rimlands are visible beyond the Moenkopi hills. The vertical walls of that rimland are Wingate Sandstone, topped with remnants of Kayenta Sandstone.

1.3 Note the abandoned loop of old road, with its steel cable guard rail.

1.5 The Pace Hill grade tops out here, offering a first glimpse of the main part of Castle Valley. Viewed from a narrow pullout at the summit, the lower end of this valley, above where the drainage narrows to the west of the Pace Hill Moenkopi deposits, was once a large, active cattle ranch. It has since been subdivided into smaller plots of land, with a growing community of private homes and businesses.

The cliffline ahead is Wingate Sandstone at the top and, in descending order, slopes of Chinle and Moenkopi, with additional near-vertical sandstone cliffs near the bottom. This lower unit of light-hued rock is an anomaly that does not appear elsewhere in this region east of the Colorado River. It is a detached deposit of White Rim Sandstone, which is an upper member of the Cutler Group. The lowest slopes in this cliffline are largely rubble, decomposed rock and wind-deposited sand, with a few exposures of dark red Cutler Sandstone showing below the massive White Rim cliffs.

Castle Valley lies above the central spine of one of the several salt anticlines in this region. It is roughly aligned with the similar anticline of Salt Valley in Arches National Park, with the remaining geologic strata between these two anticlines somewhat distorted, as described on the road log for Utah 128. See the sidebar, SALT ANTICLINES.

1.8 A pullout on the right provides a good view of lower Castle Valley, the La Sal Mountains to the left front, and the southern end of towering Parriot Mesa. Just ahead, the main road into the lower Castle Valley development goes right. From here, more of the exposure of the anomalous White Rim Sandstone in the cliff to the right becomes visible. Note that it extends across the valley floor at its lower end, phasing into the adjacent Moenkopi deposits, and helping to create a narrows in the Castle Creek drainage that goes on to the Colorado River. To the right of that narrows, but not visible from here, the anomalous White Rim Sandstone tapers out between layers of the dark red sandstones that are normal for the Cutler in this area, thus defining the easternmost extension of this detached deposit.

In late 1765, Juan María Antonio de Rivera, a Spanish explorer out of Santa Fe, was the first white man to enter what is now the State of Utah. He crossed the present Utah-Colorado border just east of the La Sal Mountains, crossed the mountains via Geyser Pass, then descended into the upper drainage of Castle Valley. From there, he continued down the valley along its western slopes toward the river, under the guidance of a native Indian familiar with the territory. In the vicinity of the developed land in the lower valley, Rivera crossed Castle Creek then climbed the red Moenkopi hills in the vicinity if this road, for his first glimpse of the "Rio Tisón," our Colorado River. This historic event was documented in Rivera's diary, as translated by an historian from his 18th century Spanish:

> *" ... after having watered the horsetrain, the guide told us that if we would climb to the top of a hill, which was on the other side of the pretty little river, we would be at the Rio Grande de el Tisón which we looked for ... "*

Rivera's *"pretty little river"* was, of course, Castle Creek, and his *"hill"* was the Moenkopi hills that Castle Valley Road climbs from the Colorado River vicinity. Passage through the lower narrows of Castle Creek would have been impractical because of flooding from the storm Rivera reported enduring before crossing the La Sal Mountains, which had not yet been named. Rivera's historic but little-recognized expedition is discussed further in the road log for Utah 128.

2.2 From a narrow pullout on the right, note the geologic features coming into view as the road continues up Castle Valley. The anomalous White Rim Sandstone is disappearing under rubble slopes below the high western cliffs. These cliffs, from top to bottom, are thin remnants of layered Kayenta Sandstone, vertical cliffs of Wingate Sandstone, then steep slopes of Chinle Formation and Moenkopi Formation deposits, with occasional exposures of dark red Cutler sandstones just above the valley floor. The valley floor itself is entirely dune sand, decomposed rock, gravels and alluvium, all still held in place in some areas by remnants of native vegetation. In other areas, the native soils have been cultivated for agricultural purposes. Straight ahead, the low, conical "mountain" in the valley is an exposed laccolith, an isolated upthrust remnant from the formation of the La Sal Mountains by intrusive lava. See the side bar, LACCOLITHIC MOUNTAINS.

To the left of the highway, Parriot Mesa still dominates, with an isolated spire of Wingate Sandstone to the right of that prominence. This spire and others will be discussed farther along the road. To the left-front, the high cliffline beyond the spire is part of Adobe Mesa, an elevated peninsula that lies between Castle Valley and Professor Valley to the northeast. The geologic formations in Adobe Mesa are the same as those in Parriot Mesa and the western cliff of Castle Valley.

3.0 From a wide shoulder here, the deposits to the left of the road most of the way up the Parriot Mesa slopes are Moenkopi Formation, with Chinle and Wingate above that. The lower ridge to the left-front is entirely Moenkopi. Straight ahead, the isolated spire of Castle Tower is set against the skyline. To the left, the standing figure of "The Priest," facing the kneeling figures of "The Nuns," are to the left of a rectangular prominence called "The Rectory." The slopes below these sheer-walled remnants of Wingate Sandstone are Chinle and Moenkopi, with the interface between these deposits being obscure.

4.0 Along this stretch of road, note the angled outcrops of older Cutler Group sandstones immediately to the left of the road. From this vicinity, the Priest and Nuns, and Castle Tower, are prominent against the sky to the left. Castle Tower is popular with rock climbers and has been ascended hundreds of times. The tower has also played a part in numerous commercial films and videos, including several with vehicles and models placed on the top of the tower by helicopter, then photographed from the ground and air. To the left-front, the walls of Adobe Mesa are on the skyline, while laccolithic Round Mountain becomes more prominent to the right-front. The upper craggy part of Round Mountain is intrusive igneous rock, while the lower slopes are a mixture of decomposed rock and sand, plus soil created by the vegetation that clings to the slopes.

Note that the western wall of Castle Valley, as well as the valley floor, are ascending toward the La Sal Mountains. Castle Valley is one of the several salt anticline valleys in this region, but its present configuration has also been strongly influenced by the much more recent creation of the La Sal Mountains by intrusive volcanic activity.

5.0 The angled ridge of Cutler sandstone continues left of the road below the higher cliffline. These ancient dark red strata were tilted by the intruding salt that formed the Castle Valley Anticline. The strata along the western cliffline slant downward away from the cliff rim for the same reason. Just ahead, the rugged contours of Round Mountain become more prominent. The valley floor continues to be relatively recent decomposed rock and alluvium deposited over a long period of time by the several drainages in the valley. The low hills developing in the upper end of the valley are geologically recent glacial gravels, deposited there on top of eroded remnants of other formations from higher in the La Sals.

7.0 Note that, to the left of the road, the dark red Cutler deposits are disappearing underground, leaving Moenkopi and Chinle slopes below the sheer Wingate cliffs, with thin remnants of Kayenta above that. The salt anticline that formed this great valley has deformed the cliffline of Adobe Mesa, to the left-front, and has left the strata in the mesa descending toward the mountains. The western cliffline in this area is gradually being lost under remnants of younger Entrada sandstones, and still younger Morrison Formation deposits, as the cliffline phases into the foothills of the La Sal Mountains. The hills in the valley are largely glacial debris on top of remnants of older strata that are now largely buried.

The gray rubble slopes in the mountain ahead are decomposed intrusive igneous rock, which resembles granite rather than extrusive lava. In the lower slopes of the La Sals, these igneous rock slopes blend into Morrison Formation and other deposits. The geologic strata along the slopes of the La Sals were violently tilted and stratigraphically displaced long ago by the salt anticlines in this region, then were distorted still further when the igneous intrusions of the La Sal Mountains were forced upward through the salt anticlines.

8.0 To the left of the road, note the relatively recent cutting into the valley floor alluvial deposits. For the next several miles, note that the geologic strata of Adobe Mesa are diving abruptly downward. The Wingate Sandstone that has dominated the cliffline is disappearing underground, as are the lower deposits, leaving the laminated Kayenta Sandstone forming most of the cliffs, with remnants of lighter-hued Navajo Sandstone above that. One such Navajo ridge is topped with glacial gravels.

8.6 The graveled road going left here ends at historic Castleton Cemetery. This road is marginal for standard highway vehicles, and is not a part of this geology road log, but is optional for those interested in this region's pioneer history.

9.0 In this vicinity, the road passes through a growing settlement. At one time a thriving community here called Castleton, that was based upon mining, logging and cattle ranching, rivaled Moab for size, and was seriously considered for county seat of Grand County. Very few remnants of the historic settlement have survived. Most of the structures in this vicinity have been built since then. Just ahead, some of the deposits to the right of the road are outcrops of Morrison Formation sediments and sandstones.

9.5 The large outcrop of eroded gray rock to the left of the road is probably Salt Wash-Morrison sandstone.

10.0 From a road junction here, note the wildly distorted strata to the left and left-front, where the Wingate, Kayenta and Navajo sandstones have been severely warped by the Castle Valley Anticline and La Sal Mountain intrusions. A detached fragment of Wingate Sandstone lies to the left-front, angled upward on the La Sal slopes. The rounded deposits that appear on the left-front skyline are Navajo and Entrada sandstones largely covered with glacial debris. This pullout provides a good last look back down the valley at Castle Tower and other nearby prominences. These features disappear from view as the road swings leftward into a tributary drainage of upper Castle Valley.

10.9 From a narrow pullout on the left here, note the remnants of an early schoolhouse in the bottomlands to the left. Only the red sandstone walls of the school remain standing. This historic site stands on public land, although much of this end of the valley was settled and homesteaded earlier. From this pullout, note that the Wingate Sandstone cliffs to the left of the schoolhouse disappear underground along with the remnants of Kayenta and Navajo sandstones. The higher cliffline to the left-front is Navajo topped with glacial deposits. The lighter sandstone cliffs straight ahead and on the skyline to the left are Entrada Sandstone. The thin layers of dark red on the open slopes are the Dewey Bridge Member of the Entrada, while the fractured vertical cliffs are the Slickrock Member of the Entrada. The other scant exposures of red rock straight ahead are also Slickrock-Entrada, with light-hued remnants of Salt Wash-Morrison above that.

11.0 The Castle Valley Road, as described, ends at this junction. The road to the right is the La Sal Mountain Loop Road, which is logged elsewhere in this book. The road ahead is the Castleton-Gateway Road, which is partly described with the geology log for the Onion Creek - Thompson Canyon - Polar Mesa ORV trail, elsewhere in this book.

Old Schoolhouse, Castle Valley.

ROAD NAME: **ISLAND ROAD**

SUMMARY: This road is essentially a branch of Utah 313 and is commonly traveled as a full day's outstandingly scenic trip together with that state road. The Island Road enters the Island-in-the-Sky District of Canyonlands National Park, and travels the meadows and terraces of the enormous sheer-walled peninsula bearing that name. The road goes to a number of breathtaking overlooks, a primitive campground and several developed picnic grounds, providing access to some very unusual hiking trails along the way. One of these reaches the rim of controversial Upheaval Dome. From the other viewpoints, the rugged, erosion-carved Maze and Needles districts of the park and equally spectacular adjacent terrain sprawl across vast distances, like an enormous three-dimensional map. These unique viewpoints are major highlights of this region, and they all offer enthralling overviews of its unparalleled geologic diversity.

ROAD SURFACE: paved, except for one short stretch of graveled road to a major overlook.

VEHICLE: any vehicle.

MAP: ***Canyon Country* OFF-ROAD VEHICLE TRAIL MAP - Island Area.**

GUIDEBOOK: ***Canyon Country* OFF-ROAD VEHICLE TRAILS - Island Area.**

MILEAGES:

TO START: 26.6 miles, from Main and Center, Moab.
ROAD: 32.8 miles, one-way, as described, to end.
RETURN: 46.5 miles, from end of described road.
TOTAL: 105.9 miles, from Main and Center, round trip.

GEOLOGIC STRATA VISIBLE FROM THIS ROAD

Stratum	Member	Page
Navajo Sandstone		Page 13
Kayenta Sandstone		Page 13
Wingate Sandstone		Page 12
Chinle Formation		Page 12
Moenkopi Formation		Page 12
Cutler Group	White Rim	Page 11
	Organ Rock	
	Cedar Mesa	
	Lower Cutler	

TIME: allow a full day for this outstandingly scenic trip, including several short hikes along the way.

ROUTE: North on Main Street (U.S. 191) to Utah 313, then west and south on that paved road to where the Island Road continues south where the state highway turns left toward Dead Horse Point State Park.

NAVIGATION: Drive north on U.S. 191, west then south on Utah 313, then continue straight ahead where the state highway turns left at its junction with the Island Road. Stay on the Island Road to the Green River Overlook and to Upheaval Dome. Then return and take the branch of the Island Road that goes to Grand View Point. Return via the shortest route to Moab, bypassing the side roads already traveled.

GEOLOGY ROAD LOG:

0.0 From this junction, 15.3 miles from U.S. 191 and 26.6 miles from Moab, Utah 313 goes left toward Dead Horse Point, while the Island Road continues south across the highlands of Island-in-the-Sky. Along this stretch of the road, the La Sal Mountains loom on the eastern horizon, the Abajo Mountains are visible to the south, and the rugged terrain below the Island-in-the-Sky is glimpsed at intervals, beginning shortly beyond this junction. The terrain adjacent to the road is almost entirely alluvium, dune sand and decomposed rock held in place by cryptogams and other native vegetation. See the sidebar, CRYPTOGAMIC SOILS.

3.0 Along this stretch of the road, the large domes of rock to the right are Navajo Sandstone. Note the red color-banding in their southern slopes. This phenomenon is unexplained by geologists. Note also that where the Navajo has been undercut toward the left end of that sandstone ridge, it has scaled away to form vertical cliffs rather than the rounded slopes on the rest of the ridge where it has not been undercut. This is an erosional characteristic of all the massive sandstones in this region, such as Wingate, Navajo and the Slickrock Member of Entrada Sandstone.

4.6 The road enters the Island-in-the-Sky District of Canyonlands National Park. Beyond here, on clear days, the distant Henry Mountains can be seen to the southwest. To the east, the La Sal Mountains loom above the immense projection of Dead Horse Point. From here on, Park Service regulations apply, rather than the multiple-use regulations of BLM land. Shortly beyond this boundary, the immense maze of the Taylor Canyon complex becomes visible to the west. The park boundary generally follows the northern rim of that canyon, which drains into the Green River gorge.

5.5 Beginning here, the spectacularly eroded lower terrain to the east of the Island-in-the-Sky becomes more visible as the road approaches a narrow place in this elevated peninsula formed by Taylor and Shafer canyons. These two massive drainage lines nearly touch at a slender isthmus called The Neck.

6.0 The Shafer Trail goes left at a junction here. This ORV trail should not be attempted by standard highway vehicles, but is generally passable in good weather to off-road vehicles and carefully-driven low-geared, high-clearance highway vehicles. It descends steeply into the lower terrain by way of a series of switchbacks which can be viewed from a pullout at The Neck. At the base of that steep grade, the trail forks, with the White Rim ORV trail going right, and the Potash ORV trail going left toward Moab. The referenced map and book depict and describe these trails. The Potash Trail is described elsewhere in this book.

7.0 Stop at the Park Service visitor center here, then continue to a large pullout on the left that gives a spectacular view to the east.

7.5 Turn left here into a pullout that offers a viewpoint and picnic site. From this viewpoint, beginning on the left, or north, the uppermost cliffs are Navajo Sandstone. The Shafer Trail winds along at the base of these cliffs on top of the uppermost layers of Kayenta Sandstone. Below this laminated rock, the sheer cliffs are Wingate Sandstone standing on a sloping base consisting of, from top to bottom, gray-hued Chinle Formation deposits partly covered with rubble, and thin red-brown strata of Moenkopi Formation. This deposit ends on wide benchlands of the White Rim Sandstone which, in this area, is the upper member of the Cutler Group. Below the White Rim, the colonnaded red sandstone is undivided Cutler Group deposits.

On toward the La Sal Mountains, Dead Horse Point dominates the skyline, with the same sequence of geologic strata. Farther to the right, another projection of Island-in-the-Sky blocks part of the mountains. Below the mountains, between these sheer walls, the northernmost tip of Hatch Point, in Canyon Rims Recreation Area, looms above the saddle of Hurrah Pass and, farther to the right, Pyramid Butte. To the right of this viewpoint, remnant Douglas fir trees cling to slopes of alluvium at the base of the Navajo Sandstone cliffs. For a better view of the Shafer Trail switchbacks and the panorama of this area, walk out onto the projection of land just below this viewpoint. A pullout at The Neck, a short distance ahead on the main road, provides another view of the Shafer Trail switchbacks.

7.8 Back at the main road, turn left. Just ahead, the road crosses The Neck, where Shafer and Taylor canyons almost meet.

8.0 The road crosses The Neck at elevation 5800 feet. Keep going to a pullout on the left at mile 8.2 for a good view of the Shafer Trail and Shafer Canyon.

8.2 For a description of the terrain below this viewpoint, see the earlier one at Mile 7.5. From this vicinity, the Taylor Canyon complex stretches off to the west. The high, rounded domes of sandstone there are Navajo standing on bases of Kayenta Sandstone. The sheer cliffs of the inner gorge are Wingate Sandstone, with some Chinle deposits visible beneath the rubble.

8.6 From a pullout here, a hiking trail descends into the upper rimlands of Taylor Canyon. In this vicinity, note the large quantities of dune sand on the terraces below the road. The wind patterns here keep these dunes active and not anchored in place by cryptogams and other vegetation. Such active dunes are relatively rare in this region. Beyond here, the road travels again through open meadows of sand and alluvium held in place by vegetation, with low ridges of Navajo Sandstone above the road level.

Notice how the grasses and other vegetation have recovered since grazing has been prohibited in the park in recent years. Most if not all of the grasses here, however, have been introduced by seeding because the native grasses were grazed beyond recovery. See the sidebar HUMANITY AS A GEOLOGIC FORCE.

10.8 Along this stretch of road, the sloping bluff to the right is Navajo Sandstone. The Navajo is exposed on the left a little farther on.

11.2 The flat-topped ridge a few hundred feet to the left of the road is Navajo Sandstone, topped with remnant deposits of a desert playa. See the sidebar PETRIFIED PLAYAS. The brown-topped ridge to the right of the road is a part of the same desert playa deposit.

12.5 Along this stretch of the road, another arm of the Taylor Canyon complex is visible. The truncated pyramidal dome to the right is Navajo Sandstone topped with desert playa sediments.

13.4 The parking area to the left is for the hiking trail to Mesa Arch, a large natural span in the rim of Island-in-the-Sky.

13.7 Here, the Island Road splits. The road on ahead travels Grand View Point and is documented as a separate stretch of road later in this road log. Take the road to the right, which goes to Upheaval Dome and a developed picnic site. This branch of the Island Road is documented next, with continuous mileage figures. A short unpaved spur of this road goes to Green River Overlook and a primitive campground. This side trip is also included in the following description.

14.0 Turn left on the graveled road to Green River Overlook.

15.1 From a junction here, the road left goes into a campground. The rounded rock to the right of the road is Navajo Sandstone.

15.5 From the Green River Overlook, at the end of the road, Grand View Point dominates the skyline to the east. On the Point, small remnants of Navajo Sandstone are visible above the horizontal layering of the Kayenta Sandstone. Below that, the sheer Wingate cliffs stand on top of terraced slopes of Chinle Formation and Moenkopi Formation down to the White Rim Sandstone. The buttes to the southwest are all of this same combination. In this area, the White Rim Sandstone is quite thick. Below it, the dark red columns of rock are the Organ Rock Shale member of the Cutler Group. This rock dominates clear to the rim of the inner gorge of the Green River, where the upper strata of the Hermosa Group appear. The identities of the various buttes and geographic landmarks beyond this overlook are given on a display here.

Note the twisting route of the White Rim Trail far below, just above the top of the White Rim Sandstone. This ORV trail travels from the Colorado River in the east to the Green River in the west, then climbs back up onto the Island-in-the-Sky peninsula. Note the old airstrip branching from that trail at one point. During the uranium boom days, before Canyonlands National Park was established, this airstrip was used to supply uranium prospectors who were searching for that mineral in the Chinle Formation below this viewpoint. Remnants of those search activities are still visible in the form of excavations and bulldozed vehicle trails on the gray terraces to the east of this viewpoint.

16.9 Back at the road to Upheaval Dome, continue left. The high promontory to the north of this junction is called Aztec Butte. It was misnamed by early cattlemen who found prehistoric rock structures in the vicinity and assumed them to be "Aztec" in origin, even though that Central American culture never extended to this part of North America. Aztec Butte is Navajo Sandstone. Its flat top was created by the highly-mineralized water-deposited sediments of a desert playa.

17.4 The pullout to the right is a trailhead for the hike to Aztec Butte. There is a prehistoric ruin visible from this trail in a cave near the top of the red-topped butte nearest the road. Federal antiquities laws should be observed near any prehistoric remnant. Beyond this pullout, the road descends into lower terrain at the interface between Navajo and Kayenta sandstones. This interface is indicated by the presence of horizontal layering separated by relatively thin sections of cross-stratified dune sand deposits. Along this stretch of road, the Kayenta is obscured under dune sands and alluvium held in place by cryptogams and other vegetation.

19.4 From a pullout on the left, note the detached butte to the left of the road, with its vertical sandstone walls. The upper knob on this butte is a remnant of Navajo Sandstone standing on a sloping base of Kayenta Sandstone. The Green River is visible far below. Along this stretch of the river, its canyon walls are entirely White Rim Sandstone. Beyond this viewpoint, the road continues to travel in the layered Kayenta Sandstone, with occasional remnants of Navajo on low ridges just above the road.

20.3 The pullout here is a hiking trailhead. The canyon beyond, to the right of the road, is one arm of the immense Taylor Canyon system. Beyond this pullout, the road climbs into Navajo Sandstone, with rounded domes of this weathered rock pushing up through sand dunes held in place by vegetation. Strong prevailing winds, thrust upward by the sheer cliffs of Island-in-the-Sky, deposit large quantities of sand just back from this rim. This atmospheric-geologic phenomenon is common in this land of vertical cliffs.

20.7 At a pullout here, a hiking trail climbs onto Whale Rock, an enormous rounded ridge of Navajo Sandstone that provides a spectacular view in all directions.

21.6 The road ends at a loop with a picnic site, restrooms and the hiking trailhead to Upheaval Dome. For a discussion of the controversy over this unique geologic phenomenon, refer to the sidebar, UPHEAVAL DOME, and a related sidebar, ROBERTS RIFT.

26.5 Back at the main road junction, turn right toward Grand View Point. The road continues out onto the Point at or near the interface between Navajo Sandstone and the underlying Kayenta Sandstone. As this leg of the Island Road heads south onto Grand View Point, it travels on top of this slender, elevated peninsula of land. From the junction, the domes of sandstone on the left for the next half mile or so are remnants of Navajo Sandstone. The road travels through rolling meadows of decomposed rock and windborne sand held in place by native vegetation. Watch just around the first bend of the road for a view to the right of the Henry Mountains in the distance, beyond a developing canyon.

26.8 While driving, note that one of the low domes of Navajo Sandstone on the left is capped by a layer of darker rock. This layer represents the harder deposits left in a desert playa, or drylake. See the sidebar, PETRIFIED PLAYAS.

27.5 From a pullout on the right, the three separate mountain peaks visible to the southwest are all part of the Henry Mountains. North of these and even more distant, the Aquarius Plateau is visible on clear days. The sheer-walled peninsula that dominates the skyline just below the Henry Mountains is known as the Orange Cliffs. This name was originally given to the Island-in-the-Sky peninsula by explorer John Wesley Powell, but has since been assigned to the more western cliffline. These cliffs lie just west of the Maze District of Canyonlands National Park. The detached buttes bear such names as Ekker Butte, Elaterite Butte and Buttes of the Cross, another name assigned by Powell.

The vertical walls of the Orange Cliffs and the nearby buttes are Wingate Sandstone topped with thin remnants of Kayenta Sandstone. Below the sheer walls the slopes are, in descending order, the Chinle and Moenkopi formations. These are difficult to differentiate visually from this distance. White Rim Sandstone is conspicuous below these darker deposits. Below the White Rim are dark red deposits of the Organ Rock Shale member of the Cutler Group. Undivided Cutler-Cedar Mesa Sandstone deposits lie below the Organ Rock, phasing into Cedar Mesa on toward the south. A short stretch of the Green River is visible from here. Beyond this viewpoint, for the next three miles, the road continues on the lower levels of Navajo Sandstone, but the rock itself is only occasionally revealed beneath the sandy soil.

29.0 The trail to Murphy Point goes right here. Only off-road vehicles and low-geared, high-clearance highway vehicles should attempt this road. The geology viewed from Murphy Point is the same as that just described.

29.8 Turn left toward the Buck Canyon Overlook. From this peninsula-rim viewpoint, the white-rimmed canyons below and to the left are upper branches of the Buck Canyon drainage. The La Sal Mountains dominate the eastern horizon. Below them, the sheer-walled cliffs of Hatch Point sprawl across the skyline. Those cliffs are Wingate Sandstone topped by layered Kayenta Sandstone, with occasional remnants of Navajo Sandstone still remaining on some parts of the complex peninsula. Below the Wingate cliffs, the usual combinations of Chinle, Moenkopi and Cutler descend to the inner gorge of the Colorado River, which can be glimpsed in a few places. The white rock rimming the western part of that inner gorge is White Rim Sandstone. Immediately below the White Rim, a thin band of undivided Organ Rock Shale-Cutler deposits phases on down into lower Cutler and Upper Hermosa deposits. These continue downward to the river level. In places the Organ Rock Shale-Cutler deposits have eroded into terraces and columns of dark red rock. The sloped red sandy deposits above the White Rim are Moenkopi Formation. Note that there is no White Rim Sandstone on the east side of the Colorado River. There, the Moenkopi, Cutler and Hermosa deposits are continuous down to the river level. The rim on each side of this projecting overlook is Navajo Sandstone that phases out into Kayenta Sandstone within a few yards in either direction.

Note the many branching off-road vehicle trails just above the level of the White Rim and angling up into the gray deposits that mark the lower part of the Chinle Formation, in which uranium is sometimes found. To the south are the Abajo Mountains and the highlands of Elk Ridge. On clear days, the Henry Mountains are visible to the southwest, as are the upper peaks of even more distant mountains.

30.0 Back at the main road, turn left. From here to the end of the road, it travels on Kayenta Sandstone, although this rock is largely obscured under sand and soil.

31.9 The road to the left goes to the Grand View Point picnic area.

32.6 Turn right here to an overlook which offers a viewpoint to the west. From this viewpoint, the Henry Mountains are visible on the skyline to the southwest. On clear days, a little of the more distant Aquarius Plateau to the right of them can be glimpsed. Still farther right, the rounded contours of Thousand Lake Mountain appear above the Orange Cliffs. To the northwest, the eastern monocline of the San Rafael Swell shows when the air is clear, as do several more distant mountain peaks. Silhouetted against the Aquarius Plateau is the high Navajo Sandstone dome of Cleopatra's Chair, in the Maze District of Canyonlands National Park. The plateau top below that is terraced Kayenta Sandstone, with the sheer cliffs being Wingate Sandstone perched above slopes of Chinle and Moenkopi deposits. The White Rim Sandstone below the soft slopes of red Moenkopi marks the inner gorge of the Green River and some of its major tributary canyons in this area. The White Rim Sandstone below this viewpoint is much thicker than the same deposit along the Colorado River gorge. The rock below the White Rim is Organ Rock Shale phasing downward into undivided Cutler-Organ Rock Shale deposits. Note that one stretch of the Green River is visible from here, and that a large expanse of bare White Rim Sandstone is exposed to the right of that. In that area, the White Rim is only thinly covered by remnants of decomposing Moenkopi deposits.

32.8 The road ends here at the southernmost overlook on Grand View Point. From this overlook, the panorama is much the same to the east as described earlier at miles 7.5 and 29.8, the Buck Canyon Overlook. The La Sal Mountains dominate the eastern skyline, with Hatch Point below them. The Abajo Mountains stand out on the southern skyline, with the high Elk Ridge country a little farther west. Elk Ridge is a northern stretch of the Monument Uplift. Between the viewpoint and Elk Ridge, countless rock spires in the Needles District of the park stand out like a forest of barren stone trees. These spires, and the canyons farther to the east, are Cedar Mesa Sandstone that was severely jointed by the northern nose of the Monument Uplift, then shaped by eons of erosion. See the sidebar, MONUMENT UPLIFT.

The viewpoint here is rimmed with remnants of layered Kayenta Sandstone, with sheer walls of Wingate Sandstone below that. The sloping rubble-covered slopes below the Wingate cliffs are Chinle Formation deposits phasing down into the Moenkopi Formation, which continues on down in broadening terraces and smooth slopes to the White Rim Sandstone. Note a segment of the White Rim off-road vehicle trail directly below the viewpoint. Below the White Rim to the southeast is Monument Basin, with its slender fins and towers of Organ Rock Shale. The top of The Totem Pole, a slender 305-foot rock tower, is visible jutting above the White Rim. In the broad lowlands to the southeast of Monument Basin, beyond the inner gorge of the Colorado River, red-hued undivided Cutler-Cedar Mesa deposits phase into Cedar Mesa Sandstone farther south. The inner gorge of the Colorado River visible to the southeast is lower Cutler deposits down to a certain point, where these phase into upper Hermosa deposits. At one time, some of these rivergorge strata were labeled the "Elephant Canyon Formation," but later research corrected this error. See the sidebar, YES -- GEOLOGISTS ARE HUMAN.

To the southwest of this viewpoint the detached islandmesa is Junction Butte, which lies almost due north of the Green-Colorado confluence. This meeting of the rivers is too deeply entrenched within ancient Cutler and Hermosa group deposits to be seen from this viewpoint. Note also to the southwest how highly eroded ORV trails climb up into the level of the Chinle Formation and skirt around the base of Junction Butte at that level. These trails are remnants of pre-park uranium prospecting booms in this area, when almost any kind of uranium ore was consider valuable and its extraction and transport to the nearest mill was subsidized by the U.S. Government. In recent years, one such old mine in Canyonlands National Park was illegally worked. This came to the attention of the Park Service when an accident with explosives killed one of the miners.

Upheaval Dome

UPHEAVAL DOME

Geologic mysteries abound in southeastern Utah's colorful and spectacular canyon country. Some of these strange phenomena are just little things that are fascinating and puzzling to regional visitors, but insignificant to professional geologists, who find them easier to ignore than to resolve.

Further, the serious study of some of these little mysteries would require geologists to venture into the fringes of other scientific disciplines, such as microbiology, paleontology or paleoclimatology, so geologists settle for labeling the curious phenomena -- like "concretions" or "bioturbation" -- then consider the matter settled.

To amateur geologists and curious canyon country visitors, this labeling without explanation is not very satisfying. It leaves too many unsettled matters, too many unexplained geological curiosities, too many unanswered questions about things they see all around them as they explore the incredibly diverse geology of canyon country, often far more thoroughly than the professionals do.

Canyon country also has some major mysteries, such as the causes of great faults, uplifts and synclines. These generally receive more attention from professional geologists because they are too obvious to ignore. However, this attention does not, as one might suppose, automatically lead to quick and incontestable resolution of the mysteries.

One such long-standing canyon country geologic mystery that is still unresolved to everyone's satisfaction is Upheaval Dome, in the Island-in-the-Sky District of Canyonlands National Park. The first scientific description of Upheaval Dome was written in 1927 by geologist T.S. Harrison. In 1940, another geologist, E.T. McKnight, wrote a U.S. Geological Survey Bulletin concluding that the strange crater had been formed by erosion of a salt dome.

Since then, studies of this immense crater in solid rock, with its sheer inner walls and concentric rings of distorted geologic strata, have produced a number of widely differing reports about Upheaval Dome, and two sharply differing hypotheses as to its original cause, plus several less likely possibilities.

The salt dome idea -- reflected in the crater's name, "Upheaval Dome" -- seemed incontestable to geologists. In a region replete with dozens, even hundreds of massive salt anticlines and salt domes, it was natural, although unscientific, to conclude that Upheaval Dome was just another example of this kind of geologic phenomenon. Hence its name, with the clear implication that the crater was just a severely eroded salt dome.

When Canyonlands National Park was established in the mid-1960s, its early interpretive brochure noted the salt dome concept. Its brochure dated 1968 described and illustrated the deposition of the deep Paradox Basin salt beds, then went on to explain how later, when the salt was deeply buried under thick and heavy layers of rock --

" .. the salt flowed upward into the anticline. It pushed and caused more doming, deforming the layers above .. turning them almost on edge."

In a stratigraphic cross-section of Upheaval Dome, this brochure showed the Paradox Formation immediately underlying the White Rim Sandstone, omitting several very thick formations in order to show the supposed doming of the Paradox salt beds. Perhaps significantly, it also showed anatomically incorrect "pterodactyls" soaring over ancient seas during the deposition of the Paradox salts, at least 150 million years before such creatures even existed.

This description of Upheaval Dome remained the official Park Service interpretation for more than a decade. Then, in 1979, in response to criticism, the park naturalist wrote a letter to a well-known geologist asking for help in upgrading the Upheaval Dome brochure, noting that --

" .. the cross-section will have to be redrawn in order to differentiate strata and incorporate names .. not in the old one. .. That degree of accuracy, of course, is neither needed nor of value to most recipients of the folder, but there is a small minority to which it is important."

-- thus once more giving scientific accuracy low priority. The letter went on to state that the naturalist was --

" .. deleting the pterodactyls .. the original artist apparently erred in his representation of the forelimbs, which may be why (they resembled) seagulls."

-- not to mention that neither seagulls nor pterodactyls were to exist yet for hundreds of millions of years. Apparently neither the park naturalist nor the geologist even considered challenging the authenticity of the salt dome hypothesis.

Meantime, another well-known geologist, E.M. Shoemaker, of the U.S. Geological Survey, was not so convinced. After performing extensive studies on known meteor impact craters, as well

continued ----

as Upheaval Dome, in March, 1984 he presented a paper titled **"Upheaval Dome, Impact Structure"** to the Fifteenth Lunar and Planetary Science Conference.

This scientific paper described in considerable detail several characteristics of Upheaval Dome that did not fit the salt dome hypothesis, but that did fit the idea of a meteorite impact crater. The report called attention to several similarities between Upheaval Dome and other known impact craters that he and other geologists had studied, and concluded that the evidence pointed to an explosive meteorite impact about 65 million years ago, rather than to the glacially-slow stratigraphic deformation caused by salt doming.

Shoemaker's report, prepared for presentation to a scientific group, was written in technical terms not within the vocabularies of most non-professionals, and is thus not very quotable in a non-technical summary such as this, but a later abstract of his report was much more readable:

"Re-examination of Upheaval Dome in the Canyonlands National Park, Utah, has shown that the structure of this remarkable feature conforms with that expected for a deeply eroded astrobleme. The structure is definitely not compatible with an origin due simply to plastic flowage of salt and other rocks in the underlying Paradox Formation, as suggested by McKnight. The most strongly deformed rocks at Upheaval Dome are bounded by a series of circumferential listric faults. Beds of the Wingate Sandstone, Kayenta Formation, and Navajo Sandstone are displaced toward the center of the dome. Near the center, these same beds have been deformed by convergent flow into tight to open folds that plunge away from the apex of the dome. Thinning of the Wingate Sandstone reported by McKnight is due entirely to faulting. At the center of the dome, beds of the Moenkopi Formation are duplicated by displacement along numerous small thrust faults. Sand from the underlying White Rim Sandstone Member of the Cutler Formation has been mobilized and injected into the Moenkopi as large massive clastic dikes.

"The convergent displacement of the rocks at Upheaval Dome corresponds to the deformation that results from collapse of a transient cavity produced by high-speed impact. Beneath the listric faults, beds are deformed in a broad structural dimple that is bounded, at least on the north, by a partial circumferential anticline. This type of structure is produced by plastic flow of rocks at great depth beneath large impact craters. Asymmetry of the development of the deep anticline suggests that impact may have been from the south. Growth of a central dome probably was due partly to early convergent flow during opening of the initial cavity, as well as to late stage collapse. It is likely that the presence of plastic beds in the Paradox Formation at depth facilitated early growth of the dome.

"From considerations of the probable depth of exposure of the impact structure and upward extrapolation of the listric faults, the final collapsed crater is estimated to have been about 8 to 10 km in diameter; the impacting body probably was on the order of 0.5 km in diameter. As much as 2 km of beds appear to have been removed by erosion since the impact structure was formed. Restoration of this thickness of rocks suggests that the impact occurred in latest Cretaceous or Paleogene time."

Was this timing just coincidental, or could the Upheaval Dome meteorite have been related to the immense meteor impact that many scientists are now convinced triggered the extinction of dinosaurs and other life at the end of the Cretaceous? Was the Upheaval Dome meteorite just another of a swarm of meteors that passed through our solar system during that period of time?

One might think that the Shoemaker studies and report would have settled the controversy over Upheaval Dome, perhaps even leading to its being given a more appropriate name, but this was not to happen. As with so many other so-called "sciences," geology is sometimes more of an art than a pure science, in which strict rules of evidence apply.

Rather, the proponents of the salt dome hypothesis continued to cling to and rationalize their beliefs, and to discount the new evidence accumulated by Shoemaker and his associates, such as Professor Peter Huntoon, Department of Geology and Geophysics, at the University of Wyoming.

Even though Huntoon wrote an article in the August, 1986 issue of FROM THE CANYONS, a publication of the Canyonlands Natural History Association, summarizing the controversy and firmly supporting the meteorite impact concept, the National Park Service decided to "sit on the fence," so to speak. Its 1990 brochure about Upheaval Dome asks --

"What caused this medley of rock to dome up in the first place? Geologists don't know for sure. As you look at the crater, it is easy to imagine that it might be the result of an ancient impact of a meteorite. Perhaps, as another popular theory states, it is the result of a more subtle process: a deep layer of salt squished up, bending and pushing the layers of rock above it."

The brochure avoided the matter of correct

continued ----

stratigraphic sequence by simply not showing the names of any strata below the White Rim Sandstone, including the Paradox salt beds. The present displays about Upheaval Dome in the Island-in-the-Sky visitor center explain and illustrate both ideas.

SUMMARY AND CONCLUSION

When all of the presently available evidence is considered, little doubt remains that Upheaval Dome and Roberts Rift (see the sidebar, ROBERTS RIFT) were formed at the same time by the impact of a meteorite or cometary body about 1,600 feet in diameter and probably composed largely of ice, as many such astronomical objects are. The shape of the disturbed geologic strata in the vicinity of the crater matches that of other known impact craters, and the immense energy required for the formation of Roberts Rift could easily have been provided by such a violent impact.

Since the time of that impact, at about the close of the Cretaceous, the impact site rose thousands of feet, leaving the present remnants about 6,000 feet above sea level after more than a mile of rock has eroded from what was the sedimentary surface at the time of the impact. This erosion has left only remnants of the geologic strata deformed by the impact far below the lower tip of the original impact crater, and has exposed in some Moab-area canyons the upper zones of Roberts Rift. See the accompanying illustration of Upheaval Dome and the map of Roberts Rift.

Since the meteorite impact took place at about the same point in geologic time as the much larger one that brought on the Great Extinction of dinosaurs and other forms of life at that time, it is probable that both of these meteors were part of a general swarm of such space debris that our solar system has encountered periodically over an immense span of time. Each such encounter has produced mass extinctions of life on Planet Earth, and has left deep scars on its surface, some of them obvious -- such as Upheaval Dome -- and others obscured by this planet's seas. Only recently have scientists found -- in the Gulf of Mexico -- the remains of the crater of the immense meteor that is believed to have caused the Great Extinction at the end of the Cretaceous.

Thus, visitors to the Moab area have a chance to marvel at two rare phenomena -- visible remnants of one of the visitors from outer space that have so devastated this planet in the past -- Upheaval Dome and Roberts Rift.

Perhaps some day, when the meteor impact origin of Upheaval Dome is no longer disputed, the Park Service will revise its brochure, interpretive displays and nomenclature to reflect the known facts. The name, "Upheaval Dome," is appropriate for a salt dome, but not for a meteorite crater. Perhaps "Impact Crater"?

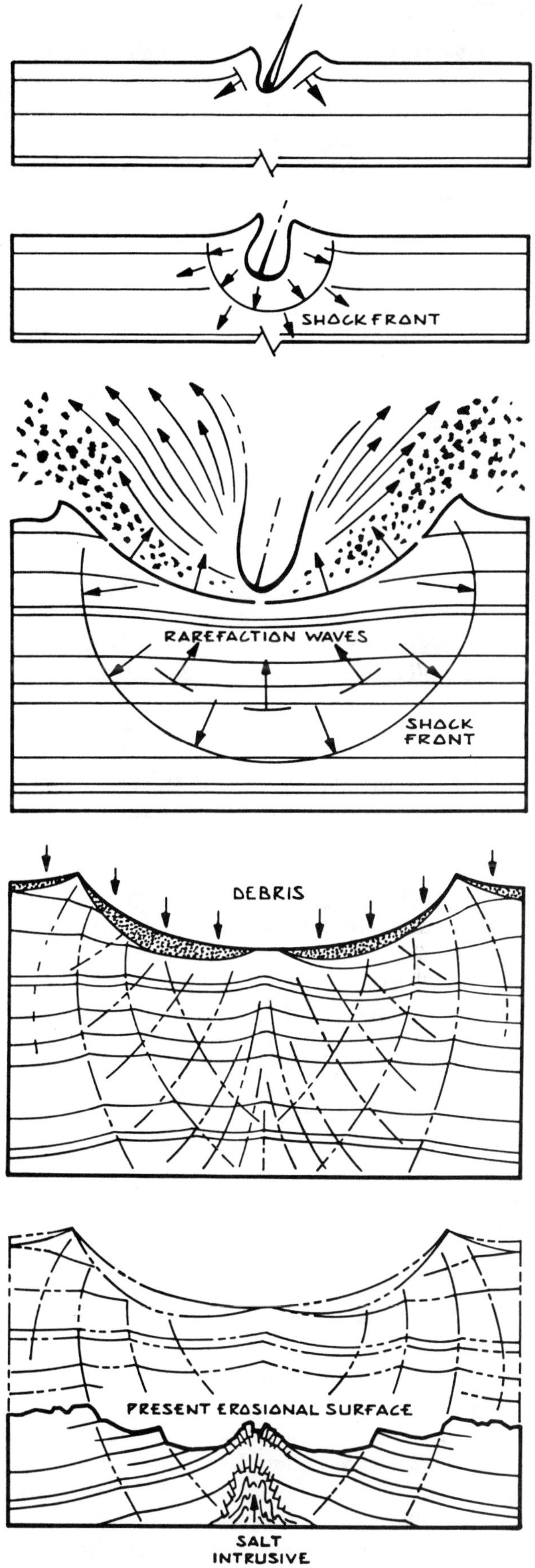

Sequence showing how the Upheaval Dome meteor crater formed, from impact to present

ROAD NAME: **NEEDLES OVERLOOK ROAD**

SUMMARY: This road offers a continuous broad panorama of immense and colorful sandstone monoliths and mesas set among rolling high-desert meadows. Along the way are a developed campground, glimpses of spectacular Harts Draw to the south, and access to Anticline Overlook Road and several off-road vehicle trails. The road ends at outstanding Needles Overlook, on a jutting western prominence of Hatch Point. The breathtaking, colorful panorama seen from this overlook encompasses most of Canyonlands National Park and the adjacent Lowlands District of Canyon Rims Recreation Area, both viewed from a lofty vantage point.

ROAD SURFACE: paved for its entire length.

VEHICLE: any vehicle.

MAPS: ***Canyon Country* OFF-ROAD VEHICLE TRAIL MAP - Canyon Rims & Needles Areas,** or ***Canyon Country* OFF-ROAD VEHICLE TRAIL MAP - Canyon Rims Recreation Area.**

GUIDEBOOKS: ***Canyon Country* OFF-ROAD VEHICLE TRAILS - Canyon Rims & Needles Areas,** or ***Canyon Country* OFF-ROAD VEHICLE TRAILS - Canyon Rims Recreation Area.**

MILEAGES:

TO START: 33.6 miles, from Main and Center, Moab.
ROAD: 22.5 miles, one-way to end.
RETURN: 56.1 miles, from end of described road.
TOTAL: 112.2 miles, from Main and Center, round trip.

TIME: a little more than 1/2 day, which allows time for walking around Needles Overlook and taking pictures along the way. Allow a full day if the spur road to Anticline Overlook is taken.

ROUTE: Main Street in Moab (U.S. 191) south to where the paved Needles Overlook Road goes west into the Hatch Point District of Canyon Rims Recreation Area.

NAVIGATION: drive south on Main Street (U.S 191) to where the paved entrance to Canyon Rims Recreation Area goes west toward cliff-walled Rone Bailey Mesa and Needles Overlook. Either return the same way, or drive the spurring Anticline Overlook Road then return, either by the same route or via Hatch Wash Road. Both of these roads are described elsewhere in this book.

GEOLOGIC STRATA VISIBLE FROM THIS ROAD

Formation	Member	Page
Morrison Formation	Brushy Basin	Page 18
	Salt Wash	
	Tidwell	
Entrada Sandstone	Moab	Page 16
	Slickrock	
	Dewey Bridge	
Navajo Sandstone		Page 13
Kayenta Sandstone		Page 13
Wingate Sandstone		Page 12
Chinle Formation		Page 12
Moenkopi Formation		Page 12
Cutler Group	White Rim	Page 11
	Organ Rock	
	Cedar Mesa	
	Lower Cutler	

GEOLOGY ROAD LOG:

0.0 This road enters the Hatch Point District of Canyon Rims Recreation Area. It ends at spectacular Needles Overlook, which is described at the end of the road-log. Along the way, several other roads and ORV trails branch off. All of these, and others in the immense recreation area, are described in the second book and map noted.

As the road heads generally west, it gradually ascends, from 5850 to 6296 feet elevation, for a net gain of 446 feet. The geological strata that make up Hatch Point ascend even more steeply, so that the younger strata visible along the first part of the road have long since eroded away toward the end of the road.

0.2 A pull-out here provides an interpretive display, and a road sign provides mileages to the principal points of interest on Hatch Point. For the next several miles, the terrain on both sides of the road is largely sand and decomposed rock held in place by native vegetation, but with occasional outcrops to the north of the road of Navajo Sandstone topped with the layered reddish Dewey Bridge Member of Entrada Sandstone.

0.6 The road crosses Tank Draw, one of a series of shallow drainages that enter Hatch Wash farther north. The alluvium deposited in this broad, flat area of Dry Valley has been cut deeply in each of these drainages since white settlers first entered this region in the late 1800s with their huge herds of cattle and sheep. Severe overgrazing brought on the cutting of native soils that took thousands of years to accumulate. See the sidebar, HUMANITY AS A GEOLOGIC FORCE.

1.0 The road that angles left here goes to Jail Rock and can also be used to intersect the Dry Valley Road, which is used as access to the Labyrinth Overlook Trail, described elsewhere in this book. For the curious history of Jail Rock, refer to the book ***Canyon Country's*** **CANYON RIMS RECREATION AREA.**

For the next mile or so, the panorama surrounding the road is dominated by three major geologic features. To the south are the Abajo Mountains. To the southwest and west, the convoluted cliffline of Rone Bailey Mesa looms high above Dry Valley. To the north, the La Sal Mountains soar above the red sandstone bluffs of the CAMEO CLIFFS recreation area. The two mountain ranges are laccolithic, intrusive-igneous in nature. See the sidebar, LACCOLITHIC MOUNTAINS. The red cliffs below the La Sal Mountains are largely the Slickrock Member of Entrada Sandstone. For more details about the geology in the CAMEO CLIFFS area, refer to the book, **CAMEO CLIFFS Biking - Hiking - Four-Wheeling,** or its matching map.

Rone Bailey Mesa is also Slickrock-Entrada, topped by a thin layer of the reddish Tidwell Member of the Morrison Formation, with rubbly remnants of the Salt Wash Member of the Morrison Formation above that. Note how the Slickrock-Entrada erodes into beautifully shaped fins and rounded abutments, and how it forms caves of various sizes wherever a seeping spring reaches the cliff surface. In parts of Rone Bailey Mesa, along its western slopes, the Dewey Bridge-Entrada has been exposed, but is not visible from this road. To the south, the isolated dome of red Slickrock-Entrada sandstone is Jail Rock, which has an immense pothole near its summit that collects rain and snowmelt. This was once a source of water for travelers between Moab and Monticello.

2.5 Dry Valley Road goes left at a four-way junction here. From here to the end of this road, wherever it passes through open brushy meadows, watch for pronghorn antelope. These graceful animals were reintroduced after having been extinct in this region, perhaps for centuries.

3.5 Along the next mile or so, the red sandstone butte of Hatch Rock dominates the northern skyline. Most of this promontory lies on Utah State land that has been leased for private development, with living quarters built in caves blasted from the solid Slickrock-Entrada Sandstone. Thin vestiges of Morrison Formation deposits remain on top of the dome.

To the left of the highway, Rone Bailey Mesa continues to dominate the skyline. Ahead and to the right, the red sandstone buttes protruding above the brushy plain are Slickrock-Entrada. The shallow drainages to the north of the road all ultimately join Hatch Wash. The one that cuts into the older strata on the other side of Hatch Rock was the site of an historic ranch settlement whose remnants are now perched precariously on a ledge of alluvium that remains after a century of severe erosion.

The road along this stretch is traveling at about the interface between the Entrada and Navajo sandstones. Decomposition of the reddish Dewey Bridge-Entrada gives the soils at this stratigraphic level their color. On ahead, some of the drainages to the right of the road cut down into light-hued Navajo Sandstone.

4.7 Along this stretch of road, note the tall evergreen trees growing in the moisture of a large seep-cave in the cliffs to the left. Some of these trees are remnants from plant communities that now thrive only in higher elevations, indicating a change in regional climate within the last century.

5.5 In the drainage to the right, the white rock in the lower slopes is Navajo Sandstone. Just above that, the dark red layered sediments are Dewey Bridge-Entrada. Above that are remnants of Slickrock-Entrada. The harder rocks lying on the Navajo Sandstone slopes are agate eroded from the Dewey Bridge-Entrada.

5.9 The road into Wind Whistle Campground goes left here. To the right of the road, white Navajo Sandstone dominates the drainage line, topped with remnants of red Dewey Bridge-Entrada. On ahead, the isolated ridge of Slickrock-Entrada that appears is Wind Whistle Rock, after which the campground was named.

6.5 The road crosses a drainage here. The red sandstone on both sides of the road is Dewey Bridge-Entrada.

7.0 Here, tributary canyons of immense Harts Draw come into view to the south. The white sandstone walling these drainages is Navajo Sandstone. Beyond that major canyon, the slopes of Harts Point dominate, with the tips of the Abajo Mountains looming above that.

7.5 Hatch Wash Road angles to the right here. That road is described elsewhere in this book. The slickrock ridge to the left at this junction is Soup Rock. This monolith and the one visible to the north are Slickrock-Entrada. Both have large seep caves in their bases. Soup Rock has an enormous pothole in the top of its western end that can be reached by hiking onto the slickrock from its lower end. The funnel-shaped pothole is 48 feet in diameter and 25 feet deep. How such deep potholes form in solid sandstone is a mystery. The elements that contribute to this are water and wind, but the exact mechanism is unclear. Beyond Soup Rock, the road travels across plains of dune sand and alluvium held in place by native vegetation, with occasional ledges and domes of Entrada Sandstone.

9.5 The major drainage coming into view on the left is Macomb Canyon, one of the side-canyons of Harts Draw, whose southern walls can be seen at the base of the distant peninsula. That peninsula is part of Harts Point, which is another district of Canyon Rims Recreation Area. The rounded, light-hued walls of Macomb Canyon are Navajo Sandstone. Note that in this area, the Navajo is variegated in color, white with red banding, indicating different sources for the dune sands that comprise the Navajo. Just ahead are the last domes of Slickrock-Entrada visible along this stretch of the road.

For more information about the Macomb Expedition after which Macomb Canyon was named, refer to the book, **CANYONLANDS NATIONAL PARK - Early History & First Descriptions.** For hiking the route of that expedition within Utah, see the book, **HIKING THE HISTORIC ROUTE of the 1859 MACOMB EXPEDITION.** The Macomb Expedition descended into Harts Draw via Macomb Canyon, but not via its upper drainage, which begins at the sandstone dome to the right of the road. In this vicinity, Harts Draw is more than 1200 feet deep.

From just ahead, on clear days, the distant peaks of the Henry Mountains are visible for a short distance, while more of Harts Point comes into view beyond Harts Draw. For the next several miles, the road goes through rounded meadows of alluvium and dune sand held in place by the sparse vegetation left after more than a century of overgrazing. The native grasses in this area that were described in historic accounts as "belly high to a horse" are now gone, leaving only arid-land shrubs.

11.9 The separate peaks of the Henry Mountains range are visible to the southwest, while the La Sals continue to dominate the northeastern horizon. The light-hued slickrock visible in the slopes to the south below the vegetated meadows continues to be Navajo Sandstone. In a few places, darker, layered Kayenta Sandstone is visible below the Navajo. The vertical walls of the inner gorge of Harts Draw are Wingate Sandstone, with slopes of Chinle Formation below that visible in a few places.

14.3 To the south, in the distance, the Needles District of Canyonlands National Park comes into view below and to the west of the tip of Harts Point. Beyond Harts Point, one of the Six-shooter Peaks soars above the lower terrain. On this side of Harts Draw, a remnant of Navajo Sandstone makes a squared-off dome. Other exposures of Navajo are visible to the left-front for the next half mile. For the next mile or so, the low ridges of reddish sandstone visible to the right are remnants of Slickrock-Entrada. The red sandstone domes in the distance between the La Sal Mountains and Hatch Point are the same rock.

15.4 Here, graveled Anticline Overlook Road angles to the right, while this road swings left. The high butte west of here is a remnant of Navajo Sandstone known as Horsehead Rock. To the northwest, the vertical cliffline of Island-in-the-Sky dominates the horizon. An ORV trail goes left here. It eventually reaches a large exposure of prehistoric invertebrate tracks within a layer of desert playa sediments in Navajo Sandstone. For a detailed description of this and other ORV trails on Hatch Point, refer to the book ***Canyon Country* OFF-ROAD VEHICLE TRAILS - Canyon Rims Recreation Area.** Just beyond this junction, both Six-shooter Peaks are visible in the lowlands to the southwest. Beyond them are the highlands associated with Elk Ridge, which is the spine of the Monument Uplift. See the sidebar, MONUMENT UPLIFT.

16.1 The low dome of rock to the right of the road is Navajo Sandstone. Note the cross-bedding that is characteristic of this sandstone.

19.0 As the road tops a final rise along this route, the extravagant panorama of Canyonlands National Park and vicinity comes into view. This panorama is described in detail from the overlook at the end of the road.

19.4 In this vicinity, the sheer walls of one tip of Hatch Point come into view. The laminated deposits on top of this wall are Kayenta Sandstone with the smooth Wingate Sandstone cliffs continuing down out of sight. Most of the rock visible between here and Needles Overlook on both sides of the road is Kayenta Sandstone, with an occasional remnant of Navajo visible in small ridges.

20.9 The ridge of rock to the right of the road is Navajo Sandstone. Note the reddish discolorations from the iron-rich layers above. This ridge and another one just ahead are topped by a deposit which contain minerals not found in the lighter-hued dune-sand. Rain runoff has carried reddish iron oxide from this deposit down onto the white dune-sand layer. In this lowest level of the Navajo, there was apparently some interlayering with the sediments from the same source that supplied the underlying red-hued Kayenta deposits.

22.5 The road ends here at Needles Overlook. Park and walk the designated trail around the protected rim of the point. Needles Overlook was so named because it provides an outstanding view of the Needles District of Canyonlands National Park and vicinity. If the distant

Henry Mountains are not crystal clear from this overlook, and the magnificent expanse of the park is not bright and colorful, the blame lies with the Huntington and Navajo coal-burning power plants to the west and southwest, both of which contribute massive amounts of air pollution to southeastern Utah.

The panorama from this overlook is described in a clockwise direction, beginning in the southeast. The convoluted cliffs of Hatch Point continue left from this viewpoint. The large gap in the meandering cliffline is the broad mouth of Harts Draw, the immense canyon system that separates Hatch Point from Harts Point, whose cliffs continue beyond the canyon. The cliffs of these two elevated peninsulas, from top to bottom, are layered Kayenta Sandstone above sheer-walled Wingate Sandstone, with Chinle Formation and Moenkopi Formation deposits forming the steep, rubble-covered slopes below. The low protruding dark red abutments are the Organ Rock Shale member of the Cutler Group. Domes of white Navajo Sandstone cap Harts Point in the near-distance.

The broad expanse of barren reddish rock below the overlook, on both sides of the meandering main drainage there, is undivided Cutler-Cedar Mesa sandstone. The developing canyon cutting into these colorful deposits is Lower Indian Creek Canyon, which drains both Upper Indian Creek Canyon and Harts Draw. For a description of the geology of Upper Indian Creek Canyon and the road into the Needles District of Canyonlands National Park, see the road log for Utah 211, elsewhere in this book. In the lower levels of the drainage below this overlook, the gray-hued limestone layer is a shallow-marine unit of the Cutler Group. The winding dirt road immediately below this overlook is the Lockhart Basin ORV trail, which is documented in the referenced books and maps.

Below the Abajo Mountains to the south, the sheer cliffs of Upper Indian Creek Canyon are cut by major drainages that originate in the mountain foothills. Below the western end of the mountains, the sharp spires of North and South Six-shooter peaks stand on top of pyramids of Chinle and Moenkopi sediments. The spires themselves are remnants of Wingate Sandstone. The terraced dark red slopes below the spires are Organ Rock Shale-Cutler deposits. The slender ribbon of Utah 211 is visible crossing the open terrain below the Six-shooter Peaks. The developed terrain beside this road in the lower end of Upper Indian Creek Canyon is part of Dugout Ranch, an historic homestead described in the book ***Canyon Country's*** **CANYON RIMS RECREATION AREA.**

The arid, red-hued flatlands on each side of the sandstone rimlands of lower Indian Creek Canyon are decomposed rock, sand and alluvium barely held in place by sparse native vegetation. The prominences on the skyline to the west of the Abajo Mountains are part of the highlands associated with Elk Ridge, the lofty spine of the Monument Uplift. See the sidebar, MONUMENT UPLIFT. Below that, and on to the west, are the complex erosional landforms after which this overlook was named. These "needles" eroded from color-banded Cedar Mesa Sandstone, which is another member of the Cutler Group.

To the west, the three separate units of the Henry Mountains dominate the horizon above the colorful Orange Cliffs. That Wingate Sandstone cliffline, with the usual Chinle and Moenkopi slopes below it, constitutes the western perimeter of the Maze District of Canyonlands National Park. The banded red and white sandstone below those cliffs is the Maze itself which is carved into Cedar Mesa Sandstone. The deep, winding sheer-walled canyon barely visible just this side of The Maze is the Colorado River gorge, whose confluence with the Green River is out of sight to the southwest. The high plateauland to the north and northwest is Island-in-the-Sky, part of which lies within Canyonlands National Park. Its cliffs are also Kayenta, Wingate, Chinle and Moenkopi in descending order, with white cliffs above that of Navajo Sandstone to the north. Below the Navajo-topped cliffs, two short stretches of the Colorado River are visible. The detached mesa to the left of Island-in-the-Sky is Junction Butte, which is almost precisely north of the Green-Colorado confluence.

The broad white-edged terrace below Island-in-the-Sky is White Rim Sandstone, an upper member of the Cutler Group. The Organ Rock Shale member of the Cutler forms vast dark red colonnades below the White Rim. The dark red spires directly below this overlook are also Organ Rock Shale-Cutler. To the northwest, in the open lowlands, the undivided Cutler-Cedar Mesa deposits on each side of Lower Indian Creek Canyon phase into similar-looking undivided Cutler deposits on north. Notice the long-abandoned dirt airstrip on the flat terrain below the viewpoint to the northwest. This was once used to supply miners who were prospecting nearby uranium-bearing Chinle deposits.

The continuing cliffline to the north and northeast is part of Hatch Point. The rounded hill of white rock to the northeast is Navajo Sandstone. In this vicinity of Hatch Point, including the surface of this overlook, the rock is Kayenta Sandstone, which descends in horizontal layers to the vertical cliffs of Wingate below. The La Sal Mountains dominate the northeastern horizon, while the distant Book Cliffs stretch across the northern skyline. The entire spectacular panorama from Needles Overlook has eroded into its present form during the last five or six million years. Prior to then, for more than 300 million years, this region was either very flat, with little topographic variety, or was under shallow seas. See the sidebar, TIME TRAVEL TOPOGRAPHY.

ROAD NAME: **U.S. 191 NORTH**

SUMMARY: This stretch of federal highway is the main route north from Moab. It provides access to Arches National Park, Dead Horse Point State Park and the Island-in-the-Sky District of Canyonlands National Park. It ends by connecting with Interstate 70. Along the way, this highway offers access to several popular scenic roads and off-road vehicle trails, some of which are described elsewhere in this book. The geology in view of the highway includes a wide variety of deposits, including several exposures of ancient marine salts, the distorted valleys produced by two salt anticlines, the severe stratigraphic offsets produced by a major fault line, and a close look at how such tectonic activities can warp normally level geologic strata. While most of this log's descriptions are made from safe pullouts, extreme caution is advisable while driving this highway and using this log, because of heavy truck traffic that generally travels at speeds exceeding the posted limits.

ROAD SURFACE: paved.

VEHICLE: any vehicle.

MAPS: ***Canyon Country* OFF-ROAD VEHICLE TRAIL MAP - Arches & La Sals Areas,** or ***Canyon Country* OFF-ROAD VEHICLE TRAIL MAP - Island Area,** or any highway map of Utah.

GUIDEBOOKS: ***Canyon Country* OFF-ROAD VEHICLE TRAILS - Arches & La Sals Areas,** or ***Canyon Country* OFF-ROAD VEHICLE TRAILS - Island Area,** or ***Canyon Country* HIGHWAY TOURING.**

MILEAGES:

TO START: Road starts at Main and Center in Moab.
ROAD: 32.2 miles, one-way to end.
RETURN: 32.2 miles, from end of described road.
TOTAL: 64.4 miles, from Main and Center, round trip.

TIME: one-half day.

ROUTE: north from the intersection of Main and Center streets in Moab.

NAVIGATION: drive north from Moab to where U.S. 191 ends at Interstate 70.

GEOLOGY ROAD LOG:

0.0 This road log begins in downtown Moab and goes north through Moab Valley. This immense northwest/southeast-trending valley is a large salt anticline. See the sidebar, SALT ANTICLINES. This anticline is exceptional in several respects. It is unusually long, its southeastern end is adjacent to the foothills of the La Sal Mountains and, for more than five million years it has been crossed by the Colorado River drainage, which has helped shape the valley that now lies above the upward bulging salt anticline.

GEOLOGIC STRATA VISIBLE FROM THIS ROAD

Formation	Member	Page
Green River Formation		Page 21
Wasatch Formation		Page 21
Mesaverde Group		Page 21
Mancos Shale		Page 20
	Ferron Sandstone	
Dakota Sandstone		Page 20
Burro Canyon/Cedar Mountain Formations		Page 19
Morrison Formation	Brushy Basin	Page 18
	Salt Wash	
	Tidwell	
Entrada Sandstone	Moab	Page 16
	Slickrock	
	Dewey Bridge	
Navajo Sandstone		Page 13
Kayenta Sandstone		Page 13
Wingate Sandstone		Page 12
Chinle Formation		Page 12
Moenkopi Formation		Page 12
Cutler Group	White Rim	Page 11
	Organ Rock	
	Cedar Mesa	
	Lower Cutler	
Hermosa Group	Upper Hermosa	Page 9
	Paradox Salts	

0.5 As the highway leaves the central part of Moab and swings to the northwest, note the low gray-hued mounds to the right of the road several places. These are exposures of the upthrust Paradox salts that formed this salt anticline valley.

To the right of the highway, just north of town, the uppermost rounded cliffs are Navajo Sandstone, with the terraced layers just below that being Kayenta Sandstone. Most of the broken rock on the higher slopes just above highway level is severely fractured Wingate Sandstone, while the deposits at and just above road level are Chinle Formation covered with rubble from higher up. The valley floor itself is an accumulation of sand, silts, alluvium and rock debris from both canyon walls, as well as from surface run-off during earlier wetter eras. The eastern valley walls are quite distorted and fractured by the original formation of the salt anticline and by faulting and slumping due to solution of the Paradox Formation salts that comprise the crown of the salt intrusion.

To the west, the giant break in the cliffline is the downriver portal, where the ancient Colorado River leaves Moab Valley. The rounded, lighter-hued rock on top of that western cliffline is Navajo Sandstone. The terraced, darker-hued rock below that is Kayenta Sandstone. On down, the short stretches of near vertical cliff are Wingate Sandstone, while the lower rubble-covered slopes are Chinle Formation. To the left of the portal, near ground level, the ridged, gray hills that are visible from the highway now and then are upthrust Paradox salts that have been uncovered and shaped by erosion. For a closer look at this unusual exposure of ancient marine evaporites, drive Cane Creek Road, described elsewhere in this book.

Note along this stretch of the highway that the cliffline to the right of the downriver portal is ascending, which places remnants of Navajo Sandstone and the layered Kayenta Sandstone high above valley level, while Navajo Sandstone is at road level as the highway passes the entrance to Arches National Park at the head of the valley. This enormous offset of geologic strata is due to the Moab Fault, which has also contributed to the formation of the canyon through which this highway leaves Moab Valley.

1.5 A large spring gushing from the broken slope to the east of the highway in this area provides a significant amount of water to the Moab vicinity. Some of the water is caught and stored in the tank above the highway, while part drains into the meadow west of the highway.

2.6 Utah 128 goes upriver here. That spectacular road is described elsewhere in this book. Just ahead, the highway crosses the Colorado River. At this upriver portal, on the far side of the river, the uppermost rounded cliffs are Navajo Sandstone above layered Kayenta Sandstone and sheer lower cliffs of Wingate Sandstone, with accumulated rubble at the base of the cliffs. Note, beyond the bridge, how these strata are sharply tilted upward to the west, again due to massive faulting and distortion by the Moab Valley salt anticline. Most of the severely fractured rock above the road to the right, beyond where it crosses Courthouse Wash, is Kayenta and Wingate sandstones, with some Chinle deposits beneath the rubble.

3.1 Here, the road crosses lower Courthouse Wash just above its confluence with the Colorado River. This drainage originates in the vicinity of Courthouse Pastures, to the north of the Monitor & Merrimac buttes, goes under this highway, then closely parallels it for a mile or so before angling east. It then enters Arches National Park via a rock-walled canyon. In the park, it continues down the deepening canyon, crosses open terrain for several miles, then travels through another canyon, emerging from the park here. Along the way, this long drainage travels part of the Courthouse Syncline, which helped shape its formation.

The sheer canyon walls at road level here are Wingate

Sandstone. Note, just beyond the drainage line, a remnant of the original vehicle road.

3.7 To the left is the site of an early uranium mill whose immense, radioactive tailings pile still occupies this end of the valley. This mill was active during each of the several booms in uranium mining in this region. While active, it occasionally spilled radioactive wastes into the Colorado River, and radioactive dusts from its piles of tailings and unprocessed ore commonly blew across the highway and into the entrance area of Arches National Park just ahead.

4.2 Utah 279, the "Potash Road," described elsewhere in this book, goes downriver here. Beginning here and continuing for a mile or so, there is Navajo Sandstone at road level on the right, while the same rock is out of sight above the left cliffline, with an offset of approximately 1,700 feet. In this vicinity, the main Moab Fault is just west of the highway, out of sight beneath sand and rock debris.

5.0 From a pullout on the right, note the immense dune of red sand to the left of the highway that has accumulated here because of prevailing winds through this narrowing part of the valley. The reddish sandstone outcroppings above the sand dune are Cutler sandstone. The Moenkopi and Chinle formation deposits above that are largely obscured by decomposed rock and larger rubble from the Wingate Sandstone cliffs on the skyline. To the right of the highway, in Arches National Park, the rounded light-hued rock just above road level is Navajo Sandstone. Above it, the layered, dark red rock is the Dewey Bridge Member of Entrada Sandstone, while the sheer cliffs above that are the Slickrock Member of Entrada Sandstone. Note how, just ahead, all three of these rock strata dive steeply downward to the west.

5.2 The Arches Road into Arches National Park, described elsewhere in this book, goes right here, then switchbacks upward along the interface between the light-hued Navajo Sandstone and the red Dewey Bridge-Entrada just above that.

5.6 Here, as the highway goes through the narrowest part of Moab Canyon, it crosses the Moab Fault. The gray layered rock on both sides of the highway is the upthrust upper deposits of the Hermosa Group. This outcrop of ancient rock is the sediments of a shallow sea and contains numerous marine fossils. Just ahead, this rock disappears underground again, leaving dark red remnants of Cutler Group sandstone visible on both sides of the road. Note just above the Cutler to the right of the road, the fractured Navajo Sandstone. That anomalous interface marks the actual line of the Moab Fault. This faultline can be observed to the right of the highway for several miles as the interface between mismatched deposits.

The high cliffs to the left of the highway continue to be Wingate Sandstone standing on terraces of Chinle, Moenkopi and Cutler deposits, in descending order, all of which are largely obscured by fallen rubble. Note, for the next mile or so, the immense boulders of Wingate Sandstone on both sides of the highway that have fallen from high above. One such fairly recent collapse can be seen by the lighter-hued nature of the cliff and rubble below.

7.6 Note along this stretch, to the right of the highway, how the colorful deposits of Cutler sandstone are angled up toward the Moab Fault line, which is bearing to the east and out of sight for now. Stretches of the old highway can be seen traveling very close to the faultline. Just ahead, the dark red, crenelated cliffs to the left of the highway are Cutler deposits. Unfortunately, the outstanding beauty of this exposure of that ancient rock has been marred by railroad tracks and a forest of utility poles and lines.

The large gap in the higher Wingate Sandstone cliffline is the easternmost extension of Little Canyon. This strange "hanging canyon" drains away from the cliffs above the highway, and eventually joins the Colorado River gorge. A large part of that drainage follows another local tectonic feature, the Kings Bottom Syncline.

9.2 From a pullout on the right at the summit of the grade, note that the western cliffline is dipping downward, leaving only scant remnants of the Wingate dominating the skyline in this part of the Sevenmile Canyon portal. The base of this lower peninsula of rock is Cutler sandstone, below thin laminations of Moenkopi sediments, then steep slopes of gray-hued Chinle deposits. To the east of the highway for the next several miles, the relatively flat expanses of terrain are largely decomposed rock and sand held in place by native vegetation, with occasional outcrops or remnants of native rock. The low bluffs of light-hued sandstone to the right-front are the Salt Wash Member of the Morrison Formation set above thin exposures of the reddish Tidwell Member of the same formation.

10.1 The Gemini Bridges ORV trail, described elsewhere in this book, joins U.S. 191 here on the left. The open terrain in this vicinity and around the mouth of upper Sevenmile Canyon is a mixture of sand, alluvium and decomposed rock, plus volcanic ash deposited within the last three-quarter million years.

11.1 The highway crosses the Sevenmile Canyon drainage. This shallow wash continues east for another several hundred yards, then plunges abruptly down into a deep canyon that is walled by Entrada Sandstone. The flat, open terrain in this vicinity is largely decomposed rock, plus sand and alluvium deposited by the Sevenmile Canyon drainage over millennia, with occasional outcrops of red and white Morrison Formation sandstones.

The main branch of the Moab Fault goes under the highway in this vicinity.

11.3 Utah 313 goes left here to Dead Horse Point State Park and the Island-in-the-Sky District of Canyonlands National Park. That road, and the geology of Sevenmile Canyon, is described elsewhere in this book. The Moab Fault now lies to the west of the highway, roughly on the alignment of the railroad tracks there for almost a mile. North of the Sevenmile Canyon portal, note that the Chinle Formation deposits below the upper Wingate Sandstone cliffs are terraced with uranium mining trails. Several mine shafts are visible in these steep slopes of gray sediments up ahead. The low outcrops of rock to the east of the highway for the next several miles continue to be Morrison Formation deposits.

12.8 Note Corral Canyon in the left cliffline. The Chinle in this short drainage has also been cut by uranium exploration bulldozers.

13.2 Stop at a small pullout here beside a shallow wash and study the western cliffline. The main branch of the Moab Fault system angles northwest in this vicinity, drastically offsetting the geologic strata in the cliff. The sheer red wall of Wingate Sandstone ends abruptly, and the Chinle Formation deposits just below the Wingate phase directly into the painted desert slopes of the Brushy Basin Member of the Morrison Formation. This same colorful deposit dominates the western skyline ahead and to the northeast. The faulting here in the western cliffline has confused many people, including some geologists, because the fault displacement has aligned the two gray-hued deposits.

To the north of this pullout, the distant red fins are Slickrock-Entrada Sandstone, topped by remnants of white Moab Member of Entrada Sandstone, with the lowlands continuing to be dominated by brush-covered sand, decomposed rock and alluvium. The distant slopes to the east are a jumble of Morrison deposits with occasional outcrops of Entrada sandstones.

14.3 Major discoveries of petrified dinosaur bone have been made in the painted desert Brushy Basin-Morrison hills to the east of the road here. Morrison deposits continue to prevail on both sides of the highway. Beyond here, however, the Morrison deposits soon disappear underground, leaving the terrain on both sides of the highway dominated by outcrops of light-colored Cedar Mountain Sandstone, topped with a few thin ledges of Dakota Sandstone.

15.8 The Monitor & Merrimac Trail, described elsewhere in this book, goes left here. Just ahead, the highway crosses upper Courthouse Wash, which it first crossed at mile 3.1.

17.0 The Klondike Bluffs Trail, described elsewhere in this book, goes right here toward the distant Klondike Bluffs. These are visible to the east of the highway for the next several miles as soaring spires of red Slickrock-Entrada topped by white remnants of the younger Moab-Entrada. The other rock outcrops to the east of the highway here are described in more detail in the Klondike Bluffs trail log, elsewhere in this book.

To the south and southwest of the highway along this stretch, the distant flat-topped buttes are Entrada sandstones, with some topped by remnants of Tidwell and Salt Wash Morrison deposits. The uppermost sheer walls beneath these rubbly deposits are Moab-Entrada, while the lower red-hued cliffs visible a few places are Slickrock-Entrada. To the northwest, the low hills of gray sediments are Mancos Shale. This ancient sea-bottom deposit dominates the landscape to the west of the highway from here to its junction with Interstate 70. To the east of the highway along this stretch, the western slopes of the Salt Valley Anticline, in Arches National Park, continue to dominate the skyline. That anticline has thrust older formations into view. These are described ahead.

17.6 The Blue Hills Road goes west here, providing access to several popular ORV trails, including the Hidden Canyon Rim Trail, described elsewhere in this book.

18.7 The road left here goes to the Grand County air terminal. In this vicinity, the Klondike Bluffs dominate the eastern skyline, with the white Moab-Entrada sandstone angled upward on the anticline slopes. Note how this rock has been fractured in a checkerboard pattern by the uplifting. The low hills to the left of the highway are Mancos Shale, topped by a thin layer of the Ferron Sandstone Member of Mancos Shale, which is rich in shallow-marine fossils. The Salt Valley Anticline continues to dominate the eastern horizon, with occasional outcrops of Dakota and Cedar Mountain deposits in the near foreground.

22.5 In this vicinity, where a road spurs left to a microwave tower, note the exposure of white Moab-Entrada slickrock to the right near the crest of the anticline slopes to the east. This slickrock is the site of an exposure of 2300 theropod dinosaur tracks. Many dinosaur tracks made by various species have been found in the uppermost levels of the Moab-Entrada, where it has been exposed on both the eastern and western slopes of the Salt Valley Anticline. Near the old mine trail that approaches that site, the first sauropod dinosaur foot tracks found in Utah were discovered and documented in 1990. These tracks are in Morrison Formation sandstone. A dirt road goes to this closer site. See the sidebar, PREHISTORIC TRACKS.

24.5 To the west of the highway, Mancos Shale

continues to dominate. The broad flatlands are dune sand, alluvium and remnants of Mancos Shale. To the east, the Salt Valley Anticline has thrust up underlying Dakota and Cedar Mountain sandstones, with the upper gray-hued sandstone ledges being the Dakota. These deposits continue to dominate the slopes east of the highway for the next couple of miles.

26.7 Here, the highway crosses a shallow drainage line that originates to the northeast. In that drainage, just beyond the tilted ridge of Dakota and Cedar Mountain sandstone to the right of the cut, are the remnants of the historic Valley City reservoir. This earthen dam was constructed between 1905 and 1908 and was used for a time to irrigate the orchards and truck gardens of nearby settlers. The dam was then destroyed by flooding, but was rebuilt several years later and used by the remaining settlers into the 1930s. The desolate area was eventually abandoned after repeated flooding made living there impractical. The tilted exposure of white sandstone east of the drainage is Moab-Entrada.

27.6 The Thompson Road goes to the right here. This was once the main road to the Thompson's Springs settlement and local depot for all railroad shipping and receiving in the Moab area. ORV trails described in the referenced book and map are accessible from this road.

North of the rocky ridges that bracket this spur road, the deposits on both sides of the highway, all the way to Interstate 70, are various units of Mancos Shale. The low, smooth ridges just ahead on both sides of the road are topped by thin ledges of the Ferron Sandstone member of this deposit. The flat, open areas are partly recent alluvium, dune sand and decomposed rock, held loosely in place by the remnants of native vegetation that survive after a century of overgrazing by domestic cattle and sheep. This severe overgrazing and denudation of the fragile soils has caused accelerated erosion in the drainage lines that go south across these flats. One of these is Salt Wash, which drains through Arches National Park and eventually reaches the Colorado River within a deep, spectacular canyon. The alluvium that built up in this long drainage over past millennia has now been deeply cut by the accelerated erosion and flash flooding promoted by the overgrazing, once again illustrating that humanity is a major geologic force. See the sidebar, HUMANITY AS A GEOLOGIC FORCE.

32.2 U.S. 191 ends here at its junction with Interstate 70. From the vicinity of this junction, the Book Cliffs dominate the northern horizon. The lower slopes of these cliffs are gray-hued Mancos Shale. Above that, the first several hundred feet of deposits are various members of the Mesaverde Group, including the shales, sandstones and coal deposits of the Blackhawk Group, the brown cliffs of Castlegate Sandstone, and several others. The higher and more distant slopes and bluffs are, in ascending order, the Tuscher Formation, the Wasatch Formation and remnants of the Green River Formation. The deposits in this line of cliffs vary to the east and west of this general locality.

The Mesaverde Group is defined as shoreline and coastal plains sandstones and shales, with some coal deposits, that overlie the deep Mancos Shale sediments left by an invading sea that covered this entire region for most of the long Cretaceous Period. The Dakota, Burro Canyon and Cedar Mountain sandstones represent shoreline and coastal plains deposits that were covered as this invading sea started its encroachment.

U.S. 191, left-center, viewed from Gold Bar Rim.

PREHISTORIC TRACKS

An enormous span of time, totaling more than 200 million years, is represented by the geologic strata exposed in the Moab vicinity, from Paradox Formation salts deposited in a shallow coastal basin, to Mancos Shale left by a great inland sea.

Over that span of time, the environmental conditions that prevailed during some periods of deposition were hospitable to life, some were harsh but bearable to a few forms of life, and others were hostile to all but microscopic life.

Paleontologists can describe the conditions necessary for the preservation of plants and animals in the form of fossils that will endure through the ages. These conditions are fairly narrow so, of the vast numbers of plants and animals that lived and died during the long time represented by exposed geologic strata, relatively few fossils exist, with the exception of one set of conditions. In shallow-marine deposits, conditions were ideal for the preservation of mollusks and other sea creatures that had hard shells or internal structures.

Land animals, however, also left other traces of their existence besides their hard bones, teeth, claws and sometimes egg shells. They left foot tracks, and the conditions required to preserve such tracks occurred fairly often. Thus, prehistoric petrified foot tracks are far more common in many geologic strata than are the fossilized remnants of the animals that made them.

All it takes to preserve visible foot tracks are the tracks themselves, in some kind of soft, wet sediments that will preserve their shapes for at least a short time. Then the tracks must be filled with somewhat different material and the entire assembly must remain undisturbed by subsequent erosion, so that millions of years later, when the petrified deposits containing the tracks are exposed to weathering, the casting material will erode away preferentially, leaving the molds of the original tracks on the harder rock surface.

In some rare cases, ledges of rock may break away and the fallen rock delaminate, exposing both molds and casts of ancient foot tracks. This author found one such specimen, and that slab of rock, bearing the casts of several dinosaur tracks, is now on display in Moab's Dan O'Laurie Museum. A photograph of the track-cast slab, as found in the field, appears in the book, Canyon Country's CANYON RIMS RECREATION AREA.

Conditions ideal for the making and preservation of tracks commonly existed along the banks of prehistoric lakes, rivers and streams, and in marine tidelands. There, wet shoreline muds, sandbars, and tideland deposits made good foot track impressions, and materials of different types borne by wind or water commonly filled the tracks, thus preserving their shapes. A few such tracks made in fine-grained sediments are so detailed that they retain the impressions of the scaly skin on the animal's feet.

Most foot tracks also compressed several layers of deeper sediments, as well as those on the surface, thus leaving not only surface tracks but less detailed impressions in lower layers. Such laminated foot track depressions become clearly visible when cross-sections of petrified tracks are exposed by vertical breaks through surface tracks.

There are a number of such cross-sections exposed at one track site in the Hatch Point District of Canyon Rims Recreation Area, near Anticline Overlook Road, which is logged elsewhere in this book. The track site is described and depicted in the book about this recreation area noted earlier.

In time, as such layered tracks are exposed by the vagaries of erosion and weathering, the surface tracks may be lost, leaving only the less distinct "undertracks." These, however, once identified as what they are, can also contribute knowledge about the creatures that made them, as well as the paleoenvironment in which they lived.

Occasionally, the several layers of sediments beneath a prehistoric foot track are so different that, once exposed to weathering, the compressed, harder layers of rock just beneath the track will resist erosion more than still deeper layers, thus allowing the surface layers of the tracks to loosen from their rock matrix. This author has found and reported several such "portable" petrified foot tracks.

Petrified prehistoric foot tracks occur in most of the geologic strata exposed in the Moab area. Some strata, especially those deposited immediately following the mass extinctions described in the sidebar, GREAT EXTINCTIONS, are virtually barren of signs of life. Other strata are rich in remnants of life, both fossils and tracks.

One stratum that some geologists and paleontologists still think was deposited in an environment virtually devoid of life, is actually rich in both fossils and petrified foot tracks, with most of the discovered sites reported by amateur geologists who live in Moab. This author has reported a number of such fossil and track sites, and there are doubtless many other sites still awaiting discovery, perhaps by readers of this book.

Following is a list of the geologic strata in the area covered by this book that are known at this time to have petrified prehistoric foot tracks. In a few cases, the tracks are at the interface between two distinct strata.

LOWER CUTLER DEPOSITS. This author has found two sites of what appear to be the foot tracks of two different types of animals in tidal-flat deposits that also contain shallow-marine fossils. Apparently, land-dwelling creatures walked across mud flats exposed at low tide, seeking food. These sites were reported in 1991, but as of the publication date of this book, neither site had been documented and studied by track specialists.

continued ----

CHINLE FORMATION. Tracks are scarce in this formation but have been found. Since the Chinle was deposited just before the mass extinction that gave dinosaurs their start, the tracks were made by the mammal-like reptiles called "therapsids" that dominated the land before that extinction. One site of four-toed tracks in Chinle deposits is in the Indian Creek Canyons District of Canyon Rims Recreation Area, and is depicted in the book about that area noted earlier.

WINGATE SANDSTONE. Since this massive aeolian sandstone was deposited following the extinction noted above, traces of life are scarce in its older deposits, but in 1990 and 1991, large exposures of small three-toed dinosaur tracks were reported and documented near the top of the Wingate, indicating that some life was returning to the region then, and that the renewed ecosystem was being dominated by the dinosaurs then evolving.

KAYENTA SANDSTONE. By the time this rock was being deposited, a major change of climate had taken place, and fresh-water sediments were being deposited beside lakes, streams and rivers. This hospitable region was frequented by a variety of larger dinosaurs, although most of the petrified tracks found in the Kayenta to date have been three-toed "theropods," or carnivorous dinosaurs. Doubtless, tracks of the animals these predators pursued are waiting to be found in Kayenta Sandstone by observant amateur geologists.

KAYENTA - NAVAJO INTERFACE. Theropod tracks have also been found at the interface between Kayenta Sandstone and Navajo Sandstone. One such site is on public display beside Utah 279, which is described in this book. Another such site, found and reported by this author in 1970, is beside Cane Creek Road, also described, but was damaged and covered by road-grading equipment before it could be studied.

NAVAJO SANDSTONE. This desert-dune deposit is rich in both fossils and petrified foot tracks, although until fairly recently its geologic age was in doubt because of a lack of fossils to use for dating. Within the last several years, however, amateur Moab geologists have found and reported a large number of fossil and track sites, most of which have yet to be studied by scientists. See the sidebars, PETRIFIED PLAYAS and PETRIFIED OASES. Doubtless there are many more such sites in this fascinating deposit. Most of the petrified tracks in this desert-aeolian sandstone are in its playa and oasis deposits, and include the tracks of therapsids, theropods and a variety of small animals. Certain paleontologists have identified some of the smaller tracks as made by "pterosaurs," or flying reptiles. Others disagree with this, claiming these tracks were made by small crocodile-like reptiles, although it is difficult to see how such a water-dependent species could survive in the shallow, transient waters of the desert playas that contained the tracks. Other such playas have myriad petrified invertebrate tracks, some large, some tiny, but all beautifully preserved. These are equally intriguing, but have yet to be studied by scientists. In any case, there was quite a variety of life, both plant and animal, at the permanent oases in the Navajo Sandstone desert, and many species of animals visited the numerous ephemeral lakes. Some of them left their foot tracks in the muds that surrounded these bodies of water. They probably came to drink, and to eat the tiny "cryptobiotic" crustaceans, insects and amphibians that hatch from desert drylake deposits when they are wetted by the scarce rain -- then and today -- or to prey on the smaller animals that ate the transient water-life.

NAVAJO-ENTRADA INTERFACE. As this book was being written, the author found what may be an extensive site of fossilized quadruped dinosaur undertracks at the interface between Navajo Sandstone and the Dewey Bridge Member of Entrada Sandstone. While the series of depressions in the exposure would be seen by the casual observer as shallow potholes, which are common in level sandstone, these particular "potholes" made patterns reminiscent of the kinds of parallel, alternating trackways left by large quadruped dinosaurs. One such set of "potholes" near Moab proved to be the first sauropod tracks found in Utah. Sauropods were large herbivorous dinosaurs, like the well-known Brontosaurus.

ENTRADA-MORRISON INTERFACE. A large number of petrified theropod and other tracks have been found at the interface between the Moab Member of Entrada Sandstone and the Tidwell Member of the Morrison Formation. While most of the tracks are in the white beach-sands of the Moab-Entrada, they were preserved by being filled with red-hued Tidwell-Morrison sediments. The Moab-Entrada eventually made harder rock, so as the interface is exposed by erosion, the less durable Tidwell-Morrison sediments quickly erode away, leaving the tracks exposed. Even though the Moab-Entrada is harder, it still weathers, and the exposed tracks are slowly destroyed. This means that at the interface where tracks are exposed, largely along the eastern and western slopes of the Salt Valley anticline, most remaining tracks are within a few feet of the distinct red-on-white interface. Those that have been exposed longer are gone. One exception to this is a large exposed area where an unusually durable expanse of Moab-Entrada has more than 2,000 well-preserved theropod tracks. Moab-Entrada beach sand and tideland deposits offered ideal conditions for the preservation of tracks, and these shorelines of one tongue of an inland sea were apparently quite popular with dinosaurs. There are a number of tracks at this interface near the Klondike Bluffs Trail, and a few near the Onion Creek-Thompson Canyon-Polar Mesa Trail, both described in this book.

MORRISON FORMATION. The periods during which the Salt Wash and Brushy Basin members

continued ----

of the Morrison Formation were deposited were rich with life of all kinds, both plant and animal, and many of these left fossil evidence of their existence. There are a number of such sites within the area covered by this book. Some animals also left petrified foot tracks. Although large expanses of Morrison Formation are exposed in the area covered by this book, track sites are not common because the conditions necessary for track preservation were relatively rare. There is one such site near Moab, however, not far from U.S. 191 North, which is described in this book. There, the first sauropod tracks to be discovered in Utah, as noted earlier, are in deposits of the Salt Wash-Morrison. There are also several theropod tracks in the same area. This site is open to the public via a short road that spurs from U.S. 191 North. In other geographic areas not covered by this book, exposed Morrison Formation deposits have more dinosaur trackways.

The chances are good that still other petrified tracks will be discovered in the general Moab vicinity by amateur geologists, in the above geologic strata and perhaps others too, where tracks are not now known to exist. Alert amateurs have long contributed greatly to areas of science that require extensive field work, because amateurs far outnumber professional scientists and are also quite often more dedicated and enthusiastic than their counterparts. Further, most amateurs pursue their interests vigorously, at their own expense, while few professionals undertake the extensive field work necessary to make such discoveries without special funding and formal sponsorship. Thus, all too often, progress in such sciences depends largely on dedicated amateurs for field discoveries. Responsible amateurs who make such discoveries in canyon country should carefully note their locations, photograph the sites, then report them to the appropriate land administration agencies. These may be either the U.S. Forest Service, the Bureau of Land Management or the State of Utah, which administers certain public lands. If in doubt about land administration, reports of discoveries made in the area covered by this book should be made to the Moab BLM.

Dinosaur tracks near the Klondike Bluffs Trail.

Typical pre-dinosaur track in Shay Canyon, near Utah 211.

Sauropod tracks at public display near U.S. 191.

ROAD NAME: **U.S. 191 SOUTH**

SUMMARY: This federal highway is the main travel route south from Moab and the access route for the Cameo Cliffs and Canyon Rims recreation areas and the Needles District of Canyonlands National Park. The highway travels through colorful high-desert terrain and offers views of a number of geologic highlights and erosional landforms. Since this popular highway is heavily traveled and offers relatively few designated pullouts where the surrounding scenery can be studied in safety, some parts of the following geologic road log are written for study while traveling along stretches of highway. Others describe, from safe pullouts, geologic features that can be spotted and studied beyond the pullouts, without being endangered by high-speed traffic.

ROAD SURFACE: paved.

VEHICLE: any vehicle.

MAPS: ***Canyon Country* OFF-ROAD VEHICLE TRAIL MAP - Canyon Rims Recreation Area** and **Map of CAMEO CLIFFS Biking - Hiking - Four-Wheeling.**

GUIDEBOOKS: ***Canyon Country* OFF-ROAD VEHICLE TRAILS - Canyon Rims Recreation Area, CAMEO CLIFFS Biking - Hiking - Four-Wheeling** and ***Canyon Country* HIGHWAY TOURING.**

MILEAGES:

TO START: Road starts at Main and Center in Moab.
ROAD: 41.1 miles, one-way to end, as described.
RETURN: 41.1 miles, from end of road as described .
TOTAL: 82.2 miles, from Main and Center, round trip.

GEOLOGIC STRATA VISIBLE FROM THIS ROAD

Formation	Member	Page
Mancos Shale	Ferron Sandstone	Page 20
Dakota Sandstone		Page 20
Burro Canyon/Cedar Mountain Formations		Page 19
Morrison Formation	Brushy Basin Salt Wash Tidwell	Page 18
Entrada Sandstone	Moab Slickrock Dewey Bridge	Page 16
Navajo Sandstone		Page 13
Kayenta Sandstone		Page 13
Wingate Sandstone		Page 12
Chinle Formation		Page 12

TIME: allow for a short day to travel this highway as described and back to Moab. If the trip is to continue on Utah 211 into the Needles District of Canyonlands National Park, the combination trip requires a full day.

ROUTE: south on Main Street (U.S. 191) to where this geology road log ends, or continue to another destination. Return via the same route.

U.S. 191 South at Wilson Arch.

NAVIGATION: drive south on U.S. 191 to just beyond its junction with Utah 211, where this geology log ends, or continue to whatever other destination is planned.

GEOLOGY ROAD LOG:

0.0 The highway heads south from Main and Center, Moab.

1.0 Park briefly here, where this federal highway begins to swing to the left. Beyond here, the highway travels the southeastern extension of Moab Valley, which is one of a series of parallel salt anticline valleys that lie on both sides of the La Sal Mountains. All of these salt anticlines are much older than the La Sals, although the valleys themselves that formed over these anticlines are at least partly the products of erosion since the mountains formed. See the sidebars, SALT ANTICLINES and LACCOLITHIC MOUNTAINS.

As the road travels southeast through Moab Valley, study the soaring sandstone walls that bound it on both sides. For the first several miles, the bare slickrock at the top of the broken cliffline on the left is Navajo Sandstone. Below this light-hued rock are slopes of Kayenta and Wingate sandstones, although these have been highly fractured by the massive faulting that borders this valley on both sides. Some of this faulting occurred during the original uplifting cycles of the Moab Valley salt anticline, some during subsequent slumping from salt dissolution as the valley formed.

The soaring cliff on the right is topped with a few visible domes and ridges of Navajo Sandstone above jagged rimlands of layered Kayenta Sandstone, with the lower Wingate Sandstone largely fractured and obscured by rock debris. A few rubble-covered slopes of Chinle Formation deposits lie at the base of the cliff and jut into the valley in places as rounded, brush-covered mounds. This cliffline, all the way to the valley end, is marked by slumping masses of rock that have been sheared from the cliff by faulting parallel to the line of the valley. The low hills that protrude outward into the valley straight ahead are examples of this. This slumping has also created stretches of detached cliffline in which the geologic strata are somewhat misaligned.

The valley floor is a mixture of decomposed rock from many formations, plus alluvium, glacial gravels and wind-borne sands that have accumulated over millions of years. In their natural form, these sandy soils were held in place by cryptogams and native vegetation, but human activities that have broken or removed the natural soil surfaces have promoted both wind and water erosion. This accelerated erosion can be observed from this road in many places as it travels Moab Valley. On windy days, great quantities of dust are lifted from such disturbed areas, but little or none from natural surfaces. See the sidebars, HUMANITY AS A GEOLOGIC FORCE and CRYPTOGAMIC SOILS.

1.5 The gap in the cliffline to the left-front is where Mill Creek enters Moab Valley. This perennial stream originates high in the La Sal Mountains at and beyond Oowah Lake, and has been used as a source of water in Moab Valley since historic times. Between the mouth of Mill Creek Canyon and South Mesa, the creek runs down a canyon behind the Navajo Sandstone fins that wall Moab Valley. The immense mass of gravel to the right of the stream portal is the remnant of a vast alluvial fan that was deposited there by heavy stream-flow during and following one of the major glaciation cycles in the La Sals. See the sidebar, GLACIERS OF THE LA SAL MOUNTAINS.

All of the water used in Moab Valley for culinary purposes comes from springs and underground aquifers that are fed by rain and snow-melt in the La Sal Mountains. Water used for agriculture comes from either Mill Creek or Pack Creek, or from underground aquifers, all of which originate in the La Sals. Without this mountain range, Moab Valley would be too arid to inhabit, even though the Colorado River crosses it. The river is too highly mineralized and too polluted by human activities to use for culinary purposes.

4.4 Along this stretch of highway, note the pyramid-shaped Navajo Sandstone fin directly to the left. Here, a break in the dominating Navajo Sandstone cliffline in that direction permits a view of the more distant rolling domes of Navajo Sandstone in the Sand Flats area, and the elevated plateaulands that jut north from the La Sal Mountains. South Mesa, Wilson Mesa and adjoining unnamed highlands are immediately below the La Sal peaks. These plateaus are rimmed with the Moab Member of Entrada Sandstone, with slopes below of the Slickrock Member of Entrada. The nearer mesa below these is called Johnsons Up-On-Top. It is walled with Navajo Sandstone, but topped with remnants of younger deposits.

In the vicinity of the pyramid-shaped fin in the valley floor to the left, a road ascends to a pass, then drops down into Mill Creek Canyon and continues upcanyon, intermittently fording the stream, then climbing out into the slickrock fantasyland of the Mill Creek Triangle. This elevated maze of Navajo Sandstone is defined by the deep canyons of the two forks of Mill Creek, plus the steep slopes of South Mesa. The valley floor continues to be dominated by a mixture of sediments and sands held in place by native vegetation, except where disturbed by human developments.

The cliff to the right of the highway is still Navajo, Kayenta and Wingate sandstones, in descending order, but as the road climbs toward the far end of the valley, the cliffline is ascending too, exposing more stretches of sheer-walled Navajo and Wingate sandstones. Large masses of these strata are detached from the cliff.

7.0 The rubble-covered ridge that walls the valley below South Mesa is Slickrock-Entrada that is largely obscured by remnants of younger Morrison deposits. The uppermost cliffs to the right of the highway are Navajo Sandstone above terraced Kayenta Sandstone, with the lower Wingate Sandstone partly obscured by rubble and slumped material.

7.9 Stop at a pullout on the right here. Opposite the pullout, a commercial operation is excavating alluvial gravels. These gravels were deposited in the valley by Pack Creek during and following various periods of glaciation, and are thus partly composed of the intrusive igneous rock now exposed in the higher peaks of the La Sals. This long-standing alluviation of upper Moab Valley by Pack Creek is more pronounced and visible where the creek first reaches the valley. The Spanish Valley Road that provides the western access to the La Sal Mountain Loop Road goes through a long stretch of this upper part of the creek. During historic times, Pack Creek flowed continuously for most of the season, but was diverted into an irrigation ditch by early settlers and is now dry most of the time.

Sheer-walled South Mesa looms above the ridge noted earlier, but still below the higher slopes of the La Sals. The road toward Kens Lake goes left just beyond this pullout. Kens Lake is an earth-fill reservoir in the valley filled by diverting part of Mill Creek through a tunnel. Beyond this stop, water can sometimes be seen cascading down the valley wall from the tunnel.

This tunnel was started decades ago by a man named Sheley, in an attempt to bring water to this part of Moab Valley for agricultural purposes. Sheley's tunnel was not completed at that time, and was abandoned until the mid-1980s, when it was completed as a part of the Ken's Lake water project. Now, a low diversion dam on Mill Creek sends part of its flow through Sheley's tunnel down a rocky cascade and into Ken's Lake. This use of the limited and seasonal flow of Mill Creek is highly inefficient, however, because a significant amount of water is lost into the softer, more permeable strata between Sheley's tunnel and Ken's Lake, and an even larger amount is lost due to evaporation from the reservoir. This, however, is typical of similar water development projects in the high desert country of southern and eastern Utah, which are constructed without regard for rational economics or environmental impact.

From the road ahead, watch the valley cliffs to the left for a large pile of whitish sediments midway up the slope. This is an exposed and eroded remnant of younger material -- probably Burro Canyon deposits -- that collapsed downward to fill a void during an early stage in the formation of Moab Valley.

9.0 The low, rounded hills appearing in the valley ahead are Burro Canyon deposits that are almost completely obscured by relatively recent glacial alluvium. Along this stretch, the highway goes through several cuts in large masses of material slumped from the right-hand cliffline.

11.0 From where an ORV trail spurs right, note the low ridge in the valley to the left of the highway. This is another remnant of older deposits, in this case Entrada and Navajo sandstones, that are almost completely buried under alluvium. Note how the geologic strata on the western cliffline have descended in this part of the valley, leaving vertical or rounded remnants of Navajo Sandstone at the top of the cliffs. The lower slopes are fractured Kayenta and Wingate sandstones.

11.7 The color-banded deposits to the left and left-front here are a mixture of Burro Canyon and Morrison deposits that have been severely faulted. The gray slopes visible through the gap in the same direction are Mancos Shale.

11.9 Note, to the right-front, just before the road begins to ascend Blue Hill, the unnatural slope there that was created during the installation of the gas pipeline that descends into the valley floor at that point. To the east, the gentle gray-brown slopes of sediments are Mancos Shale, displaced to this level by salt valley faulting and surface subsidence.

12.2 As the highway ascends Blue Hill, it climbs through Morrison Formation deposits. The blue-gray color in one level of these deposits gave this grade its name. As the highway tops the grade, the rock walls on both sides are the Salt Wash Member of the Morrison Formation.

13.3 The Cane Creek Canyon Rim ORV trail goes right here, while Black Ridge Road goes left. Both of these are described elsewhere in this book. Beginning here, the terrain to the left of the road is part of the Cameo Cliffs recreation area, described in the book **CAMEO CLIFFS Biking - Hiking - Four-Wheeling.** The terraced rock ahead on both sides of the road is Salt Wash-Morrison. As the road swings left and descends into the Cane Creek drainage, note where the gas pipeline noted earlier has cut through terraces of Morrison Formation deposits.

As the highway descends into the Cane Creek drainage about 1.5 miles ahead, note to the left where the red Entrada-Slickrock that appears there has been aligned with the Morrison Formation by a fault. Above the road to the right, as it descends, remnants of the original vehicle road can be seen in the form of rock retainer walls. This first motor vehicle road between Moab and Monticello was completed about 1920. Remnants of an earlier wagon road can be glimpsed as a narrow, rocky terrace to the left of the highway as it first starts descending. This eroded historic trail is more easily spotted while ascending the grade.

15.0 The fault noted earlier is in the steep slope to the left here.

15.5 The highway crosses the Cane Creek drainage.

15.6 Viewed from a pullout here, the soaring red cliffs are Slickrock-Entrada topped in most places with remnants of Salt Wash-Morrison. To the right of the Kane Springs Roadside Rest, note the hills of gravel carried down by Cane Creek from the La Sal Mountains over a long period of time, but stranded as the creek cut more deeply. For more information about this drainage, refer to the log for Black Ridge Road. The spring water available at this roadside rest was used historically by travelers between Moab and Monticello, and by ranchers on cattle drives. For more information about the colorful early history of this general area, refer to the books, ***Canyon Country's*** **CANYON RIMS RECREATION AREA,** and **CAMEO CLIFFS Biking - Hiking - Four-Wheeling.**

NOTE: The original name for this drainage was "Cane Creek" and the springs in the vicinity of the roadside rest were "Cane Springs." Both were named after the native water-loving cane-grasses that once thrived in the area, and still do farther down the drainage. When the State of Utah decided to develop Cane Springs as a roadside rest, it applied the catchier name "Kane Spring" to the development. Unfortunately, this has led to confused nomenclature in subsequent literature, including U.S. Geological Survey maps. Earlier U.S.G.S. maps, dated in the 1950s, show "Cane Springs," while more recent ones use the incorrect "Kane." Canyon Country Publications uses the original spelling, here and in its other guidebooks and maps.

As the road continues beyond here, the terrain to the left is all part of Cameo Cliffs recreation area, while the terrain to the right, and on to the end of this description, is in Canyon Rims Recreation Area. As the road travels for the next several miles, the red sandstone cliffs continue to be Slickrock-Entrada, sometimes topped with Morrison deposit remnants. Along some exposures, the layered, dark red Dewey Bridge Member of Entrada is exposed below the smooth Slickrock-Entrada cliffs. Most developing drainages cut down into lighter-hued Navajo Sandstone.

16.6 Stop here at an ORV trail junction on the right. The dome of bare rock to the right-front is Slickrock-Entrada and Dewey Bridge-Entrada, resting on a terraced slope of Navajo Sandstone that continues down into lower Muleshoe Canyon, which is visible from the road on ahead, and which ultimately joins Cane Creek Canyon west of the highway.

To the left of the highway, the rounded cliffs are Slickrock-Entrada, while the higher rubble-covered slopes are Morrison Formation topped with a thin remnant of undivided Dakota-Burro Canyon sandstone. Note the microwave tower on top of the promontory that marks the end of Black Ridge Road. These Morrison Formation slopes have been heavily prospected for uranium, with bulldozed mine trails visible here and ahead. The flat-topped hill straight ahead is also a remnant of Morrison deposits on top of Slickrock-Entrada.

18.5 The highway crosses a bridge over the Muleshoe Canyon drainage.

20.0 The sharp-peaked dome to the left-front along this stretch of the road is called The Nipples. This feature was probably named by cowboys on the early cattle drives that traveled this route before any roads existed.

21.2 The Cane Creek Canyon Overlook Trail goes right here, just beyond a gas pipeline pumping station. This ORV trail is the main access route into the Flat Iron Mesa District of Canyon Rims Recreation Area.

21.3 Browns Hole Road, described elsewhere in this book, goes left here. Beyond this junction, the highway enters an open area of sand, soil and alluvium that is mostly vegetated, with a few exposures of Navajo Sandstone in the distance to the right of the road. The bulldozer-slashed hill to the left is largely Morrison Formation deposits that have been heavily prospected for uranium, with the tailings of one old mine visible. The red sandstone at the base of the cliff is Slickrock-Entrada.

22.8 Utah 46, described elsewhere in this book, goes left here.

23.3 The highway crosses the drainage of West Coyote Creek, which has formed a spectacularly deep and narrow gorge several miles west of the highway. That gorge eventually joins equally deep Hatch Wash Canyon.

23.6 Hatch Wash Road, described elsewhere in this book, goes right here.

24.0 Along the next stretch of road the bare clifflines on both sides are again Slickrock-Entrada topped with remnants of Morrison Formation deposits. Note the brush-covered soils in this vicinity. Depletion of the native grasses by grazing has promoted the spread of sage, blackbrush, saltbush and other arid-land shrubs in the lower terrain ahead.

26.1 Stop at a wide pullout on the right here for a view of Wilson Arch. This immense span is in Slickrock-Entrada sandstone. Note the light-hued cliffline beyond the arch to the left-front. One ORV trail described in the book, **CAMEO CLIFFS Biking - Hiking - Four-Wheeling,** ends on that ridgeline with a magnificent view of the arch from the other side, and the colorful surrounding terrain. For information on how such arches form, refer to the book ***Canyon Country* ARCHES & BRIDGES.**

Beyond Wilson Arch, the highway drops into the main part of Dry Valley. For more than a century, this immense valley has been heavily grazed by domestic livestock, to the point where its native grasses are completely gone. The sparse grasses found there now are exotic species that do not thrive in this high desert climate. This destruction of the natural ground cover has promoted dramatic erosion in every drainage line in the valley, and in the accumulated alluvium in the major canyons that drain the area. All of the terrain to the right of the highway continues to be part of Canyon Rims Recreation Area.

The cliffs to the left of the highway as it crosses Dry Valley continue to be Entrada Sandstone -- the Cameo Cliffs after which the recreation area was named -- with the Slickrock Member uppermost, and the layered, dark red Dewey Bridge Member below that. In many locations, lighter-hued Navajo Sandstone is exposed below that. Some cliffs are topped with remnants of Morrison Formation deposits. Note in the vicinity of drainage lines how the alluvium that took thousands of years to accumulate is now being cut by erosion, leaving raw banks of exposed soil. The immense flat-topped plateau to the right of the highway as it crosses Dry Valley is Rone Bailey Mesa. For the story behind this mesa's curious name, see the book, ***Canyon Country's* CANYON RIMS RECREATION AREA.**

As the road continues south, it heads for the Abajo Mountains, which dominate the southern skyline. Like the La Sals, this range is laccolithic. See the sidebar, LACCOLITHIC MOUNTAINS. In the lower terrain ahead, cottonwood trees grow in drainage lines where there is permanent subsurface moisture.

29.7 The highway crosses Hook and Ladder Draw, which drains an immense area to the east. Much of this complex drainage has cut down into white Navajo Sandstone below the red-hued Entrada cliffs.

31.3 Pull off right onto a fenced-off road, just before a paved road goes east and the main highway crosses Hatch Wash. The county road east marks the southern boundary of the Cameo Cliffs recreation area, and goes to an area where uranium was mined in the past, and gas and oil resources are currently being developed. The fenced road that goes right here is part of an older highway alignment that provides access to several highlights in Canyon Rims Recreation Area. Hatch Wash drains an immense area to the east of the highway beyond this point, including East Wash Canyon, which will be to the east of the highway where this geology log ends in another 9.8 miles.

To the west of the highway here, Hatch Wash continues to wind through open terrain for several miles before dropping into a deeper canyon. Hatch Wash Road, described elsewhere in this book, crosses Hatch Wash just above this developing canyon. Eventually, Hatch Wash Canyon joins Cane Creek Canyon. The sandstone near the road in this vicinity and walling Hatch Wash is Navajo Sandstone. In some locations, red-hued ledges of Dewey Bridge-Entrada top the light-hued Navajo slopes. Where the highway goes near such a slope just ahead, note the sharply defined dune-sand cross-bedding of this aeolian deposit.

The 1859 Macomb Expedition crossed Hatch Wash and continued west just a few hundred yards beyond this junction. Men from this expedition were the first to see and describe the fantastically eroded terrain now in Canyonlands National Park, ten years before the well-known 1869 Powell Expedition, which boated through the area on the Green and Colorado rivers. The Utah part of the Macomb Expedition route is mapped and described in the book, **CANYONLANDS NATIONAL PARK - Early History & First Descriptions.** Its companion book, **HIKING THE HISTORIC ROUTE OF THE 1859 MACOMB EXPEDITION,** provides detailed instructions for hiking the Utah portion of that route.

32.5 The shallow drainage to the right of the road along this stretch is historic Tank Wash, a tributary of Hatch Wash. This drainage got its name from natural potholes in the Navajo Sandstone farther up the drainage that held water that could be used by early stockmen. A small concrete reservoir was added to hold even more water. The red-hued rim of the drainage is Dewey Bridge-Entrada.

33.7 Needles Overlook Road, described elsewhere in this book, goes right here.

34.0 Along this stretch of highway, notice to the left in the distance, remnants of Entrada Sandstone standing above the valley floor. Beyond these, note how the strata are tilting upward to the east. The lower slopes there are Navajo Sandstone. The layered mid-slopes are Kayenta Sandstone and the upper slopes and exposed sheer cliffs are Wingate Sandstone. This entire tilted ridgeline is the western monocline of the Lisbon Valley salt anticline.

The major canyon system to the left-front is East Wash Canyon, down which the 1859 Macomb Expedition descended from the higher terrain farther east. The bare, white Slickrock-Entrada knobs at the left side of the canyon portal were noted in the official journal of that expedition, which traveled between them and the main canyon wall, then due north to the large, reddish Slickrock-Entrada butte there that was named "Casa Colorado" by earlier Spanish explorers.

The lower walls of East Wash Canyon are also Slickrock-Entrada sandstone. The uppermost rim of the mesa is Dakota-Burro Canyon sandstone, while the terraced slopes on down to the sheer cliff are the three members of the Morrison Formation, with the Brushy Basin Member at the top, the Salt Wash Member in the middle, and the reddish Tidwell Member just above the Slickrock-Entrada cliffs. The upper two members of the Morrison there have been heavily prospected for uranium. In the immediate vicinity of the highway, the sloped sandstone ridges and buttes are all Slickrock-Entrada topped with remnants of Tidwell-Morrison.

39.1 The graded dirt county road that goes left here enters East Wash Canyon. It can be used as access to the canyon to study the route of the 1859 Macomb Expedition, referred to earlier.

39.9 The isolated red sandstone dome fairly near the road to the right is Sugar Loaf Rock. For a curious vignette of history associated with this rock, refer to the book, ***Canyon Country's* CANYON RIMS RECREATION AREA.** The rope ladder that provided access to the large pothole in the top of this dome of Entrada Sandstone is still there, but is too weathered for safe use.

40.7 Utah 211, described elsewhere in this book, goes right here. This road log ends just ahead at a wide pullout that offers a beautiful view of Church Rock to the left of this junction.

41.1 This geology log ends here, at the Church Rock viewpoint. This immense Entrada Sandstone monolith received its name for its startling resemblance to a church with its welcoming front door open. To the east, the Slickrock-Entrada walls of East Wash Canyon lie at the base of lofty Deer Neck Mesa. To the right of the canyon is South Canyon Point. South Canyon is the tributary of East Wash Canyon down which the 1859 Macomb Expedition descended from the higher "sage plains" to the east. These promontories, above their Slickrock-Entrada walls, are Morrison Formation topped with Burro Canyon and Dakota sandstones. To the right-front, these same deposits form Peters Point Ridge. On to the right, Shay Mountain, of the Abajo Mountains range, looms against the skyline. The yellowish sandstone butte to the west of this pullout is also Slickrock-Entrada. For more information about the geology west of here, refer to the road log for Utah 211.

Between here and the town of Monticello, the geology is largely what is visible from here. This highway climbs through the narrowing canyon between South Canyon Point and Peters Point Ridge, ascending through the geologic strata just described, to the higher plains that lie to the east of the Abajo Mountains. These vast "sage plains" of historic accounts are now largely under cultivation, with that segment of the Macomb Expedition route obscured and inaccessible to the public. The historic route down East Wash Canyon, and on west to where it ends in what is now Canyonlands National Park, is almost entirely on public land.

Church Rock

HUMANITY AS A GEOLOGIC FORCE

The planet on which humanity evolved and lives -- Gaea, Earth, Sol III -- is subject to a broad spectrum of natural geologic forces, all of which work separately or in concert to change the face of the planet. Most of the time such natural changes take place slowly, over millions of years, but once in a while the changes are rapid, not only on the leisurely geologic time scale but on the subjective human time scale -- within minutes, hours, days, weeks or years.

The basic natural geologic forces that drive all the lesser forces are heat and gravity. The heat deep within planet Earth comes from the original accretion of the planet's matter during its creation from an immense mass of space debris, and from the decay of radioactive elements within the planet since its formation.

Within the formed planet, heat and gravity create massive convection currents of molten rock that cycle slowly upward from near the planet's intensely hot nickel-iron core, to beneath its relatively thin and light-weight surface crusts, its continents and ocean floors. There, some of the heat is lost, and the somewhat cooler and heavier plastic rock heads back down for more heat, in a never-ending cycle.

These glacially-slow convection currents of molten rock provide the energy that triggers changes on the planet's surface, including continental drift and its attendant secondary forces -- volcanism, earthquakes, changes in ocean currents and atmospheric circulation, global climate and erosional processes. Each of these forces in turn activates still other geologic forces, some major, some minor, such as landslides, floods and glaciers.

Still other forces act to bring changes on the Earth in more subtle ways, including periodic reverses in magnetic polarity, shifts in axial tilt, the slow accumulation of space debris, variations in atmospheric composition, and changes in the solar radiation received on the planet's surface. In contrast to these comparatively minor changes, at least one outside force brings catastrophic change -- the impact of large meteors.

Altogether, these natural geologic forces, from within and elsewhere, bring about natural changes on planet Earth, most of them gradual, a few rapid. When the changes to the planet's climate, atmosphere, lands and oceans are gradual, life can adapt. When such changes are fast they are said to be "catastrophic," and this term is appropriate. Many forms of life cannot adapt to quickly-changing conditions and thus go extinct, leaving the surviving species of plants and animals to evolve slowly to fill the newly-emptied ecologic niches. Such major extinctions are geologic forces in themselves, because life, especially plant life, is a major factor in slowing, stopping or even reversing, the naturally destructive geologic force of erosion produced by weather and gravity.

The cyclic process of extinction and rebirth to which life on the planet has been subjected many times over the past 350 megayears is also natural and has always produced new and more complex forms of life, but is nonetheless disastrous to the individuals and species that have borne the brunt of the catastrophic geologic changes.

Thus, it can be said, on an objective, philosophical level, that geologic forces are neither good nor bad -- they just are. But on a personal, subjective level, all too many geologic forces are devastating, murderous, calamitous, fateful, catastrophic, ruinous and cataclysmic. To individual life forms, and to plant and animal species, these geologic forces are not a good thing. Which brings up the matter of humanity as a geologic force.

Humanity is, indeed, a geologic force -- one of the most rapid and destructive geologic forces on the planet right now. Few of the noted natural geologic forces can compare with the destructiveness, the scale and pace of change, wrought by humanity. Only the relatively rare earthquakes and more violent volcanoes create similar havoc on the planet's surface. All the other natural geologic forces are too slow-acting to compete with humanity, which is destroying the planet's natural environment at blinding speed on the geologic time scale.

Within a few thousand years, mostly in the last couple of centuries, humanity has wrought more change on the lands and waters and atmosphere of Earth than all other geologic forces combined, including continental drift, changing climates, shifting ocean currents, meteor impacts, volcanoes and earthquakes.

Humanity is a major, almost explosive geologic force -- no matter how you look at it! Only the impacts of large meteors that occur every few tens of millions of years exceed humanity in geologic and biologic destructiveness.

How does this work? What are the prime factors that make humanity a geologic force? The basic factor is simple -- humanity's inability, or unwillingness, to control its population growth. This in turn has led to humanity's development of and uncontrolled use of agriculture, animal husbandry, mineral development and destructive forestry to support that growth. It has led to modern technology -- with its pollution of the lands, lakes, rivers, oceans and atmosphere -- again in support of explosive population growth. Within a couple of centuries, mankind is returning to Earth's atmosphere the carbon dioxide and other gases that it took nature hundreds of millions of years to lock away in coal and petroleum. This will inevitably affect the atmosphere, and the global biosphere and climate that it controls, and hence general erosion rates.

Above all else, the invention and universal use of earth-moving machinery -- primarily plows, bulldozers, graders and power shovels -- has been a major factor in humanity-as-a-geologic force. These basic human geologic mechanisms have

continued ----

worked to change the planet's surface in many and complex ways, too many to discuss in detail here, but basic examples of how these factors work as geologic forces are readily observable, in southeastern Utah's canyon country and throughout the world.

As human populations have grown, all too often ahead of the growth of their food supplies, pressures to cultivate more land, graze more domestic livestock and develop more natural resources, have grown correspondingly. In just the last century or so, the rapid advancement of technology has enhanced the growth of agriculture, animal husbandry, logging and mineral extraction enormously, converting vast areas of natural grassland, forest and mountainous terrain to human use.

The devastation of plant and animal communities that have evolved over megayears does happen naturally, but with rare exception quite slowly, slowly enough that the survivors can adapt to the changing conditions. But when humanity changes the environment, the changes are too fast, biologically explosive. Plant and animal species go extinct. This, of course, is not in itself a geologic force -- but the way humanity goes about wiping out native ecosystems certainly does have geologic consequences.

All too often, the conversion of -- woods to tobacco fields, grassy prairies to endless ranks of corn, deserts to rows of irrigated cotton, forests to eroding slopes, swamps to theme parks and vacation homes, high-desert pinyon-juniper woodlands to rocky barrens, flowing rivers to polluted reservoirs -- is done carelessly, done where the changes do more damage than good, done mindlessly, thus creating rapid changes in natural water flow and absorption, and drastically accelerating erosion, a normally slow geologic process.

One of mankind's greatest geologic contributions is to create new erosion, drastic erosion, either directly or by destroying ecosystems that slow or even reverse erosion, thus vastly accelerating this naturally very sedate geologic process. Forest clear-cutting creates drastic erosion. Land cultivation for agriculture all too often creates gullies and minicanyons in fertile soils that took nature millennia to build by slow alluviation. Countless megatons of soils in the sandstone canyons of southeastern Utah and vicinity have already eroded away as a result of the destruction of the region's native grasses by massive overgrazing by sheep, cattle and horses.

All over the world, humanity's activities have accelerated erosion of the land to the point of disaster -- to the land itself, to its native plant and animal life, to the life in and adjacent to the land's streams and rivers, and even in the oceans, as erosion products and pollution have devastated river estuaries, their ecosystems and the marine life that depends on the nutrients that are brought from the land by its rivers.

Along with the explosive growth of agriculture, animal husbandry, forest destruction and the erosion these produce, the creation of homes and other civilized amenities for the use of humanity's burgeoning population have had their effects, some also leading to vastly accelerated erosion, some to less obvious geologic effects. Air pollution can and does change climates and precipitation patterns. These in turn affect erosional rates, fresh water flows, water tables, aquifers and even sea levels, through the growing and shrinking of snow-packs and glaciers.

A recent world resource investigation revealed that, just since World War II, as much as 11% of Earth's land surface has been so badly damaged by such human activities as overgrazing, deforestation and bad agricultural practices that it has lost most of its biological productivity.

The construction of great reservoirs for the purpose of providing culinary water and for agriculture also changes land, river and marine ecosystems, atmospheric humidity, natural sedimentation processes, water tables, and even geologic fault stability. These in turn create still other secondary geologic effects, in a wide-spreading pattern that echoes throughout the entire planetary ecosystem and its effect on natural geologic processes.

The mineral industry that supports most of what else humanity does in this technological age also promotes massive erosion and changes to native plant and animal communities. Strip mining is perhaps the worst culprit, but conventional mining and its support operations also contribute. Both approaches move great amounts of native rock, sediments and soils, thus promoting erosion. Both bring to the surface and add to the nearby water tables, streams and rivers great quantities of minerals in concentrations poisonous to the plants and animals subsequently exposed to them, thus bringing about more extinctions and erosion.

The region described in this book exhibits the results of virtually all of humanity's geologic forces. Endless miles of bulldozed trails to speculative uranium "prospects" and low-grade mines have promoted erosion and mineral contamination of nearby land and water -- to say nothing of the aesthetic damage. For another example of massive destruction in support of mineral extraction, see the sidebar, POTASH.

More than a century of extreme overgrazing, in a high-desert region where even a little grazing is overgrazing, has drastically changed plant communities and thus promoted rapid erosion. Failed attempts to cultivate the region's high-desert meadows has also led to faster erosion.

Chaining -- the uprooting of native trees and shrubs by dragging enormous anchor chains or steel rails between great bulldozers -- to "enhance" domestic livestock grazing has largely failed to achieve that goal, but has greatly accelerated natural erosion and wiped out entire plant and wildlife communities. The construction of thousands of miles of dirt roads, with little thought given to preserving natural erosional surfaces and following natural land contours, has also produced surface erosion, massive rockfalls and landslides, and has created new surface drainages that cut deeply into the sparse native soils.

The list of global, continental, regional and

continued ----

local geologic destructiveness of humanity goes on and on. But the bottom line is inescapable -- humanity is, indeed, a major geologic force. Our activities are playing a greater role in changing the ecosystems, planetary climate, land erosional rates and biological extinction processes than are all natural geologic forces put together. We are rapidly driving countless other species to extinction. One study indicates that the present extinction rate is 1,000 times the normal rate, thanks to mankind.

Is this good or bad? As with changes to the planet by natural geologic forces, whether gradual or catastrophic, the changes humanity is wreaking are neither good nor bad -- from an objective viewpoint. Change is the name of the game in nature, both slow and rapid change. But from humanity's very subjective and personal viewpoint, drastic change is bad. By destroying the ecosystems of which we are a part, by changing the very lands, waters and air upon which we depend for our lives, by fouling our one and only nest, so to speak, we are definitely damaging -- seriously threatening -- our own chances of survival as a species.

At the rate we are presently going, humanity may turn out to be one of the shortest-lived species on the planet. Sure, humans, if broadly defined, have been around now -- for what? -- two or three million years. But what is that on the general biologic time scale, where even other mammals that have carefully adapted to their environment, rather than adapting their environment to their whims, have already been around for 40 or 50 million years -- like the cetaceans, for example.

And how do a couple of megayears compare to the hundreds of millions of years that many insect species have survived, in some cases virtually unchanged? What about the many plant species that have existed for similar lengths of time? Or the anaerobic bacteria and blue-green algae that have been a part of the Earth's biosphere since the very beginning of life on the planet? Biologically, humanity is a Johnny-come-lately species that has yet to prove its durability, has yet to demonstrate that its much-vaunted "intelligence" and its penchant for changing the environment rather than adapting to it, are long-term survival traits.

Would anyone like to wager that humanity will still be around even a mere one million years from now, given its proclivity for self-destruction and its proven ability to be a major geologic and life-extinguishing force? An objective observer might easily conclude that humanity is clever enough to create forces that could lead to its own destruction -- but is not intelligent enough to avoid using those forces.

Enjoy the beauty of canyon country while it endures. Both natural and human geologic forces are still working to destroy it, but humanity is the greater threat. Some of the signs of this are noted in the geologic road and trail logs in this book.

Human governmental, commercial and private developments continue to destroy the natural protective surfaces of this high-desert region, thus accelerating water and wind erosion.

Domestic livestock continue to overgraze public land, further decimating the sparse remaining native plants, this with the full cooperation of federal and state land administration agencies. The myriad hooves of sheep, horses and cattle continue to crush and destroy the fragile cryptogamic soils that are the basis of this arid region's entire biological community, thus greatly accelerating erosion and causing species extinctions.

These same organizations, and ranchers, continue to destroy vast areas of native pinyon-juniper forest by chaining, all in the name of improved grazing, thus severely damaging all associated plant and animal communities and promoting drastic wind and water erosion.

Hardrock mineral claims based on the archaic 1872 mining law continue to appear on public land, with the attendant threats of assault-by-bulldozer, the destruction of large areas of terrain, and the introduction of human and mine wastes and unnatural mineral concentrations into biological communities, all of which in turn increase erosion and life extinction.

Seismograph operations in support of petroleum exploration continue to bulldoze unspoiled terrain, and further damage its natural surfaces with explosives. Then drillers for oil and gas bulldoze still more roads and large areas, drill dry holes, then leave the scarred land to erode for decades, while slow-growing arid-land shrubs and cryptogams attempt to heal the damage.

Yes -- humanity is a geologic force of major magnitude.

GEOLOGIC STRATA VISIBLE FROM THIS ROAD

Formation	Member	Page
Burro Canyon/Cedar Mountain Formations		Page 19
Morrison Formation	Brushy Basin	Page 18
	Salt Wash	
	Tidwell	
Entrada Sandstone	Moab	Page 16
	Slickrock	
	Dewey Bridge	
Navajo Sandstone		Page 13
Kayenta Sandstone		Page 13
Wingate Sandstone		Page 12
Chinle Formation		Page 12
Moenkopi Formation		Page 12
Cutler Group	White Rim	Page 11
	Organ Rock	
	Cedar Mesa	
	Lower Cutler	
Hermosa Group	Upper Hermosa	Page 9
	Paradox Salts	

Cashin Mine

ROAD NAME: **UTAH 46 - COLORADO 90**

SUMMARY: This highly scenic and geologically varied road travels the vast alluvial southern slopes of the La Sal Mountains, crosses lofty Pine Ridge, then descends through lush highland meadows watered by mountain streams before plunging steeply into verdant La Sal Creek Canyon with its beautifully eroded walls of colorful Entrada and Navajo sandstones. Abandoned uranium mines in Morrison Formation deposits are visible from the road, while communities of tall pine and fir trees lend contrast to the surrounding arid high-desert pinyon-juniper plant zone. Beyond this lovely, terraced canyon, the road enters Paradox Valley, another of the several immense salt anticline valleys that surround the La Sal Mountains. The description ends near where the Dolores River enters Paradox Valley through a gap in its cliffline, crosses the valley, then leaves through another such portal.

ROAD SURFACE: paved.

VEHICLE: any vehicle.

MAP: ***Canyon Country* OFF-ROAD VEHICLE TRAIL MAP - Arches & La Sals Areas.**

GUIDEBOOKS: Utah 46 - Colorado 90, as described in this log, is mentioned as an access route for roads and ORV trails in the La Sal Mountains area in the book, ***Canyon Country* OFF-ROAD VEHICLE TRAILS - Arches & La Sals Areas,** but no details are given. A general description of the terrain surrounding the first stretch of this trip appears in the book, **CAMEO CLIFFS Biking - Hiking - Four-Wheeling.**

MILEAGES:

TO START: 22.7 miles, from Main and Center, Moab.
ROAD: 37.8 miles, one-way to end.
RETURN: 60.5 miles, from end of described road.
TOTAL: 121.0 miles, from Main and Center, round trip.

TIME: this trip can be taken in a short day, leaving adequate time for short hiking or driving side-trips along the way, such as to the lovely waterfall in the upper end of the La Sal Creek narrows. Those with off-road vehicles can return from Paradox Valley via Buckeye Reservoir, Geyser Pass and the La Sal Loop Road, as described in the referenced book. This very scenic loop route requires a long day.

ROUTE: Main Street (U.S. 191) south, Utah 46 - Colorado 90 east.

NAVIGATION: Drive south from Moab on the main highway for the indicated distance, then east on Utah 46 and Colorado 90 into Paradox Valley. Return to Moab via the same route. If returning via Buckeye Reservoir and Geyser Pass, use the referenced guidebook and map for detailed instructions.

GEOLOGY ROAD LOG:

0.0 As this road angles east from U.S. 191, it enters the CAMEO CLIFFS recreation area described in one of the referenced guidebooks, and begins traveling over terrain that is largely glacial debris and sand, soils and decomposed rock from a variety of geologic strata. The heavily mined hill to the north of the highway is Morrison Formation deposits, with the Brushy Basin Member of the Morrison in the upper levels and the Salt Wash Member of the Morrison on down, with occasional outcrops of the Slickrock Member of Entrada Sandstone along its base and as separate ridges or low cliffs near the road. As the road continues, the barren rock cliffline to the south is also Slickrock-Entrada topped with remnants of Salt Wash-Morrison. These strata dominate the first several miles of this road, with more and more Salt Wash-Morrison, with fewer and fewer exposures of Slickrock-Entrada, and with increasing alluvium covering the Morrison.

As the road travels for the first couple of miles parallel to West Coyote Creek, note how this stream is now cutting into the alluvium that it built up over a long period of time. This accelerated erosion is due to the extreme overgrazing of this entire region by domestic livestock for more than a century. This has drastically reduced the grasses that hold soils and restrain runoff, thus promoting accelerated erosion.

5.8 From a paved pullout on the right just beyond the entrance to a defunct uranium mining operation, study the La Sal Mountains to the north. Note how all the nearby terrain is a series of vast interlocking alluvial fans that descend from the steeper mountain slopes. These gently sloping rocky deposits, with thin coatings of native soils, dominate both sides of the road between here and the small settlement of La Sal, on ahead. See the sidebars LACCOLITHIC MOUNTAINS and GLACIERS OF THE LA SAL MOUNTAINS. This predominance of alluviation on the southern slopes of the La Sals contrasts with the Morrison Formation that prevails on the high plateaulands on the north side of the range, as described in the log for the La Sal Mountain Loop Road, elsewhere in this book.

9.2 The road goes through the settlement of La Sal. For a summary of the curious human history of this community and of Old La Sal, on ahead, refer to the book **CAMEO CLIFFS Biking - Hiking - Four-Wheeling.** Beyond La Sal, the road climbs steeply into the higher terrain of Pine Ridge, which is the uncollapsed spine of a salt anticline that is continuous with the Moab Valley salt anticline on the other side of the La Sal Mountains and the Gypsum Valley salt anticline to the southeast. See the sidebar, SALT ANTICLINES. The slopes in this area are largely Morrison deposits, with the summits of the major ridgelines topped with Burro Canyon Formation, although all of these deposits are obscured by dense vegetation. Watch for occasional old mines in the Brushy Basin-Morrison.

10.2 Along this stretch of the road, and ahead for a mile or so, the light brown rock exposed along road cuts and the deposits above it are Burro Canyon Formation.

12.6 The Forest Service off-road vehicle trail that goes right here travels the lofty spine of Pine Ridge. The tall trees that dominate the Burro Canyon spine of the ridge are Ponderosa pines set among scrub oak and other brushy vegetation. The lower, more open terrain ahead beyond the ridge is again dominated by alluvial deposits from the La Sals. As the road descends into the lower terrain, it goes through the community of Old La Sal. Another early settlement in this area was called Coyote.

While traveling this stretch of road as it descends toward the southeast, note to the left-rear a broad sloping valley between major peaks of the La Sal Mountains. The summit of that valley is La Sal Pass, at elevation 10,400 feet. The drainage line is La Sal Creek, which starts high in the mountains, travels the length of the lower valley parallel to this road, then descends with the road into a deep canyon system. To the south, Pine Ridge walls this open ranching community. This entire valley was alluviated by La Sal Creek and its many tributary drainages over the last two million years.

18.0 At a junction here, just beyond a large reservoir to the right of the road, a graded dirt road spurs right to ascend steeply to Ray Mesa, or Wray Mesa, depending on which map is being used. The sandstone ridgeline to the left of the road here and for another half mile is Burro Canyon Formation. The "painted desert" deposits in the road cut just ahead are Brushy Basin-Morrison. The tall evergreens in this vicinity are Ponderosa pines. Just ahead, La Sal Creek flows through a stretch of swampland that is frequented by water-loving birds, although golden eagles have been seen there, too.

Beyond this junction, the road begins a very steep descent into La Sal Creek Canyon. As it drops into lower terrain, the road cuts down through older geologic strata in the following order: Brushy Basin-Morrison, Salt Wash-Morrison, then barren slopes of Entrada Sandstone. As the road descends a series of switchbacks, note the several long-abandoned mines in the steep slopes of the canyon. These mines are in the Morrison Formation, which was heavily mineralized by volcanic ash fallout over a long period of time. Later, subsurface water concentrated these minerals into pockets of ore that were sometimes economically feasible to mine. See the sidebar, URANIUM.

The tall trees at the base of the grade are a mixture of Ponderosa pine and Douglas fir. Ponderosas dominate the slopes of the canyon for most of its length, thriving on the spring water seeping from its steep sandstone cliffs and terraces. Douglas firs again appear where the canyon narrows and the road leaves the main canyon.

20.7 Beginning at the base of the switchback grade and continuing along the next two miles of road, the red sandstone exposed on both sides of the canyon is Slick-rock-Entrada.

22.3 Utah 46 crosses the Utah-Colorado state line and becomes Colorado 90.

22.8 Here, beginning at the mouth of a large tributary canyon to the left, the first outcrops of Navajo Sandstone appear just above road level on the left. This rock continues on down the canyon as a sandstone wall slowly growing in height above the road, although it is generally difficult to distinguish from the overlying and continuing Entrada Sandstone that is now higher above the road. In this area, the Navajo Sandstone that is so prominent in much of southern Utah is diminishing in thickness. Morrison Formation deposits continue to dominate the cliffline above the bare slopes of the red-hued Slickrock-Entrada.

As the road continues, note how the sandstone slopes to the left of the road, and continuing for a mile or so, are beautifully streaked with vertical draperies of dark desert varnish, lichens, mosses and minerals from higher strata. This is caused by intermittent water seepage from the higher soils and terraces in this well-watered canyon. See the sidebars DESERT VARNISH and LICHENS. Note, too, how the sandstones that dominate the lower canyon walls display a spectacular variety of erosional landforms.

26.7 An ORV trail spurs right here to descend the lower narrows of La Sal Creek Canyon. This trail goes to several old mining operations and is closed to the public beyond about 3-1/2 miles, but provides an intimate look at lower La Sal Creek and beautiful erosional forms in the Wingate Sandstone that walls the inner gorge. About 0.9 miles down this trail it reaches a lovely waterfall in a sandstone grotto, where a tributary stream joins La Sal Creek. Carefully driven highway vehicles can travel that far on this trail when it is dry. To avoid confusing this road log's mileages, this spur could be explored on the return trip.

Just ahead, as the road winds along the steep canyon slopes, note the sharp interface, just above road level,

between the light-hued Navajo Sandstone and dark red Kayenta Sandstone. The road continues to travel near this interface for the next couple of miles.

27.2 Stop at a narrow pullout to the right of a curve for a look on down the narrows of La Sal Creek Canyon. About eight creek-miles from here, La Sal Creek joins the Dolores River, where both are in deep and spectacular gorges. The waterfall noted earlier is directly below this overlook.

East of here, the highest light-hued rock in the ridgeline is a remnant of Navajo Sandstone. The lower red terracing is Kayenta Sandstone, while the inner gorge of La Sal Creek is Wingate Sandstone. Note the tall Douglas fir trees on the slopes across the canyon.

Beyond this pullout, the road ascends the steep slopes of the tributary canyon below the overlook. Navajo and Kayenta deposits continue to dominate both sides of this road as it climbs toward the high wooded ridge that is the southwestern wall of the Paradox Valley salt anticline. This immense anticline has tilted all the geologic strata in this area sharply upward. The summit of the ridge ahead is badly fractured by a series of faults that were a part of the formation of that salt anticline and the subsequent formation of Paradox Valley.

27.7 Note near the top of the cliffline ahead the distinct red and white color-banding in the Entrada Sandstone. In the drainage to the right of the road, the deep alluvium is being cut and destroyed. This is caused by grazing and other human activities in this steep tributary canyon that have disturbed the water-holding surface of the native soil. The erosion begins where the road swings right around a small ranch. Beyond that ranch, the road climbs steeply through cuts in the remnant Navajo Sandstone.

29.3 Stop at a pullout just beyond the summit of the steep grade, where the road starts descending into Paradox Valley. This valley was the type locality for the Paradox Formation, which was first studied here and named after the valley. There are large outcrops of Paradox salts in the valley floor near the settlement of Paradox, and even larger ones on down the valley. Colorado 90 crosses one such exposure of this ancient marine deposit.

Note how the geologic strata in the upper end of this immense salt valley are massively faulted and tilted, although this is somewhat obscured by dense vegetation except for a wall of bare red and white Slickrock-Entrada sandstone. This color-banded deposit is the stratigraphic equivalent of the lofty two-toned ridge noted at mile 27.7, even though it is at a much lower elevation due to salt anticline faulting, tilting and slumping.

The summit of the far wall of Paradox Valley is dominated by remnants of Morrison Formation deposits. A thin cliffline of color-banded Slickrock-Entrada is visible below the ridge summit, with more of this sandstone appearing toward the right. Below the Slickrock-Entrada, terraced Kayenta Sandstone appears above sheer cliffs of Wingate Sandstone. Below the Wingate, the rubble-covered slopes are deposits of Chinle and Moenkopi formations, in descending order. Along the base of the far cliffline, a few outcrops of dark red Cutler Sandstone jut outward. Note that in the far rim of the valley, Navajo Sandstone is missing from its usual position just above the Kayenta.

The valley floor is alluvium that has been developed into ranches and farmland. The break in the cliffline toward the east is where the Dolores River leaves Paradox valley. As with Moab Valley, a river enters this valley through one narrow sandstone portal, crosses the valley, then leaves it through another narrow portal. This strange phenomenon is discussed in the sidebar, PARADOXICAL VALLEYS, and was responsible for the name, Paradox Valley.

The rocky bluffs adjacent to this road cut and pullout are thin remnants of Navajo Sandstone. Beyond the pullout, the road descends steeply into Paradox Valley through a variety of geologic strata that are too tilted, fractured and faulted to identify in detail, although they are mostly Kayenta, Wingate and Chinle deposits. The road then continues down the valley along its southwestern flanks to the Dolores River, where this description ends.

32.1 The road reaches the valley floor at a junction. The road to the left goes to the ranch community of Paradox, which offers no traveler facilities or services. A branch of that road goes to Buckeye Reservoir, high on the eastern slopes of the La Sal Mountains. The distant wall of the valley was described earlier. Beginning just ahead, and continuing to the Dolores River, the lower slopes above the road to the right are gray and red Chinle Formation deposits. The jumbled slopes above the Chinle are faulted, fractured and tilted Wingate and Kayenta deposits topped with thin remnants of Navajo Sandstone.

As the road travels southeast down Paradox Valley, note in the far cliffline how the Wingate and Kayenta sandstones that were obscured under rubble in the upper end of the valley rise higher and become readily apparent, with older Chinle and Moenkopi deposits appearing below the sheer Wingate cliffs. Dark red Cutler Sandstone appears in a few places at the base of the cliff.

37.2 The road reaches the tiny settlement of Bedrock, with its picturesque general store, just ahead nearer the river.

37.5 From the pullout in front of the general store, the downriver portal is visible as a gap in the cliffline in the opposite side of the valley. The relatively level geologic strata in the lefthand wall of the downriver portal are Salt Wash-Morrison on the skyline, with a band of red and white Slickrock-Entrada sandstone just below that. Navajo Sandstone, which is thin to the west of Paradox Valley, is missing entirely in its eastern rim. Terraced Kayenta Sandstone lies just above the sheer cliffs of Wingate Sandstone. Below those cliffs, steep slopes of Chinle and Moenkopi deposits descend almost to the valley floor, with outcrops of dark red Cutler Sandstone jutting from the rubble slopes just above the valley floor. To the right of the downriver portal, the upper cliffline is Wingate topped with remnants of Kayenta, with the same sequence of strata on down to the valley floor.

The cliffs adjacent to the upriver portal were more sharply tilted by the Paradox Valley salt anticline. While the lower narrows of La Sal Creek Canyon began traveling through Wingate Sandstone, here, four miles downriver from the La Sal Creek - Dolores River confluence, the Paradox Valley salt anticline has tilted the strata upward, leaving the Wingate high above the drainage line. The rubbly uppermost cliffs on both sides of the portal, and on top of the distance cliffs visible through the portal, are Morrison deposits. The red and white banded sandstone cliff below the Morrison is Slickrock-Entrada. Below that, a relatively thin band of Navajo

Sandstone lies above steeply terraced Kayenta layering. Wingate sandstone cliffs continue down to slopes of Chinle deposits, which then descend to the river, with a little Moenkopi exposed on the right side of the portal just above ground level. The broken cliffs high above the left side of the portal are also Wingate.

The dirt road angling right from this pullout can be safely traveled by any vehicle. It goes to a take-out ramp for rafters floating down the Dolores River from Slickrock, Colorado. There are several shaded picnic pavilions near the ramp that can be used by anyone.

37.8 Here, Colorado 90 crosses the Dolores River, just below its upriver portal into Paradox Valley. Although this geologic road log ends here, as Colorado 90 continues to travel the southeastern extension of Paradox Valley, the geology of both clifflines remains much the same for many miles. Beyond the Dolores River, large outcrops of Paradox salts appear in the center of the valley as low, sparsely-vegetated gray hills, with the first appearing just beyond the river. After several miles, the road crosses one such exposure of ancient marine salts.

As the Dolores River crosses Paradox Valley it picks up great quantities of salts dissolved from these outcrops, which it carries on into the Colorado River. This salt has caused problems with the use of Colorado River water for domestic and agricultural purposes on downriver in Arizona, California and Mexico. Over the years, various impractical schemes have been considered for avoiding the Paradox Valley salts, but to date none of them have been implemented. Instead, the federal government has constructed immense desalinization plants near the lower Colorado River for removal of salts introduced here and elsewhere along the Colorado River and its many tributaries, especially where they cross or drain any of the many salt anticline valleys in the eastern part of the Colorado Plateau.

The salt added to the Colorado's water by the Paradox Formation in this region has even had international repercussions. Before the desalinization plants were built, the Colorado River water that reached Mexico violated an international treaty requiring that the river water be usable for agricultural purposes, which the excessive salt prevented.

Aerial photograph of Upheaval Dome. See sidebar page 55.

PARADOXICAL VALLEYS

The Colorado River enters Moab Valley through a portal -- a gap in a high cliffline. It crosses the valley, then leaves through another sandstone portal. The Dolores River, a major tributary of the Colorado River, enters Paradox Valley through a gap in a cliffline, meanders across the valley, then leaves it via another portal in a great wall of sandstone.

These two "paradoxical valleys" stand in stark contrast to conventional valleys, which have been formed by rivers or streams flowing their lengths, with one end of each valley higher than the other. Moab Valley and Paradox Valley are higher at both ends, with drainages flowing toward where the rivers cross.

To understand this odd geologic phenomenon, it is necessary to retreat million of years into the past and observe earlier stages of this region's present phase of erosion.

From a vantage point of perhaps eight million years ago, visualize what is now the general Moab region. It is a vast, flat, arid plain, relieved only by the rounded hills of what will eventually become the La Sal Mountains, and by the low spine of the Uncompahgre Uplift, which looms on the skyline to the east.

A much smaller Colorado River meanders across the flat terrain to the north of the La Sal hills as a shallow stream. A major tributary, the stream that will become the Dolores River, is winding northward between the Uncompahgre ridge and the La Sal hills to join the Colorado, after crossing the still-buried Paradox Valley salt anticline. Between the two highlands, the Dolores is joined by the creek that will eventually become the Gunnison River, which is slicing across the northern nose of the Uncompahgre, slowing cutting what will later become Unaweep Canyon.

From the vicinity of the La Sals, the ancestral Colorado River, now joined by the Dolores, wanders across the flat, gently sloped, arid terrain toward the southwest, meandering in great leisurely loops, slowly cutting downward into sediments that took more than 300 million years to accumulate. There is no trace of the salt anticline that later becomes Moab Valley, nor is there any sign along the meandering course of the ancestral Dolores River of the similar anticline that eventually becomes Paradox Valley.

Now, jump ahead to perhaps four million years ago, to when the Colorado, fed by water from the Rocky Mountains, is flowing more vigorously and has cut its channel more deeply into the flat terrain. Note, however, that the deepening river gorge has few tributary canyons, because the vast region west of the Rockies is still relatively arid. Only the deepening, meandering river gorge gives a hint of the canyon country to come.

Look more closely, and see that where the Colorado crosses the still-buried Moab Valley salt anticline, it has now cut deeply enough to reach the two underground sandstone ridges that mark its eastern and western monoclines, even though these ridges are still not exposed above the general ground level. The same thing has happened where the Dolores cuts across the Paradox Valley salt anticline.

Visualize the same terrain two million years later, after the entire region has lost additional deposits from surface erosion, and the Colorado has cut to depths of 800 feet in the vicinity of the La Sal hills. Now the parallel monoclines of these two salt anticlines are exposed as eroding ridges of sandstone on each side of the valleys-to-come and, as the surface of the land has eroded away, the two rivers deep within their entrenched meanders have continued to cut into these ridges in order to continue flowing. The two sets of river portals are developing, although the land between the parallel anticline ridges is still barely lower than the surrounding terrain.

Take a look in another million years. Cyclic episodes of wetter climate have produced a series of glacial pulses in the La Sals. The drainages between the sandstone portals, fed by local rain water and melting snow and ice from the La Sals, have deepened the depressions between the two sets of parallel sandstone ridges. Moab Valley and Paradox Valley are starting to form. See the sidebar, GLACIERS OF THE LA SAL MOUNTAINS.

Up to this point, the slow deepening of the two valleys has come from surface runoff, from tributary streams, melting glaciers and flash floods that carried surface materials into the rivers as they cross the shallow valleys. But soon, a critical stage of valley formation will be reached.

For a long time, surface water has percolated downward through sediments and rock debris to the soluble marine salts that formed the two salt anticlines. Once there, the water has dissolved great quantities of these minerals. But until the rivers cut down to these subterranean salt-saturated lakes, the lakes could not drain.

Once the two rivers cut deeply enough to encounter these underground salt lakes, however, they began draining them, carrying their marine salts back to the sea, and new downward seepage of surface water dissolved and carried still more salts into the rivers as they continued to deepen their channels. This process literally undercut the terrain between the parallel monocline ridges, allowing it to sink and promoting massive slumping along the faulted valley walls, thus accelerating the deepening of the valleys.

Now take a fresh look at Moab Valley and Paradox Valley as they are today, and study the results of those millions of years of intricate and unique erosional processes. To form these two paradoxical valleys, the rivers had to be there first, cutting ceaselessly downward, long before the present valleys existed or their formidable sandstone walls were exposed. The rivers had to have been active well before any traces of the two salt anticlines were visible above the surface.

These two rivers had to have time, millions of years of time, to cut their deep meanders. These, in turn, forced the entrenched rivers to cut through the parallel sandstone ridges of the buried salt anticlines, then eventually reach and drain the subterranean lakes of saturated salts, carrying these away to the sea and lowering the valley floors, thus creating the present deep, sheer-walled valleys -- with their strange rock portals through which the two rivers enter and leave.

Of course, the Colorado and Dolores rivers also contributed to the formation of other salt anticline valleys in this region, but they just happened to cross the still-buried Moab and Paradox anticlines at the correct angles to create valleys that were higher at each end while, by the same vagaries of chance, other salt anticlines were not crossed by the rivers. These other salt anticline valleys lost their sediments and salts through their ends, or through single portals out of their sides. Their valleys are not paradoxical.

For related information, see the sidebars, ENTRENCHED MEANDERS, SALT ANTICLINES, LACCOLITHIC MOUNTAINS, UNCOMPAHGRE UPLIFT and TIME TRAVEL TOPOGRAPHY.

GEOLOGIC STRATA VISIBLE FROM THIS ROAD

Mancos Shale	Ferron Sandstone	Page 20
Dakota Sandstone		Page 20
Burro Canyon/Cedar Mountain Formations		Page 19
Morrison Formation	Brushy Basin Salt Wash Tidwell	Page 18
Entrada Sandstone	Moab Slickrock Dewey Bridge	Page 16
Navajo Sandstone		Page 13
Kayenta Sandstone		Page 13
Wingate Sandstone		Page 12
Chinle Formation		Page 12
Moenkopi Formation		Page 12
Cutler Group	White Rim Organ Rock Cedar Mesa Lower Cutler	Page 11

ROAD NAME: **UTAH 128**

SUMMARY: This road is one of the scenic and geologic highlights of the Moab area. The road closely parallels the Colorado River as it flows through a deep and narrow gorge that has been cut over millions of years into ancient and colorful rock strata. The erosional land forms visible from the road are outstandingly varied and beautiful, with soaring cliffs, sheer-walled gorges, broad valleys, severely warped rock strata of ancient salt anticlines, skyscraper-tall towers of red sandstone, cliffs streaked with draperies of desert varnish, large seep-caves, natural arches and gigantic rockfalls. Several roads and ORV trails described elsewhere in this book branch off along the way, offering further unique geologic and scenic highlights.

ROAD SURFACE: paved.

VEHICLE: any vehicle.

MAP: ***Canyon Country*** **OFF-ROAD VEHICLE TRAIL MAP - Arches & La Sals Areas.**

GUIDEBOOK: ***Canyon Country*** **OFF-ROAD VEHICLE TRAILS - Arches & La Sals Areas.**

MILEAGES:

TO START: 2.6 miles, from Main and Center, Moab.
ROAD: 35.2 miles, one-way to end, as described.
RETURN: 37.8 miles, from end of described road.
TOTAL: 75.6 miles, from Main and Center, round trip.

TIME: one-half day, plus time taken for stops along the way. Allow for a full day if side trips to the Fisher Towers and up Castle Valley are planned.

ROUTE: Main Street (U.S. 191) in Moab, then Utah 128 east to where this description ends, several miles before the state highway joins Interstate 70.

Historic Dewey Bridge

NAVIGATION: Drive north on U.S. 191 from Moab for the indicated miles, then turn upriver on Utah 128 just before crossing the bridge over the Colorado River.

GEOLOGY ROAD LOG:

0.0 Utah 128 leaves U.S. 191 just south of the highway bridge across the Colorado River. The road enters the Colorado River gorge and closely parallels the river for most of its described length. The geologic stratigraphy along most of this stretch of highway is relatively uniform, but has some interesting tectonic features and erosional land forms. The first of these is at the road beginning. Note that as the road enters the upriver portal through the cliff of Moab Valley, the geologic strata are tilted, sloping down in the upriver direction. This is due to distortion of the strata by the Moab Valley salt anticline. See the sidebar, SALT ANTICLINES. Such ancient upthrusts of deeply buried Paradox Formation salt deposits have also affected the geologic strata along the river farther on.

At the river portal, the uppermost round-topped cliffs on the north side of the river are Navajo Sandstone. Below this, layers of Kayenta Sandstone extend downward until they meet the relatively unstratified Wingate Sandstone that continues down to river level. On the righthand side of the portal, the uppermost rock visible from the junction is fractured Wingate Sandstone with a trace of Chinle Formation deposits at road level.

0.2 At a pullout on the right here, Matrimony Spring can be seen gushing from a pipe at the base of the cliff. Folklore has it that couples drinking the water from this spring will be forever bonded in matrimonial bliss. Beyond this pullout, the Wingate Sandstone, now on both sides of the river at road level, continues to descend.

0.5 From a pullout on the left, straight ahead and high above the road in the Navajo Sandstone cliffline, the vertical opening of Updraft Arch is visible just below the skyline. As the road continues closely parallel to the river, note that the river meanders continuously in broad, sweeping curves. See the sidebar, ENTRENCHED MEANDERS, for an explanation of this.

0.8 In this vicinity, the Wingate Sandstone disappears underground on both sides of the river, leaving ledgy Kayenta Sandstone just above road level. About a half mile ahead, the Kayenta also disappears underground, leaving Navajo Sandstone cliffs on both sides of the river.

1.3 Note the large, arching seep-cave high in the opposite cliff wall. Relatively soon, geologically speaking, the seeping water will form a hole in the cave roof and thus create a new arch. For details concerning the formation of natural rock openings, refer to the book, ***Canyon Country* ARCHES & BRIDGES.** Just ahead, note how rainwater and snowmelt runoff have created beautiful draperies of desert varnish in the sheer cliff walls. See the sidebar DESERT VARNISH.

1.5 At a pullout here, there is another seep-cave at the base of the cliff to the right behind the vegetation. During the colder months of winter, this cave is curtained with immense icicles from snowmelt and spring water dripping down from higher up. Beyond this pull-

out, note the numerous seep-lines in the cliff on the other side of the river, where water in the relatively porous Navajo Sandstone reaches a less permeable layer and is forced laterally to the surface of the cliff. Such seeping water wetting the surface rock causes exfoliation, thus creating deepening caves and horizontal crevices. See the sidebar ROCK EROSION.

1.9 Beginning about here, note on the benchlands on the opposite side of the river, the great masses of gravel abandoned by the river as it cut more deeply. Beginning at the river portal and continuing to the Salt Wash-Colorado River confluence, the land on the opposite side of the river is in Arches National Park. At one time, there was an attempt to placer-mine gold and certain unclaimable minerals from these deposits, even though they were in the park. After considerable controversy, the placer-gold claims were disallowed.

At intervals along the river, depending upon its seasonal level, sand bars appear. Some of these sand bars are transient and move gradually during higher water. Others along the riverbank are more permanent. Some of these become heavily vegetated by native river willow and exotic tamarisk shrubs and thus create a permanent embankment. Only during occasional periods of exceptionally high water does the river flood into these brushy river margins. On rare occasions of high water, the river has flooded stretches of this road and Utah 279, which goes downriver from Moab Valley.

3.1 The pullout to the right provides parking for hiking up Negro Bill Canyon. This canyon extends many miles into the foothills of the La Sal Mountains and provides excellent hiking. It has an intermittent stream for about the first six miles during wetter seasons, less during the drier seasons. The lower walls of that canyon for the first several miles are Kayenta Sandstone, with round-topped Navajo Sandstone cliffs above that.

Note that, on the opposite side of the river, there is still some ancient river gravel on the rounded Navajo Sandstone clifftop. Beginning here, the geologic strata start to ascend gradually from the river level. This tilting is due to the somewhat continuous salt anticlines of Salt Valley, in Arches National Park, and Castle and Fisher valleys some miles ahead. These three salt anticlines are connected by areas of uplifted geologic strata. About one half mile ahead, Wingate Sandstone once again rises above the river level and soon dominates the canyon walls. The broken horizontal strata above the Wingate cliffs are Kayenta Sandstone. The lighter bluffs above that are remnants of the Navajo Sandstone which, on the distant horizon ahead, has eroded almost completely away. The combination of Wingate, Kayenta and Navajo sandstones is known as the Glen Canyon Group because this sequence dominates Glen Canyon, in southern Utah, which is now largely inundated by Lake Powell.

3.8 Note the large rock column of remaining Navajo Sandstone to the left. Note also that, above the cliffline ahead, the Navajo is almost entirely gone. As the Wingate Sandstone climbs steeply above the river level, the slopes below are Chinle Formation that is largely obscured by rubble fallen from above.

5.6 Stop at a pullout on the left of the road. Along this stretch of highway, note the many immense boulders that have tumbled down from the Wingate Sandstone cliffs high above. The slopes below the vertical Wingate cliffs continue to be Chinle Formation down to the river level, but very little of the Chinle is exposed beneath the massive amounts of debris from the Wingate cliffs. Where exposed, the Chinle deposits take the form of layered, reddish sandstones and blue-gray silts that have been colored by the minerals in volcanic-ash fallout. Just ahead, the road enters an immense sweeping curve known as Big Bend. The lofty peninsula formed by Big Bend and lower Castle Valley is known as Mat Martin Point.

Just ahead, the upper strata of the Moenkopi Formation appear but are largely obscured by immense masses of rubble from the cliffs above. As the strata continue to climb, more and more of the thin-layered Moenkopi is exposed but still partly concealed by rubble. Below the uppermost sheer Wingate Sandstone cliffs, the rubble-covered slopes are Chinle, then Moenkopi on down to the road level, with the interface between these two being virtually impossible to see here because of the rubble.

7.0 From a pullout on the right, note how the tip of Mat Martin Point straight ahead is isolated by collapse of the upper Wingate cliffs, leaving a snaggle-toothed gap. The Moenkopi Formation is exposed in a low, layered bluff of red dark deposits just below the spire in this gap in the Wingate.

7.8 A short road to the left enters a BLM picnic area, with designated camping sites nearby. To the right of the road, below slopes capped by harder rock, the Moenkopi Formation is exposed. There are more of the same red sediments to the right of the road as it continues for the next mile, but they become largely obscured by rubble just beyond the picnic area. As the road continues around the immense meander of Big Bend, the geologic strata once again descend, with the Moenkopi going underground and the Wingate Sandstone cliffs dropping to a lower level. The Chinle Formation deposits just below the Wingate cliffs continue to be obscured by collapsed rubble from the Wingate and layered Kayenta sandstones above.

9.1 For a short distance, the Chinle Formation is exposed to the right of the road cut. Chinle deposits along this stretch of the road contain more reddish sandstone than in other areas.

9.8 From a narrow pullout on the right here, note about forty feet above the river level on the far side, a harder red-gray layer sandwiched between layers of red Chinle sandstone. This is the first outcrop visible of an anomalous deposit in the Chinle that appears on both sides of the river gorge beginning here, and continuing beyond where the river gorge widens into Richardson Amphitheater at the mouth of Castle Valley. Along the next three-fourths of a mile, this anomaly is visible intermittently in the Chinle as a strange lighter-hued striated rock. It then becomes a continuous mass of lighter material that erodes in a different fashion from the Chinle above it.

10.5 Beginning at a pullout on the left here, note the anomalous rock layer mentioned earlier that appears just above river level on the left, and watch for the first appearances of it in the rubble slopes to the right of the road, continuing almost to where the river gorge widens into Richardson Amphitheater. Note in the high cliffs straight ahead that there are remnant domes and ridges of Navajo Sandstone above the layered Kayenta Sandstone that tops the vertical Wingate cliffs.

10.8 Beginning here and extending on toward the mouth of Salt Wash, the anomalous deposit is clearly visible above the road to the right in the form of strangely eroded gray cliffs stained with reddish runoff from above, with occasional detached boulders of the same material. Notice how this anomalous stratum contains angled layers in its exposures on the opposite side of the river.

11.1 The pullout on the left of the road here is perched on this anomalous layer, which is also exposed in the rubble-covered slopes to the right of the road. Straight ahead is the confluence of the Colorado River and Salt Wash, a major drainage that goes through Arches National Park. The lower part of the Salt Wash drainage cuts through the anomalous rock layer. Note across the river from this pullout, the angled strata of deposits in the anomalous layer that are topped by a level layer of the same material.

Geologists differ about the nature and identity of this strange intrusion in the Chinle. There is general agreement that the anomaly is a conglomerate, that is, a mixture of sands, silts and gravels cemented together with various minerals. A 1985 geologic map of Arches National Park, by H. H. Doelling, shows the anomaly as Chinle Formation, as does a 1964 Utah State Land Board geologic map by Hintze and Stokes, although in a personal conversation with the author of this book, Stokes admitted that the deposit was anomalous. A 1933 report by geologist A. A. Baker both described and identified the anomaly:

> "*The Shinarump Conglomerate .. is absent or thin north of Lockhart Canyon.* (The anomaly here in the river gorge) *consists of rounded quartz pebbles 1-1/2 to 2 inches in diameter embedded in a matrix of buff to reddish-brown quartz sandstone, but ... grades into a nondescript indurated clay, which in turn grades into the thin gray zone at the base of the Chinle Formation* (in Richardson Amphitheater). *Detailed mapping by the writer has shown that the normal sequence of beds is present in the canyon above and below this series of beds, and they are therefore interpreted as a local development of the Shinarump Conglomerate* (a member of the Chinle Formation otherwise rare in this vicinity)."

On the other hand, a 1983 Grand Junction Geological Society report by Chenoweth and Daub states unequivocally that:

> "'*The Shinarump and Moss Back members* (of the Chinle Formation) *are not present this far to the northwest.*"

The readers of this book are welcome to believe whatever suits them, since professional geologists do not seem to agree. What is certain, however, is that during the Chinle period of deposition, a river and its tributary streams gathered gravels, silts, minerals and assorted sediments from elsewhere and dropped them in a broad river-delta, perhaps where it drained into a large lake, leaving an anomalous deposit in this vicinity. The unusual coloration of this deposit is probably due to the river's concentration of the silts and soluble minerals from the volcanic ash that frequently blanketed the region at that time. One group of geologists reported finding the fossilized casts of burrows of some kind of mud-dweller in this deposit. They believed the burrows were made by lungfish, but another geologist, in a later report, thought they were made by crayfish. Either way, the river delta was inhabited by some kind of water-dwelling species.

Beyond the Salt Wash-Colorado confluence, the anomalous deposits continue to be apparent on both sides of the river just above the river and road level. Along this stretch, the Chinle continues upward to the base of the sheer Wingate Sandstone cliffs that are topped with Kayenta deposits, with a few remnants of Navajo Sandstone above that.

12.2 About here, the anomalous layer disappears underground on the opposite side of the river, but continues to be visible to the right of the road. This strange deposit reappears on the other side of the river about one-half mile ahead, but is largely obscured by rubble for another half mile. It then reappears as distorted lighter-

colored layers beneath the rubble along several stretches.

13.5 The anomaly appears on the other side of the river here as a vertical cliff, sharply tilted with respect to the Wingate and Chinle above it, yet shortly disappears again under the rubble. This exposure is one of the best and most obvious along this stretch of the river. The anomaly disappears under rubble as the river swings around the wide bottomland occupied by a ranch at the mouth of Castle Creek. Beginning here, the major drainages from the La Sals, Castle Valley, Professor Valley and Fisher Valley have widened the Colorado River gorge. This broad cliffwalled valley is called Richardson Amphitheater.

14.4 The road crosses Castle Creek, with a cattle ranch on the left. The lower narrows of Castle Valley extend right from here, with the lower slopes being Moenkopi Formation, the mid slopes the Chinle Formation, and the upper cliffs Wingate Sandstone topped with remnants of layered Kayenta. To the north of the river, the ascending geologic strata form a lofty prominence called Dry Mesa. This elevated plateau is topped with ledgy Kayenta Sandstone, with occasional remnants of Navajo Sandstone above that. Its sheer cliffs are Wingate Sandstone with slopes below being Chinle and Moenkopi. As the road continues, the low hills and bluffs to the right of the highway are Moenkopi Formation. Note that these deposits are multiple thin layers of siltstone, with occasional thicker ledges of sandstone, that are characteristic of this formation in this area. The deep red color of the Moenkopi indicates that it is rich in iron minerals.

15.9 Castle Valley Road goes right here. The geology along this road in Castle Valley is described elsewhere in this book.

Recent historic research has strongly indicated that the Spanish explorer, Juan María Antonio de Rivera, reached the Colorado River, known to the Spanish as the Rio Tizón, in this vicinity in late October, 1765. Rivera was thus the first white man to enter what is now the State of Utah. He was guided by a native Indian to a river ford just upriver of this location that was used by the Indians at that time for trading with the tribes that dominated the region beyond the river. To reach this ford, the Indians from the north used a trail between Cache Valley, which lies beyond the cliffline above the river, and Richardson Amphitheater. Allegedly, this trail was later used by settlers in Richardson Amphitheater to bring supplies from the railhead at Thompson's Springs. Rivera's expedition was discussed in issues 9 and 10 of the *CANYON LEGACY,* the quarterly journal of the Dan O'Laurie Museum in Moab. These back issues may be available at the museum.

Beyond this road junction, to the south note the several detached buttes of Wingate Sandstone standing on bases of Chinle and Moenkopi deposits. These spectacular buttes have been formed by erosion promoted by the salt valleys in this vicinity. Castle Valley and Fisher Valley are both the product of erosion above salt anticlines. Professor Valley was formed by a major drainage that originates near the base of the La Sal Mountains, Mary Jane Creek and its several tributaries.

Across the river from this junction, note the remnant deposits of gray-hued river gravel. There are other such deposits ahead on both sides of the road. Some of the rocks in these gravels were carried here from elsewhere by the river, but some came from relatively recent local glacial activity. See the sidebar, GLACIERS OF THE LA SAL MOUNTAINS.

In the cliffs beyond the river, note the exposures of the anomalous deposit described earlier, appearing at intervals as a whitish band in the red Chinle deposits, below the sheer Wingate cliffs.

16.5 Just ahead, watch to the right of the road for the spires of Castle Rock, The Rectory, the Priest and Nuns, and other unnamed spires in a nearby ridge. These are prominent from both this road and Castle Valley Road. Note ahead, to the right of the road, the remnants of abandoned river gravel on the red Moenkopi slopes. More such gravel deposits appear on the Moenkopi slopes on the far side of the river.

17.5 From this vicinity, park where convenient and study the cliffline to the north of the river. Note in a saddle in the cliffline, how the rock strata there are severely distorted. This distortion between Cache Valley, just south of Arches National Park, and lower Fisher Valley, was caused by severe faulting and a small syncline that were related to the nearby salt anticlines. Note that east of the distorted area, these same geologic strata are undisturbed and relatively level. The historic trail between the river gorge and Cache Valley climbs through this saddle in the cliffline.

Dry Mesa ends at this distorted saddle. The lofty cliffs east of the saddle comprise an immense prominence known as Dome Plateau, which slopes generally downward to the north. Its upper cliffline is largely Wingate Sandstone. The slopes below continue to be Chinle and Moenkopi, with Moenkopi bluffs dominating the sloping plains just above river level. Several miles ahead, remnants of lighter colored Navajo Sandstone can be seen above the layered Kayenta Sandstone. As the road continues upriver, the low hills of Moenkopi to the right are topped by gray river gravel deposits.

18.4 As the road goes through a cut, notice the warped and tilted Moenkopi strata. This is part of the stratigraphic distortion that lies between Cache Valley and Fisher Valley.

18.8 The road to the right enters Professor Valley. Beginning here, the La Sal Mountains are prominent to the south. To the east, the lofty rimlands of Fisher Mesa loom above Richardson Amphitheater, while Dome Plateau continues to dominate beyond the river. To the left of Fisher Mesa, the pair of jagged domes are more of the distorted strata between Cache Valley and Fisher Valley. Still farther left, some of the strange spires that loom above the Onion Creek labyrinth are visible against the skyline. The tallest spire there and the mass of rock to its right are known as Little Red Riding Hood (standing) and Grandma (sitting in a chair). Still farther left, the higher distant cliffline of Wingate Sandstone is known as Top-of-the-World. An ORV trail described in the referenced book ends at that breathtaking rim viewpoint. To the left of Top-of-the-World and a little lower, some of the soaring spires of the Fisher Towers are visible from here and from the road as it continues upriver. The uppermost parts of the Fisher Towers are Moenkopi Formation, while the lower parts are Cutler Group. Cutler sandstones also appear in the base of the clifflines just above the river to the northwest.

19.5 The road crosses Professor Creek, which is a combination of Professor Creek and Mary Jane Creek, with Mary Jane Creek contributing most of the water. The broad meadowlands on each side of the road ahead are dune sand and decomposed rock barely held in place by native vegetation. The lush native grasses that formerly bound these soils against erosion have been destroyed by a century of overgrazing by domestic livestock.

20.7 The road to the right here goes toward Fisher Valley through the Onion Creek labyrinth. It can be traveled by off-road vehicles and by carefully driven, low-geared, high-clearance highway vehicles. The road fords the creek many times before reaching an exposed upthrust of Paradox Formation salts. This road, and the ORV trail beyond its end, are described elsewhere in this book.

21.2 The road crosses Onion Creek. Note in this vicinity the great masses of alluvium and abandoned river gravel perched on ledges of dark red Cutler sandstone.

21.5 The road to the right goes to the Fisher Towers. This road can be driven by any vehicle, although the last few yards of it are steep and usually rough. At the end of the road, there is a parking area, a developed picnic site, and a spectacular hiking trail that climbs among the Fisher Towers. The tallest of these is The Titan, which soars 900 feet above its base. The road north from this junction soon enters private land.

22.0 From a pullout on the left, note the thin ridge of rock toward the river. This ornately eroded remnant of Cutler sandstone is known as Locomotive Rock. From this pullout, note just below the high Wingate cliffs of Dome Plateau, how the rubble on the steep slopes has funneled through narrow gaps in the lower Moenkopi cliffs, then spread into triangular alluvial fans. Note also the isolated remnants of Navajo Sandstone on the skyline above the Wingate cliffs. The road continues to travel through Cutler sandstone, but this dark red rock is only exposed occasionally.

23.9 The road left toward the river goes to an historic site and a popular launch ramp for one-day raft runs on the river. In this vicinity, the uppermost sheer cliffs continue to be Wingate Sandstone topped by remnants of Kayenta Sandstone. The slopes below the Wingate are Chinle Formation, and the lower crenelated bluffs are Moenkopi Formation with alluvial debris fanning out through gaps in that cliffline. To the right-front, the lowest bluffs just above ground level, with spires and balanced rocks, are Cutler sandstone.

From this junction and the road ahead, note the four power poles on the skyline above the cliff, high to the right-front, where the powerlines that are paralleling the road climb upward in steep, catenary curves to continue on north. An ORV trail that spurs from the Entrada Bluffs Road goes to those power poles for a spectacular view of Richardson Amphitheater. Beyond here, Richardson Amphitheater ends where the river emerges from a narrow gorge, with the high cliffline beyond the river continuing to be Dome Plateau. Within a mile or so, the strata on both sides of the river begin a gentle decline.

25.7 Note, on the other side of the river, how the low Moenkopi bluff is disappearing beneath the river level, leaving the Chinle Formation dominating the slopes below the Wingate cliffs. Beyond here, all the strata continue to descend, so that most of them visible here will be underground by the time this road reaches Dewey Bridge.

26.9 Note on the detached boulder to the left of the road, the words KINGS TOLL ROAD. Samuel King developed the first "road" along this stretch of the river-gorge, and established a ferry at Dewey, in the late 1800s. A small remnant of Navajo Sandstone appears above the Wingate cliffline ahead.

27.5 At a pullout on the left here, note that the base of the Wingate Sandstone is now at river level, with a wall of Navajo Sandstone above the layered Kayenta on the skyline straight ahead. Just beyond the bend ahead, the strata climb briefly once again as the river curves in a new direction, then descend again as the river goes around a wide sweep to the left.

29.0 The bottom of the Wingate is once again at river level. As the road continues, the Wingate dives more steeply downward and the terraced Kayenta Sandstone descends to a lower level.

20.5 From a pullout adjacent to a small bridge, search the low cliffline to the left-front for remnants of an old wooden ladder that ascends the cliff in two stages. Although no record of the ladder's construction was found, it was probably built to enable prospectors in the higher terrain to reach the river for water.

29.9 From a pullout on the right, note that the Wingate Sandstone has largely disappeared beyond the river and that the lower cliffs are now dominated by layered Kayenta Sandstone, with one remnant dome of Navajo Sandstone to the left-front. The Navajo continues beyond a gap, then descends steeply toward the river level to disappear beneath the ground shortly beyond the river bridge ahead. The rock that dominates the skyline straight ahead is Entrada Sandstone, with the uppermost thick layer being the Slickrock Member. At the base of the spires that project from that cliff, the darker mudstone is the Dewey Bridge Member of the Entrada, which was first studied one mile east of that historic structure, then named after it. Geologists call such a first-study site a "type locality." Note that in this vicinity, the Navajo Sandstone is much thinner than it is farther west. In this part of the state, the Navajo diminishes to nothing in the general vicinity of the Utah-Colorado border. The rubble above the Slickrock-Entrada is lower Morrison Formation deposits.

30.3 From a pullout on the left here, note that the Navajo Sandstone has almost disappeared beyond the river and that the low slopes below the Slickrock-Entrada cliffs are topped by immense quantities of abandoned river gravel. From here, the Morrison Formation deposits above the Slickrock-Entrada dominate the northern skyline. The cliffs above the road to the right are Navajo Sandstone.

30.6 The Entrada Bluffs Road, noted earlier, goes right here, while this road continues across the new bridge. Historic Dewey Bridge, just upriver, was constructed in 1916. For more information about this historic structure, see the sign beside the bridge. The sheer cliffs that dominate ahead and to each side in this vicinity are Slickrock-Entrada topped with Morrison Formation deposits. There is a thin layer of the reddish Tidwell Member just above the Slickrock-Entrada, then the Salt Wash and Brushy Basin members of the Morrison appear above that in ascending order. Note that, in this area, the Slickrock-Entrada does not erode in exactly the same manner as Wingate Sandstone. It tends to erode away into vertical cliffs only when severely undercut. Otherwise, it takes on the form of more rounded, gentle slopes. The rock at river level here is Navajo Sandstone, with slopes of river gravel above that. Beyond the bridge, the riverbottom land on each side of the road is private property. This area once hosted a small but thriving settlement of cattle ranchers. The rounded cliffs to the left of the road are Slickrock-Entrada.

31.7 Note the large spring-seep cave to the left of the road, with its seepline of water-loving vegetation and beautiful run-off streaks in the cave roof. Note that the geologic strata are still descending as the road continues.

continued on page 93

ENTRENCHED MEANDERS

"Entrenched meander" is a term geologists apply to a meandering watercourse that has existed for so long during a period of erosion that it has entrenched itself, or cut its looping path deeply into the terrain through which it flows.

Rivers and streams in general either cut downward into the land through which they flow, or they "alluviate," that is, they build up the surrounding terrain by depositing silts in their beds until they overflow and change course, then repeat this process endlessly, gradually building the depth of the alluvium, or deposited sediment, that covers the surrounding land.

When conditions are wrong for alluviation, rivers and streams cut into the sediments and rock below their beds, gradually entrenching themselves more and more deeply, and carrying their burdens of sediments on to the sea, where they are deposited to create immense deltas of salt-marsh that eventually becomes dry land, thus slowly but steadily expanding the area of a continent that is above sea level.

There are several major factors that control whether a river or stream builds or cuts the terrain through which it flows. These include the general slope of the terrain, the local climate and weather, the variability of the flow and, in recent centuries, how the watershed feeding the flow has been affected by human activities. Such common practices as forest clearcutting, the conversion of prairies and brushy plains to dry-farming, poor plowing techniques, and the destruction of native soils and plant communities by grazing and grazing-support operations, all drastically accelerate water runoff and hence cause flash-flooding and rapid cutting by streams and rivers.

This is another way that humanity affects geologic processes. See the sidebar, HUMANITY AS A GEOLOGIC FORCE.

In canyon country, the present erosional phase was started by the regional uplifting that occurred about 10 million years ago. This general rise in the region's elevation brought increased precipitation to its interior mountains and highlands and to those surrounding it. This, in turn, gradually increased the flow of the few rivers and streams in the region at that time. The region's generally arid climate and relatively sparse vegetation allowed the scant precipitation that did fall on its lower terrain to run off quickly, thus causing rapid erosion from flash-flooding, and the cutting of numerous arid sidecanyons. This also created violent fluctuations in the flow of such rivers as the Colorado, Green, San Juan and Dolores, which then cut more deeply into their channels, creating the present system of spectacular canyons and gorges that are major topographic features of the Colorado Plateau. See the sidebar, COLORADO PLATEAU.

The general flow of a river or stream, and the way in which this flow varies, is obviously related to the regional climate and the local weather that produces the flow. But how the general slope of the land affects watercourses is not so obvious. This curious effect has been demonstrated clearly by a reduced-scale experiment with water flow. Arid land crossed by a river was simulated on a long table with built-up sides. Variable terrain and rock strata were simulated by adding many thin, fairly uniform layers of soils, sands, clays and gravels to the table, compacting each layer so as to closely simulate various types of harder and softer geologic strata and the ways in which they are eroded by a flowing river.

Once this miniaturized terrain was ready, the table was tilted slightly from level and a trickle of water was run lengthwise, simulating the flow of a river. The experiment was repeated several times, with various static degrees of tilt, and also while slowly increasing the tilt, and the patterns of river flow and erosion of the simulated rock strata were recorded.

Several curious general patterns related to river flow became readily apparent, as did many interesting details about rock erosion. If the tilt of the miniature land was gentle, the tiny river created widely looping meanders, gradually filling the river's channel in the lower part of its course with sediments from the higher part, and constantly changing its course where this siltation occurred.

If the simulated terrain was given considerable tilt, the little river cut a much straighter course. It did much less meandering, cutting quickly and deeply into the simulated rock strata. Virtually none of the silts it picked up along the way were redeposited lower in its course. They were, in effect, carried on to the "sea" beyond the lower end of the table.

The most interesting effect was observed when the table was tilted very little at the beginning of the tiny river's flow, allowing it to construct a looping, meandering course, then the tilt was gradually increased while the river continued to flow. As the tilt slowly increased, the river's flow cut its meandering course more and more deeply, thus creating the kinds of "entrenched meanders" so common to canyon country.

As time passed, some of the large loops in these entrenched meanders were bypassed, as the flowing water eventually cut through the narrow necks of the loops, thus creating "abandoned meanders." These erosional phenomena are fairly common in canyon country. Most of them occur in remote areas that are not easily reached, but in the area of this book there is a large abandoned Colorado River meander called Jackson Hole that is visible from the end of Utah 279, and from the Potash ORV trail that continues beyond that paved road. Both of these routes are described elsewhere in this book.

There are at least two major abandoned meanders in the narrows part of deep Harts Draw, the spectacular canyon that is paralleled by

continued ----

Needles Overlook Road, also described in this book, and there are several such abandoned loops in Lower Indian Creek Canyon that can be seen with binoculars from Needles Overlook.

As the simulated terrain on the table was tilted still further, its tiny river cut across more and more of its original meanders, reducing the meandering nature of its course, and continuing to deepen it, until eventually the river was flowing through a miniature gorge that was almost straight.

This relatively simple experiment revealed much about the history of the Colorado, Green, San Juan and Dolores rivers, as they entered the canyon country region, joined their waters, then continued southwestward across the elevated Colorado Plateau. It demonstrated that when the rivers first began flowing from their sources -- three in the Rocky Mountains, one from the Wind River range in Wyoming -- then across the broad open terrain that was to become canyon country, that this open terrain had very little overall slope. This allowed the young rivers to wander across the land in broad sweeping turns, doubling back on themselves in many places, creating and deepening countless meanders.

It also showed, however, that there was enough slope, at least in the Colorado Plateau, to prevent the sediments the rivers were carrying from being deposited until the Colorado River reached the sea, although there is strong evidence that the Colorado did not at first actually reach the sea, but instead dumped its waters and their sediments into an enormous inland lake in what is now Arizona.

The experiment also suggested that canyon country terrain and climate were such that for millions of years these rivers were stable, that they were able gradually to deepen their meandering courses, cutting slowly into older rock strata, creating entrenched meanders that could last for additional millions of years, with relatively few such meanders being abandoned. These same general conditions and this same process also allowed the region's countless flowing tributary rivers and streams to meander and become entrenched.

The simulated river also revealed why the canyons of rivers and streams that flowed steadily developed deeply entrenched meanders, while most of the tributary drainages cut only by violent, intermittent flash-flooding, developed relatively straight canyons, thus further increasing the complexity of the system of canyons and gorges of the Colorado Plateau that make the region so spectacular, and such an open geologic book.

Just around the bend ahead, the Slickrock-Entrada goes underground on both sides of the river. The gap in the bluffline beyond the river indicates where the Dolores River joins the Colorado River. Beyond here, the rubbly slopes are dominated by the Salt Wash-Morrison and, above that, the painted desert deposits of the Brushy Basin-Morrison.

32.6 Stop at a pullout on the left to see several prehistoric petroglyphs on a large boulder about 10 feet above the pullout.

32.9 Here, the Yellowcat ORV trail goes left, climbing the rocky slopes toward the Yellow Cat mining area. In this vicinity, the geologic strata have descended to the point where the Brushy Basin-Morrison is at road and river level, with the stratigraphically higher deposits being the Cedar Mountain Formation. Cedar Mountain is the equivalent of the Burro Canyon Formation east of the Colorado River. Some geologists consider the Cedar Mountain to be a lower unit of Dakota Sandstone. Notice in the Cedar Mountain slopes ahead above the road and in the equivalent Burro Canyon across the river, the reddish color-banding caused by unusual mineral content.

34.2 From a pullout on the right here, above a bend in the river, the low gray hills ahead and on to the north are Mancos Shale. To the left, Morrison and Cedar Mountain deposits extend to the horizon, with remnants of still younger Dakota Sandstone. The panorama to the east and northeast, beginning to the right-rear, is the northern nose of the Uncompahgre Uplift in the distance, with more vast expanses of Morrison, Burro Canyon, Dakota and Mancos Shale deposits in the foreground. See the side bar, UNCOMPAHGRE UPLIFT. In the distance, the river emerges from high terrain, with brown Dakota sandstone and varicolored Burro Canyon deposits forming high cliffs above the river.

35.2 The road crosses a bridge over a shallow drainage. This geologic road-log ends at the bridge because, from here to Interstate 70, the geologic scenery is unrelieved Mancos Shale. Note the old cattle corral to the left of the road. The bottomlands along the river in this area are private land, some of it under development.

From here, the way back to Moab is longer by continuing to Interstate 70, then going south on U.S. 191. The return via Utah 128 will permit viewing the many outstanding geologic features along that route from a different angle and with different lighting, making the return trip over the same route well worthwhile. This return route also permits exploring Fisher Towers and the Castle Valley Road, if time allows.

ROCKHOUNDING

Rockhounding in most mountainous and desert areas, where the geologic strata are jumbled and distorted, is usually rewarding but generally takes the form of lengthy, random searches over large areas of broken terrain, with satisfactory results being more a matter of luck than planning.

In the high-desert terrain of Utah's canyon country, however, the geologic strata are easily identified and generally stacked neatly in layers, with minimal distortion by tectonic forces, thus making rockhounding far easier and more rewarding, although the collectible specimens may be quite different from those found in western mountain ranges and lower deserts.

The last chapter of the book, ***Canyon Country* GEOLOGY - For the Layman & Rockhound**, describes how rocks, fossils and other collectible mineral specimens sought by rockhounds are found in canyon country in specific geologic strata, and lists what kinds of specimens are common within each of the rock formations that occur in the canyon country region. The principles summarized in that book also apply to the general Moab area.

As described in the section in this book on geologic formations, the strata in the general Moab area range from the ancient marine deposits of the Paradox Formation to the deep marine deposits of Mancos Shale, with several younger formations visible in the Book Cliffs to the north of Interstate 70. The following chart indicates what collectible specimens are most apt to occur in the most promising geologic strata in the area.

The oldest strata exposed in the Moab area are not listed because they are found only in stretches of the Colorado River gorge that are not accessible to most rockhounds. The deposits that appear only in the Book Cliffs are also not listed, because they are not accessible from the area covered by this book. Formations that contain few if any rock or fossil specimens are also omitted from the chart.

SPECIMEN SOURCE CHART

Lower Cutler Group. Profuse marine fossils in the uppermost limestone layer and at intervals in lower strata, plus occasional petrified pre-dinosaur foot tracks.

Organ Rock Shale, Cutler Group. Agate, chert and jasper.

White Rim Sandstone, Cutler Group. Iron-rich concretions.

Moenkopi Formation. Ripple rock, mud casts, and occasionally reptile tracks and fossilized bone.

Chinle Formation. Petrified wood, calcite, agate, barite geodes, jasper, aragonite, selenite, fossil horsetail, ferns, bone and tracks, coal, jet, carbonized plant life and aquatic-life fossils.

Wingate Sandstone. Ripple rock and dinosaur tracks.

Kayenta Formation. Ripple rock, mud casts, petrified wood, iron concretions and dinosaur tracks.

Navajo Sandstone. Chert, iron concretions, petrified wood, algae and tracks.

Dewey Bridge Member, Entrada Sandstone. Agate, chert, jasper, iron concretions, manganese nodules and petrified wood.

Slickrock Member, Entrada Sandstone. Manganese nodules.

Moab Member, Entrada Sandstone. Copper and other concretions, rare marine fossils, dinosaur tracks.

Tidwell Member, Morrison Formation. Agate, chert, jasper, gypsum, ripple rock, mud casts and petrified tracks.

Salt Wash and Brushy Basin members, Morrison Formation. Agate, chert, petrified wood and bone, plant and invertebrate fossils, calcite, barite, geodes, celestite, selenite, gastroliths, coprolite and dinosaur tracks.

Cedar Mountain - Burro Canyon Formations. Petrified wood and other plant fossils.

Dakota Sandstone. Agate, chert, fossilized bone, plants and coal, with abundant petrified wood in some locations.

Mancos Shale. Calcite crystals, aragonite, marine fossils, coal, jet and carbonized plant life, with some fossil plant impressions in upper layers. The Ferron Sandstone Member is rich in shallowmarine fossils.

To look for any of the listed specimens, rockhounds need only seek out exposures of the above-listed rock formations, many of which are noted in the geology road and ORV trail logs in this book. For a more comprehensive overview of places where these strata occur, geologic maps are available for sale at some Moab area visitor centers and from offices of the U.S. Geological Survey.

In addition to these rockhound "supermarkets," where certain kinds of specimens occur in specific geologic layers, there are also places in the Moab area that are giant "grab-bags" of mineral specimens. At intervals along the banks and terraces above the Colorado River, ancient river gravel deposits have been left high and dry as the river has cut deeper. These deposits contain durable rock specimens from the entire geographic area drained by the Colorado and its numerous upper tributaries. To reach such deposits, rockhounds need only drive any of the roads and ORV trails that closely parallel the river, watching for piles of rounded river-rocks that are often blackened with desert varnish.

The best roads and ORV trails for this purpose are Utah 128, Utah 279, the Potash Trail and the Chicken Corners Trail, all described elsewhere in this book. The Potash Trail continues downriver from the end of paved Utah 279. The Chicken Corners Trail goes downriver from a three-way trail junction at the western end of the Hurrah Pass Trail. A number of ancient river gravel deposits lie beside or near this trail where it closely parallels the Colorado River's inner gorge. The Jackson Hole Trail, described in the referenced book and map, heads upriver from the same junction and goes to an immense abandoned river meander. There, countless tons of river gravel fill the meander wall-to-wall with rounded, durable rock specimens transported there by the river over many thousands of years, from outside of this region.

Rockhounds should note that there are legal limits to the kinds and amounts of prehistoric fossils that collectors are permitted to take from public land. For current regulations applicable to Moab area public lands, including the Canyon Rims and Cameo Cliffs recreation areas, contact either of the two Moab offices of the Bureau of Land Management. For current regulations governing specimen collecting in National Forests, contact the U.S. Forest Service office in Moab. Collecting of any kind is prohibited in Canyonlands and Arches national parks.

ROAD NAME: **Utah 211**

SUMMARY: This popular highway, together with its access route, U.S. 191, offers spectacular scenery and considerable geologic variety along the way. It is the only route into the Needles District of Canyonlands National Park that can be used by standard highway vehicles. Between U.S. 191 and the park, this road travels through three districts of beautiful Canyon Rims Recreation Area, each of which offers different combinations of geologic strata, tectonic features and erosional landforms. These include deposits ranging from relatively young Morrison Formation to ancient Cutler Group sandstones; faults, grabens and laccolithic mountains; a magnificent canyon system; and a variety of colorful sandstone buttes, spires, ridges, fins and domes.

ROAD SURFACE: paved.

VEHICLE: any vehicle.

MAPS: ***Canyon Country* OFF-ROAD VEHICLE TRAIL MAP - Canyon Rims & Needles Areas,** or ***Canyon Country* OFF-ROAD VEHICLE TRAIL MAP - Canyon Rims Recreation Area.**

GUIDEBOOKS: ***Canyon Country* OFF-ROAD VEHICLE TRAILS - Canyon Rims & Needles Areas,** or ***Canyon Country* OFF-ROAD VEHICLE TRAILS - Canyon Rims Recreation Area,** or ***Canyon Country's* CANYON RIMS RECREATION AREA.**

MILEAGES:

TO START: 40.7 miles, from Main and Center, Moab.
ROAD: 35.3 miles, one-way to end.
RETURN: 76.0 miles, from end of described road.
TOTAL: 152.0 miles, from Main and Center, round trip.

GEOLOGIC STRATA VISIBLE FROM THIS ROAD

Morrison Formation	Brushy Basin	Page 18
	Salt Wash	
	Tidwell	
Entrada Sandstone	Moab	Page 16
	Slickrock	
	Dewey Bridge	
Navajo Sandstone		Page 13
Kayenta Sandstone		Page 13
Wingate Sandstone		Page 12
Chinle Formation		Page 12
Moenkopi Formation		Page 12
Cutler Group	White Rim	Page 11
	Organ Rock	
	Cedar Mesa	
	Lower Cutler	

TIME: allow for a short day to reach and travel this route, then return to Moab. A full day is required if any hiking is done, or any side roads or ORV trails are explored, along the way.

ROUTE: south from Moab on Main Street (U.S. 191), then west on Utah 211 toward the Needles District of Canyonlands National Park.

NAVIGATION: drive south from Moab on U.S. 191 for the indicated distance, then west on Utah 211 into the park. Return via the same route to Moab. Those with off-road vehicles who are prepared for primitive camping could return via the Lockhart Basin - Chicken Corners - Hurrah Pass - Cane Creek Road route. Of these, the last three are described elsewhere in this book, but traveling in the opposite direction. The Lockhart Basin Trail is described in the referenced guidebooks, and should not be attempted without such guidance.

GEOLOGY ROAD LOG:

0.0 As this road heads west toward the Needles District of Canyonlands National Park, the nearby slickrock domes and ridges are the Slickrock Member of Entrada Sandstone topped with remnants of Morrison Formation deposits. The peaks of the Abajo Mountains are visible to the left-front. This intrusive igneous mountain range was named by early Spanish explorers. For a description of its geology, see the sidebar, LACCOLITHIC MOUNTAINS.

For the first three miles, the road crosses open brushy meadowlands from which the native grasses have long since been removed by more than a century of overgrazing. The sparse grasses seen are an exotic species introduced in an attempt to restore better grazing. The meadows consist of decomposed rock, soils, alluvium and sand held in place by the remaining vegetation. The notch in the ridgeline ahead is known as Photograph Gap because of the panoramic view that opens up at that point. As the road travels toward Canyonlands National Park, it goes through the Dry Valley, Indian Creek Canyons and Lowlands districts of Canyon Rims Recreation Area. See the book ***Canyon Country's* CANYON RIMS RECREATION AREA.**

0.3 The road to the right goes to Sugar Loaf Rock. The curious history of this colorful butte with a large pothole in its top is described in the above book.

0.5 The abandoned structures to the left of the highway, beginning here and continuing to Photograph Gap, are remnants of a 1930's religious community. For more details, see the sidebar HOME OF TRUTH in the above-noted book.

1.7 The abandoned structures to the left of the road are more of the Home of Truth colony.

3.2 From a road junction pullout on the left, at the summit of the pass through Photograph Gap, more structures from the Home of Truth colony appear to the right of the road. The sandstone slopes on both sides of the road are Slickrock-Entrada. The red layer exposed at the top on the righthand bluff is the Tidwell Member of the Morrison Formation.

As the road continues, the flat-topped domes of sandstone ahead to the right are all Navajo Sandstone topped with vegetated remnants of the Dewey Bridge Member of Entrada Sandstone. Just beyond Photograph Gap, the northernmost peak of the Abajo Mountains is Shay Mountain, which is one of several central stocks in this intrusive igneous range. Farther to the right are the Elk Ridge highlands on the northern nose of the Monument Uplift.

3.5 Dry Valley Road, described in the second two of the three referenced guidebooks, comes in from the right. Beyond here, Utah 211 drops into a broad valley that is crossed by one major drainage and contributes to another. Harts Draw, which originates high in the Abajos, crosses this valley, then plunges deeply into the rising terrain to the northwest. The Needles Overlook road described elsewhere in this book, parallels that awesome gorge and provides a few glimpses of its 1200-foot high sandstone walls. Farther on, another drainage contributes to the formation of Upper Indian Creek Canyon.

5.0 The long ridge to the left along this stretch of road is Slickrock-Entrada topped with red-hued Tidwell-Morrison deposits. As the road continues, the sandstone base of the terraced cliff south of the road is the same rock.

6.0 The white rock exposed in the lower drainages to the right is Navajo Sandstone.

7.4 The road crosses the Harts Draw drainage, which is going to the right.

8.0 The low slope of slickrock to the right of the road, and to the left on ahead, is Navajo Sandstone largely covered with the pinyon-juniper plant community.

9.8 Upper Harts Draw Road goes left here and, just ahead, Harts Point Road goes right. These roads are described in the referenced books on Canyon Rims Recreation Area, but not in this book. Beyond here, Utah 211 leaves the Dry Valley District and enters the Indian Creek Canyons District of the recreation area.

10.6 From a pullout on the right, the slickrock slopes on both sides of the road are Navajo Sandstone. Note on the left, in the rock nearest the road, how the cross-bedding that is typical of this deposit has been accented by surface weathering. The flat-topped slopes to the right-front of the road are topped with the limestone deposits of an ancient desert playa. See the sidebar, PETRIFIED PLAYAS. The immense dome straight ahead is also Navajo Sandstone. As the road drops down into Upper Indian Creek Canyon, it continues to descend through more light-hued Navajo Sandstone.

11.1 Along this stretch of road, the terraced reddish rock ahead and to the left is Kayenta Sandstone. Note that the Kayenta Sandstone on the left is aligned with light-hued Navajo Sandstone to the right of the road. This misalignment is due to a major fault that runs approximately along the drainage at the base of the Navajo Sandstone cliff. Note the small remnant community of Ponderosa pine at the base of the Navajo cliffline.

11.7 From a narrow pullout here, the layered reddish rock straight ahead is Kayenta Sandstone, while the lighter rock on each side of the road is younger Navajo Sandstone. Note to the right, where several cottonwood trees are growing, how the red Kayenta strata are misaligned several yards by a spur fault.

11.8 Note the seep caves to the right of the road beyond the drainage.

12.1 From a narrow pullout on the right, on a curve in the road, Wingate Sandstone is appearing above ground level to the right. The layered rock just above the smooth Wingate cliff is Kayenta Sandstone, and the high dome straight ahead is Navajo Sandstone. The sloping cliff to the left of the road here is also Navajo Sandstone. This misalignment of the strata is due to the fault noted earlier that runs down the drainage to the right of the road, then crosses it about where the road angles right around the cliff ahead, then continues through the cliffs to the left. There is another major fault parallel to this one farther southeast, with the land between these two faults being displaced somewhat downward. Such depressed areas between faults are called "grabens," a German word meaning "grave." These parallel faults continue for many miles in a southwesterly direction around the base of Shay Mountain, hence the name "Shay Graben." Note the old stock corral at the base of the Wingate cliffs just ahead.

12.4 Indian Creek enters the canyon from the left. This road parallels the creek for a number of miles before emerging into more open terrain.

12.6 Pull into the parking lot for the petroglyph panel at Newspaper Rock State Historical Monument. These prehistoric petroglyphs are in desert varnish on a wall of Wingate Sandstone. See the sidebar, DESERT VARNISH. As the road continues beyond here, it drops more deeply into this beautiful canyon. The Wingate Sandstone walls, topped with layered red Kayenta Sandstone, become higher and higher, with occasional remnant domes of Navajo Sandstone looming against the skyline.

13.4 Indio Arch is in the Wingate Sandstone cliff to the left, about midway up.

14.6 Stop at a wide pullout to the left of the road. Shay Canyon comes in from the left, making a break in the canyon wall. On the righthand abutment of that sidecanyon, the base of the vertical cliff has hundreds of prehistoric petroglyphs, some of which can be viewed from here with binoculars. Well-watered Upper Indian Creek Canyon was heavily used by prehistoric Indians, who left their primitive graphics in many places in the dark brown desert varnish that forms on the Wingate Sandstone.

There are dozens of petrified animal tracks in a Chinle Formation sandstone exposure in the lower drainage of Shay Canyon. These tracks were made by pre-dinosaur mammal-like reptiles known collectively as "therapsids." Most of these mammal predecessors went extinct during the stratigraphic unconformity between the Chinle Formation and the base of the overlying Wingate Sandstone. See the sidebar, GREAT EXTINCTIONS. If Indian Creek is not flowing too vigorously, it is practical to hike from here into lower Shay Canyon to see the petroglyphs and tracks at and near its confluence.

15.7 From a pullout on the right just beyond a cattleguard, note the high sheer-walled dome of Navajo Sandstone to the left, looming above the terraced Kayenta Sandstone and colonnaded cliffs of Wingate Sandstone. As the road continues to descend Upper Indian Creek Canyon, note how the Wingate Sandstone walls are fractured into vertical columns. From here on down the canyon, Chinle Formation deposits appear as rubble-covered slopes at the base of the sheer-walled Wingate.

16.5 Remnants of two historic ranch buildings constructed with native stone and logs are visible in the trees to the left of the road. This canyon was the site of considerable settlement and development during historic times. For more information about this early canyon country history, see the sidebar, INDIAN CREEK RANCHES in the book ***Canyon Country's* CANYON RIMS RECREATION AREA.** During this period of settlement, most of the lower canyon was homesteaded, although these homesteads have since been consolidated under one ownership as Dugout Ranch. Along the next several miles of road, many of the canyon's brushy meadows have been cleared of native vegetation and cultivated for livestock support. Note that these historic and current human uses have promoted alluvium cutting by Indian Creek. This is visible from the road in a number of places. See the sidebar, HUMANITY AS A GEOLOGIC FORCE.

In this vicinity the canyon has now cut deeply enough that the layered Chinle Formation is now prominent below the Wingate Sandstone cliffs as rubble-covered slopes. Here, the Chinle deposits are largely red-hued rather than the gray that dominates elsewhere.

17.0 The mouth of Titus Canyon appears on the left.

18.2 Note the relatively fresh rockfall to the right of the road that occurred in the spring of 1992, while this book was in preparation.

19.1 From a narrow pullout here, at a trail junction entering the public-land canyon to the right, note the spectacular panorama opening up beyond the mouth of Upper Indian Creek Canyon. The two spire-topped pyramids there are the Six-shooter Peaks. The lofty peninsula across the distant skyline is the Island-in-the-Sky. The detached prominence to the left of that is Junction Butte. The high terrain to the right is all part of Harts Point.

19.9 County 104, the Cottonwood Canyon ORV trail, goes left here through some of the older structures of Dugout Ranch, then continues up Cottonwood Canyon and into higher terrain. That trail is described in the referenced books and maps. Note ahead, the conical towers of the two Six-shooter Peaks. The base of the cliffs in Cottonwood Canyon are Chinle Formation, and the sheer walls are Wingate Sandstone topped with layered Kayenta Sandstone. The white rock domes above that are Navajo Sandstone. This combination of rock strata applies to everything visible from this point except the low red terraces in the distance ahead. Note the reservoir to the right of the road which collects natural drainage from the canyon to the right. The high terrain to the right is more of Harts Point. The road and ORV trails that penetrate that lofty, complex peninsula are described in the referenced guidebooks.

21.0 The sediments exposed along drainage lines near this stretch of the road are red-hued Moenkopi Formation topped with remnants of gray Chinle deposits.

23.6 Stop at the junction where the Lavender Canyon ORV trail, described elsewhere in this book, goes left. Lavender Canyon is the broad drainage going south. The canyon is walled with cliffs of sheer Wingate Sandstone topped with layered remnants of Kayenta Sandstone and one white dome of Navajo Sandstone. Below the Wingate cliffs, steep slopes of Chinle Formation sediments descend to a ridge of harder sandstone, which is also Chinle. The smooth, rubble-covered slopes below that are Moenkopi Formation. The crenelated bluffs that descend below that to the canyon floor are the Organ Rock Shale member of the Cutler Group. The deposits in the nearby Six-shooter Peaks are this same combination, except the Navajo and Kayenta are gone. The broad drainage to the left of the Six-shooter Peaks is Davis Canyon. The ORV trail that enters that beautiful drainage is described in the referenced guidebooks.

25.0 Along this stretch of road, the western end of immense Harts Point dominates the skyline to the right. Harts Point is topped with Navajo Sandstone or, in some places, with layered Kayenta Sandstone. Below that, sheer walls of Wingate Sandstone and slopes of Chinle, Moenkopi and Organ Rock Shale descend into the lower terrain. The low bluffs on both sides of the road along this stretch are Organ Rock Shale.

26.9 The road crosses the Indian Creek drainage. The low redrock bluff walling the drainage is Organ Rock Shale. The creek is dry during the warmer months because its water is diverted for ranching. The drainage continues in a northwesterly direction for several miles, then cuts deeply into older Cutler Sandstone deposits before finally plunging into the Colorado River gorge.

27.1 The Davis Canyon ORV trail goes left here. That trail is not described in this book, but the geology of Davis Canyon is generally very similar to that of Lavender Canyon, which is described.

27.4 From a paved pullout here, the high cliffs to the north, beyond the broad mouth of Harts Draw, are part of immense Hatch Point. This elevated peninsula has the same geologic strata as Harts Point, which were described at mile 25.0. The low terrain in this vicinity, from Harts and Hatch points to the park boundary, is part of the Lowlands District of Canyon Rims Recreation Area. All of Hatch Point and Harts Point are also part of this public land recreation area, which is larger than the national park it surrounds on the south and east.

continued on page 101

DESERT VARNISH (ROCK VARNISH)

Desert varnish is a kind of thin patina or coating on rock that appears in desert areas throughout the world. It ranges from light red-brown with a matte texture to coal black with a lustrous, almost metallic sheen. It occurs on almost every type of rock that has a surface durable enough for the slow-growing varnish to become established.

Since desert varnish also occurs to some extent in climate zones and locations that are not desert-like, scientists prefer to call it "rock varnish," although it is far more common in arid regions because these provide the most ideal conditions.

In the canyon country heartland of the Colorado Plateau, rock varnish of all shades is common. The soaring cliffs of massive sandstones that dominate the region -- Wingate, Navajo, Entrada, Cedar Mesa and White Rim -- are coated with varnish of one shade or another, with some of it streaked, some smooth and uniform, some red-brown and some coal-black. Fresh rockfalls that have detached from such cliffs are made conspicuous by the absence of colorful rock varnish on the newly-exposed surfaces.

Less apparent exposures of rock varnish are also found on other durable sandstone deposits, such as Cutler, Kayenta, Morrison, Dakota, and Mesaverde, as well as on the harder sandstone layers within virtually all other formations in the region.

It is also found on detached boulders and on hard, water-rounded rocks and gravels that have been imported into the region by its rivers, then abandoned as the rivers have cut deeper. Such varnish-blackened river gravels are common along the banks and terraces of the Colorado River, as described in several geology logs in this book.

Rock varnish was first noticed by explorers who remarked about it in their journals, sometimes with suggested explanations for the strange substance. Of recent decades, scientists have taken an interest in the subject for several reasons, but their investigations have most often been spurred by the idea that understanding rock varnish might lead to its use in dating archaeological sites and artifacts.

For example, if the new varnish that has formed inside the chipped areas of a petroglyph could be age-dated with any degree of accuracy, that would indicate when the petroglyph was made. Some prehistoric artifacts, such as stone points or tools, also have varnish on their worked surfaces that might be dated.

Early conjectures about the nature and origin of rock varnish were wrong, largely because until very recently the scientific tools available for investigation were inadequate, leaving the investigators little recourse but to make their best guesses on the basis of limited data. Even as recently as the early 1980s, some leading investigators were claiming that rock varnish was the result of natural inorganic processes, although their explanations of the suggested processes left much to be desired.

With the development of such scientific tools as electron microscopes, however, and various techniques for using these instruments, the mysteries of rock varnish were largely solved, although there are still a few minor points unresolved. Following is a summary of present knowledge.

The main components of rock varnish are about 60% clay and 20% to 30% iron and manganese, in the form of oxides or hydroxides, with traces of as many as thirty other elements. The clays are largely composed of common minerals of aluminum and silicon. Varnishes that are light-hued, usually pale red-brown, are rich in iron but poor in manganese. Varnishes that are dark or black are heavy in manganese, with intermediate shades indicating proportionate mixes of the two oxidized elements.

The iron hydroxide and manganese oxides in rock varnish are deposited by bacterial action, although under the right conditions iron-based varnish can be deposited without the help of oxidizing bacteria. It has long been known that bacterial action forms the manganese nodules that are found on the sea floor in many locations. Similar bacteria create the black manganese oxides and reddish iron hydroxides that give most rock varnish its color. In darker rock varnishes, both iron and manganese are hundreds of times more concentrated than they are in the surrounding environment, due to this bacterial activity. Although the general proportion of iron to manganese in rock is about 5 to 1, the bacterial preference for manganese increases the ratio in rock varnish to about 1 to 1.

The varnish-making process is very slow. One scientist estimates that under normal conditions, rock varnish grows at the rate of only 2 to 5 micrometers per thousand years. A micrometer is one-millionth of a meter. With this extremely slow growth rate, most dark varnishes are tens of thousands of years old.

The minerals that make up rock varnish created by bacterial action come from the environment of the rock, not from the rock itself, although the iron in the red-brown iron-based varnish that forms without bacterial action probably comes from within the parent rock. This process, however, has yet to be fully described. The most likely explanation is that seeping water in the porous sandstones on which this type of varnish forms dissolves minute quantities of the iron within the rock. The iron-laden water then travels by osmosis to the rock surface, where evaporation deposits the iron as a relatively uniform reddish patina of iron hydroxide.

The bacteria that create rock varnish need wet-dry cycles to produce it. Once the appropriate bacteria have inhabited the rock surface,

continued ----

they grow when the surface is wetted, either by rain, snow-melt or transient flowing water, as down a cliff face or in the bed of an intermittent or seasonal stream. As the bacteria grow, they absorb the minute quantities of soluble iron and manganese picked up by the water from other deposits. They then secrete an enzyme that oxidizes these elements, deriving metabolic energy from the process and leaving the insoluble brown or black oxides as a coating on the rock surface.

Minute clay particles in the water or airborne dusts then stick to the wet manganese oxide, and in turn provide another source of iron and manganese. The bound clay also provides other mineral nutrients for the bacteria during wet cycles, and both the clay and precipitated oxide, or rock varnish, provide physical protection from damage to the dormant bacteria during dry cycles. The deposited rock varnish also protects the rock surface from weathering, thus further preserving the environment favorable to the bacteria. This wet-dry cycle is essential to the production of rock varnish by bacterial action, and is the reason that arid environments are most favorable to its creation.

In caves, beneath rocks, or in rock cracks and fissures, where there is not enough sunlight for bacteria to thrive, no black manganese varnish grows, but some reddish iron varnish can form by a non-biological process if conditions are right. In canyon country, varnish streaking is prominent on sandstone cliffs that are exposed to intermittent water runoff, although some of this darkening is due to lichens. Undercut sandstone cliffs that receive little or no wetting are generally fairly uniform red-brown in color, indicating that their varnish was not generated by bacterial action and that the iron mineral in the varnish probably came from within the rock itself.

Of course, the growth of rock varnish is not always a simple process. Environmental factors can slow varnish formation or even destroy it partially or completely. Major climate changes affect its growth rate. Rock surfaces close to the ground are abraded by wind-borne grit, and flowing water carries abrasive rock particles. The fungal components in lichens growing on varnished rock surfaces produce acids that dissolve rock varnish, as do the decay products of plants that may reach the varnish in runoff water. This is another reason why most rock varnishes grow in regions that are generally arid, with relatively little vegetation in the environment.

This is especially important to the varnish that is produced on water-rounded river gravels that are lying on the ground or on bare sandstone. Most such gravels that are lying on soils have very little varnish because that environment is made too acidic by the plants growing in the soils, but gravels lying on bare sandstone take on extremely dark coatings of varnish, with the iron and manganese minerals brought to their surfaces by splashing and surface osmosis from rainfall and snow-melt.

Recent research using electron microscopes has produced clear images of rock varnish structures, which are sometimes quite complex and revealing. Just as originally hoped, researchers have developed techniques for dating rock varnish that are useful in determining the ages of prehistoric artifacts. The most successful technique uses a complex process called "accelerator mass spectrometry," which can perform carbon-14 dating using less than one milligram of sample. Such tests use the bacterial and other organic matter in the varnish. The minute quantities needed leave the prehistoric artifacts being tested virtually undamaged.

At best, however, such tests can only tell that an object is "at least" a certain age, because the test reveals only the age of the varnish, which may not have started forming for some time after the artifact was exposed to varnish-forming conditions. Using these techniques, certain petroglyphs in Australia have been dated as at least 36 thousand years old, making them the oldest so far tested.

Other complex testing techniques involving rock varnish can also provide valuable information about past climates, vegetation communities and other environmental factors.

It should be noted that atmospheric acids created by human air pollution also attack rock varnish. Unfortunately, the prevalence of such air pollution could destroy the effectiveness of rock varnish for artifact dating and other research purposes, just as this valuable archaeological and paleontological tool has become available.

Rock varnish analysis has revealed another curious factor. Varnishes near highways have lead content as high as 3 to 4 percent in their surface layers, but only 0.1 percent a few micrometers deeper, with the extra lead coming from the additives used in automotive fuels over recent decades.

Another use of the research into the nature of rock varnish has produced a way to create the coating artificially and very quickly by precipitating manganese and iron oxides on a rock surface chemically. This method has been used to "varnish" conspicuous graffiti placed by vandals on panels of petroglyphs by chipping or scratching through the natural varnish.

An alkaline solution is applied to the raw rock in the graffiti, then a solution of iron and manganese is applied. The alkaline surface precipitates the two elements as iron hydroxide and manganese oxide, the same compounds found in natural rock varnishes. By controlling the ratio of iron to manganese in the solution used, the artificial varnish can be made to match the color of the original natural patina exactly, thus completely obscuring the visual aspect of the blemishes. It is hoped that this artificial rock varnish will encourage re-establishment of the bacteria that produce natural varnish.

Thus, the seemingly esoteric research into the nature of rock varnish has produced knowledge and practical techniques that are useful in many other fields of study, and has resolved the long-standing mysteries about the nature and origin of this strange rock patina that is found so many places, including the spectacular rock exposures of canyon country.

Note the beautifully eroded crenelated cliffs of Organ Rock Shale to the left of the road ahead. The lighter-hued red sandstone layers at the base of that ridge are a first appearance of undivided Cutler Group sandstone. Beyond where the road passes the end of that ridge, the exposed rock on both sides of the road is the same undivided Cutler Sandstone for about three miles.

29.8 The Lockhart Basin ORV trail, not described in this book, joins the road here. In the rock mass to the left of the road here, note the clearly discernible junction between the darker Organ Rock Shale and the lighter-hued undivided Cutler Sandstone. This distinct interface is visible in the bluffs on each side of the road for the next mile or so.

32.2 From a paved pullout here, the panorama ahead provides a first glimpse of the spectacular erosional landforms of the Needles District of Canyonlands National Park. The nearer ridges of light-hued sandstone are undivided Cutler deposits. The higher ridges to the left and right are topped with dark red Organ Rock Shale. In the more distant red-and-white banded cliffs to the west, the undivided Cutler phases into the Cedar Mesa Sandstone member of the Cutler Group. The white sandstone that dominates the lowland there is Cedar Mesa Sandstone.

32.7 The road enters Canyonlands National Park. Although Utah 211 technically ends at the park boundary, this road log continues into the park to the visitor center.

34.2 The road to the right here goes to a commercial development that offers automotive fuel and camping supplies, plus a snack bar, campground and air strip. All of the red slickrock on both sides of the road here is undivided Cutler phasing into Cedar Mesa Sandstone. The mass of sandstone to the left of the road here offers good slickrock hiking, as described in the book ***Canyon Country* SLICKROCK HIKING & BIKING.**

35.3 Turn right into the Park Service visitor center and register. This road log ends there, although the main paved road continues into the park to end abruptly at Big Spring Canyon. This road-to-nowhere was once part of a political boondoggle. The road was supposed to go all the way to the confluence of the Green and Colorado rivers, but died considerably short of that goal for lack of further porkbarrel funds.

The geology along that undescribed stretch of road is entirely Cedar Mesa Sandstone, although upper layers of still older Cutler sandstone beds are visible in the bottom of Big Spring Canyon just down the drainage from where the road dead-ends at the canyon. These older Cutler deposits were once called the Elephant Canyon Formation, but that designation is now obsolete. See the sidebar, YES, GEOLOGISTS ARE HUMAN.

Newspaper Rock

GEOLOGIC STRATA VISIBLE FROM THIS ROAD

Navajo Sandstone		Page 13
Kayenta Sandstone		Page 13
Wingate Sandstone		Page 12
Chinle Formation		Page 12
Moenkopi Formation		Page 12
Cutler Group	White Rim	Page 11
	Organ Rock	
	Cedar Mesa	
	Lower Cutler	

ROAD NAME: **UTAH 279**

SUMMARY: This road is one of the few anywhere that closely parallel the Colorado River in its long and deep gorge. It offers lovely views of the complex erosional phenomena associated with the Glen Canyon Group of rock formations, especially Navajo Sandstone, and provides insight into the way a river cuts deeply into the land through which it flows.

ROAD SURFACE: paved for its entire length as described.

VEHICLE: any vehicle.

MAP: ***Canyon Country* OFF-ROAD VEHICLE TRAIL MAP - Island Area**

GUIDEBOOK: ***Canyon Country* OFF-ROAD VEHICLE TRAILS - Island Area**

MILEAGES:

TO START: 4.2 miles, from Main and Center, Moab.
ROAD: 15.7 miles, one-way to the end.
RETURN: 19.9 miles, from end of the described road.
TOTAL: 39.8 miles, from Main and Center, round trip.

TIME: a little more than 1/2 day, which allows time for hiking to Corona and Bowtie arches, and viewing the roadside petroglyphs and dinosaur tracks along the way. Allow for a full day if the Long Canyon Trail and some of the roads beyond it are to be explored.

ROUTE: Main Street in Moab (U.S. 191) north to where Utah 279 spurs left. As an alternate, after traveling Utah 279 to its end, those with off-road vehicles can continue beyond the end of the pavement, where this description ends, on the

Potash Trail - Shafer Trail loop, then return to Moab via Utah 313 and U.S. 191. As another alternate, from the end of the paved road, return to its junction with the Long Canyon Trail and take it to higher terrain, then explore Utah 313 and the Island Road beyond there. Use the referenced map and guidebook for navigating these alternate return routes.

NAVIGATION: drive north on Main Street (U.S 191) to where Utah 279 spurs left, just north of the uranium-vanadium mill site in the northern end of Moab Valley.

GEOLOGY ROAD LOG:

0.0 Utah 279, a designated Utah Scenic Byway, turns west from U.S. 191 and immediately crosses Moab Canyon Wash. It then continues southwesterly along the base of the cliffs of Moab Valley toward The Portal, where the Colorado River leaves the valley. Upriver, the entry portal is visible on the opposite side of the valley. See the sidebar, PARADOXICAL VALLEYS. The sheer cliffs above the road are Wingate Sandstone with a little visible remnant of Navajo Sandstone above that at The Portal, with intervening layers of Kayenta Sandstone between the two. The heavy desert varnish on the vertical cliffs indicates that they have been exposed to the weather for a long time. See the sidebar, DESERT VARNISH. The rubble slopes below the Wingate cliffs cover the Chinle Formation, the Moenkopi Formation, and the Upper Cutler Formation, in descending order, with very little of these showing through the fallen debris. Notice that the Wingate Sandstone cliffs have a general downward slope toward the river, caused by the massive faulting that took place at this end of Moab Valley.

The first stretch of the road curves around an enormous tailings pile which was constructed to hold the radioactive wastes from the uranium-vanadium mill-site beside U.S. 191.

0.3 Above the road to the right, the railroad tracks that come from the north enter a long tunnel. The other end of this tunnel appears farther down this road. The layered rock strata to the right of the road are Moenkopi Formation. Just beyond here, the full panorama of the La Sal Mountains looms above Moab Valley. Below them, the round-topped cliffs are Navajo Sandstone. The distant benchlands closer to the mountains are Wilson Mesa and South Mesa, which are rimmed with the Moab Member of Entrada Sandstone, with Slickrock Member of Entrada slopes below. The broad swamplands adjacent to the Colorado River are alluvium, or soils that have been deposited there by the river over a long period of time. The large segment of swampland on the south side of the river has been dedicated as the Scott Matheson Wetlands Preserve.

2.9 The road has entered the downriver portal of Moab Valley. From a pull-out on the left, note that the strata on each side of the river were steeply tilted by the Moab Valley salt anticline. The high domes with vertical walls visible on each side of the portal are Navajo Sandstone. The layered rock below that is Kayenta Sandstone. The vertical cliffs at the road level at and near The Portal are Wingate Sandstone set above rubble-covered slopes of Chinle Formation.

From here on to the end of this road as described, it closely parallels the Colorado River. Both riverbanks are heavily vegetated, providing habitat for a variety of wildlife. During early morning and late evening, and during the cooler months, animals such as deer, geese, ducks, porcupines, beaver, ravens and other birds are frequently seen. Drivers should be alert for deer crossing the road, especially at dusk.

3.9 The Portal Hiking Trail leaves the road to the right here at a pull-out. Notice from here the beautifully eroded cliffs and streaks of desert varnish on the opposite side of the river, with several seep-caves in the vertical cliffs. A seep-cave not visible from this road but above it to the right here, has joined the pothole that formed it and created Little Arch. To view and photograph this huge span, drive down Cane Creek Road.

From here on downriver, note on both sides of the river on terraces and slopes, the great quantities of river gravel. These water-rounded rocks were deposited before the river cut down to its present level. They originated from everywhere that the Colorado River and its tributaries drained, and thus make interesting rock-hounding because there are specimens in the gravel that do not occur naturally in this region.

4.1 The pullout to the right goes to a public campsite. About here, the geologic strata level off from the tilting caused by the Moab Valley salt anticline.

4.8 The cliffs along this stretch of the river are popular with rock-climbers because the Navajo Sandstone that is at road level here, on both sides of the river, erodes into near-vertical cliffs, with narrow cracks useful for climbing. Such massive sandstones are virtually unclimbable without the cracks and crevices that appear when they fracture on near-vertical planes from undercutting, as along canyon bottoms. Note for the next mile or so the fins and domes of Navajo Sandstone on the opposite side of the river. These were produced by eons of weathering and erosion.

5.1 Beginning here, and continuing for the next half-mile, there are hundreds of prehistoric petroglyphs in the cliffs to the right of the road. These were chipped into the rock through the dark desert varnish by prehistoric Indians of the Anasazi culture, using sharp quartz rock shards and perhaps hammerstones as well. Note the slender spire of Navajo Sandstone directly across the river. All of the rock visible on both sides of the river here is Navajo Sandstone, although straight ahead there are angled slopes below the Navajo cliffs beyond the mouth of Cane Creek Canyon, where the rubble covers ledges of Kayenta Sandstone. Note the heavy deposit of river-gravel on the terraces on the opposite side of the river here. Although the top of Navajo Sandstone erodes into rounded domes that resemble the sand dunes from which the stone formed, the shapes of these domes do not conform to the shapes of the original dunes. Their present shapes are determined by a number of complex factors involving the erosion modes of Navajo Sandstone.

5.8 Pullouts to the right here enter more primitive campsites. On the other side of the river, the lower narrows of immense Cane Creek Canyon reaches the Colorado River. Cane Creek Road, described elsewhere in this book, enters that canyon. Ahead to the left-front, near the top of the cliff, the farthest of two large alcoves is open to the sky, hence forming a natural arch. At road level on both sides of the river, the layered rock below the vertical cliffs is Kayenta Sandstone.

6.0 The Poison Spider Mesa ORV trail angles off to the right here through a gate. Park at the pull-out just beyond the gate and look for the dinosaur tracks on a loose rock slab on the rubble slopes above the ORV trail, and on a much larger slab at the base of the upper cliff. These tracks were made in muds at the interface between the Navajo and the Kayenta sandstones, and were later exposed on slabs of rock as the cliff collapsed. See the sidebar, PREHISTORIC TRACKS. The large track-covered slab at the base of the vertical cliff was exposed when a section of the cliff collapsed in the winter of 1970/71. The lofty span of the arch on the other side of the river is clearly visible from here. The rock at river level from here on for some distance is Kayenta Sandstone, with Navajo Sandstone cliffs above that.

6.8 The vertical cliffs appearing on both sides of the road are the upper levels of Wingate Sandstone.

7.5 Note that as the river gorge curves to the right here, the geological strata on each side of the canyon are sloping downward, with the Wingate Sandstone disappearing and the ledgy Kayenta Sandstone appearing at road level on both sides of the river.

8.9 Stop at a pullout on the left. Here, the descending rock strata have brought the Navajo Sandstone cliffs almost to road level. Note the mounds of ancient river gravel above the far river bank. The immense sandstone figure on the skyline to the left-front is known as The American Eagle. It has its wings spread and its tail flared. The streaks of desert varnish descending from its wings resemble feathers.

9.9 In the canyon to the right, note the railroad cut which has been blasted through solid Navajo Sandstone. The tunnel from which the tracks emerge is out of sight around the bend. As the tracks descend into the river gorge and continue down the river closely parallel to the road on the right, they are subject to periodic damage from rock falling from the cliffs above. See the sidebar, ROCK EROSION. Some such massive rockfalls have severely damaged the tracks and road, and have required enormous amounts of work to drill, blast and clear away. This demonstrates that canyon-forming is an ongoing geologic process that takes no heed of transient mankind.

10.1 The parking area here is for the hike to Corona, Bowtie and Gold Bar arches. Gold Bar Arch is visible from the road ahead, just below the saddle in the skyline to the right. All three of these immense arches have formed in Navajo Sandstone. It is well worth the one-mile hike to see magnificent Corona Arch, also known as Little Rainbow because of its resemblance to Rainbow Bridge on a somewhat smaller scale. Even so, Corona Arch is large enough that small airplanes have been flown through its opening. From here, note straight ahead the two large seep caves in the upper levels of the Navajo. The terraced rock at road level and up to the vertical cliffs of Navajo Sandstone is Kayenta Sandstone. Note on ahead that the cliffs above the railroad track have been drilled and blasted to create the trackbed. This artificial disturbance of a natural rock surface often leads to accelerated collapse of the artificial cliff. The canyon visible to the left-front is Gold Bar Canyon, a drainage line that extends clear to the rim of Moab Valley.

10.5 Park here at the pullout to hike up Culvert Canyon, so named because the hike begins through the big tubular culvert beneath the railroad tracks. For a description of this and several other hikes from this stretch of the Colorado River gorge, refer to the book, ***Canyon Country* SLICKROCK HIKING & BIKING.**

11.7 Here, the descending railroad track reaches road level. Deep and narrow Day Canyon angles right. Hiking up this spectacular canyon is made difficult by dense vegetation at its lower end. The entire Bull Canyon system drains into Day Canyon over a magnificent pouroff about 1 mile up the canyon. There is usually a large pool of water at the base of this pouroff. One spur of the Bull Canyon Trail, described elsewhere in this book, reaches the top of this pouroff. The rock on both sides of the road here is Navajo Sandstone, with collapsed rubble at its base.

12.6 The feathery-leafed tamarisk bushes that dominate both banks of the river along this stretch of road are not native to this continent. They were imported into the Unites States earlier in this century from North Africa for the purpose of stabilizing the banks of irrigation canals in the western states. From there, the exotic plant has spread throughout the west along riverbanks and in other wet locations. In the Colorado River gorge, where sandbars once shifted seasonally, allowing the river to change its course between the cliffs, tamarisk, native willow and other shrubs have stabilized many such shifting sandbars. Here, between the far cliffs and riverbank vegetation, this stabilization has created a permanent lake that is not visible from the road because of the shrubs. It can be reached only by using a boat, then crawling through dense vegetation.

13.0 Note the giant boulder beside the railroad tracks, and other nearby rubble on both sides of the road. These are remnants of a cliff collapse that took place during the winter of 1990/91. Above the tracks here, note that the older cliff faces are heavily patinaed with desert varnish and lichens, while the surfaces exposed by the railroad construction have little desert varnish and no lichens. See the sidebars, LICHENS.

13.7 From a pullout on the left, note magnificent Jughandle Arch in the cliff to the right of the road. The cliffs at river level on both sides here are Wingate Sandstone. The layered rock above the sheer cliffs is Kayenta Sandstone. The lighter uppermost sheer cliffs with rounded tops are Navajo Sandstone. The smooth, heavily desert-varnished cliff opposite Jughandle Arch is known as The Billboard.

13.8 At the road junction here, the Long Canyon ORV trail, described elsewhere in this book, goes right. This trail can at times be traveled for some distance by standard highway vehicles, but is often closed by rockfalls or damaged by rain run-off, and its upper stretches are always very rough and steep. Highway vehicles who venture up this canyon should turn around and return at the first signs of road damage or steep grade. The vertical cliffs above the canyon are Wingate Sandstone above rubble-covered slopes of Chinle and Moenkopi deposits. There are two arch openings in the left abutment of the canyon. One is visible from the trail junction, and the other from the trail a little beyond the junction. The cliffs on both sides of the river here are Wingate Sandstone.

14.4 From a pullout on the left, note that the geologic strata are again ascending from the river level, forced upward by another salt anticline. In this vicinity, the Colorado River has cut through the Cane Creek Anticline. This is described further under the Hurrah Pass geology log, elsewhere in this book. Here, the vertical cliffs on both sides are Wingate Sandstone, with traces of layered Kayenta Sandstone above that. The rubble-covered slopes below the Wingate are Chinle Formation. The rounded mounds just above the river on the other side are ancient river gravel. Straight ahead is the rounded dome of a mill for refining various Paradox Formation salts mined from deep beneath the crust of the earth here, then concentrated by evaporation in ponds that the Potash ORV trail circumnavigates several miles beyond the end of this road. The Potash Trail is described elsewhere in this book.

15.0 The high cliffs recede to the right. On both sides of the road the uppermost cliffs are Wingate Sandstone. The slopes immediately below are Chinle Formation. On the right, there are some banded strips of red-brown Moenkopi Formation showing below the Chinle rubble, with a little dark red Cutler sandstone at and just above road level. The Cutler is marked by the white banding and rounded surfaces of its exposed cliffs.

15.7 The pavement ends here at the refining mill. Highway vehicles should not continue beyond this point, although low-geared, high-clearance highway vehicles, such as some pickups and vans, can continue on the Potash ORV trail for a considerable distance, but generally not all the way into Canyonlands National Park. The lower dark red, crenelated cliffs beyond the mill are Cutler sandstones. For more information about the mill here, see the sidebar, POTASH.

Beyond the river in this area is an isolated section of cliff. Before the Colorado River cut to its present depth, it flowed around the looping canyon behind that isolated promontory. As the river cutting continued, it broke through the narrow part of the loop here and abandoned that part of its course, taking the new shorter route. Such a forsaken loop is known as an "abandoned meander." This particular one was named Jackson Hole after an early local character. The abandoned meander of Jackson Hole contains countless tons of river-rounded gravel, making it a rockhound supermarket. The cliffs beyond Jackson Hole are Wingate Sandstone, with the slopes below being rubble-covered Chinle Formation. At river level, the reddish sandstone is more Cutler, covered with mounds of gray river gravel.

From here, return to Moab via the same route or, if traveling in an off-road vehicle and time permits, continue downriver on the Potash Trail, or return to the junction near Jughandle Arch and drive up the Long Canyon Trail.

POTASH

Visitor displays and literature at Dead Horse Point State Park and Anticline Overlook in Canyon Rims Recreation Area attempt to "interpret" the conspicuous ponds and chemical mill visible in the otherwise spectacular redrock panoramas below these lofty viewpoints. They carefully avoid, however, telling the full story which was, and continues to be, a human tragedy and an environmental crime of major magnitude.

The story began almost three decades ago, in 1964, when a large chemical firm decided to mine potash, a mineral found in the Paradox Formation deep beneath most of this region that is used primarily as a chemical fertilizer. As usual, state and county officials enthusiastically supported the project with public funds, in this case by constructing the present downriver paved road to the mining site, Utah 279. Both state and federal land administration agencies expedited the construction of the railroad spur that parallels U.S. 191 and Utah 279, a spur that was needed to haul the mineral to market.

At first, the big evaporation ponds did not exist, but the mill was visible from Anticline Overlook, from the summit of Hurrah Pass and at the end of Utah 279, and a large area behind the mill was also devastated by bulldozers and cluttered with assorted industrial equipment and junk. The potash was removed from a conventional mine shaft that was almost 3,000 feet deep, and that eventually had many miles of horizontal shafts. The raw product was then refined at the mill and shipped out on the railroad. The mill waste -- thousands of tons of common salt -- was dumped behind an earth-fill dam out of sight behind the mill, in a natural canyon beneath the soaring redrock cliffs.

Soon, however, this conventional form of mining encountered severe problems. It was found that the deep strata of potash were not neatly horizontal, thus conventional horizontal shafts produced too much waste salt and too little potash. Constant flooding from seeping water, and heat up to 100 degrees F, made the deep mine difficult and dangerous to work. Underground explosions killed an appalling number of miners. After five years of operation, an alternate method for removing the soluble mineral salts was proposed, but the new approach was economically unproven and quite expensive to initiate, and would require additional large areas of nearby public land.

Since the mining corporation was faced with having to pay for many years yet on the contract it had signed in order to get the railroad spur to the mine, the decision was made in 1970 to try the questionable new approach, in an attempt to mitigate that loss.

In a land exchange that would be illegal today, and was highly questionable even then, the area now occupied by the great solar evaporation ponds and their supporting equipment was acquired by the mining corporation. At that time, the area was public land, a broad and colorful expanse of incomparably beautiful and unspoiled redrock desert with a single narrow off-road vehicle trail winding through it. Without public knowledge or input, the federal Bureau of Land Management that administered the land transferred it to Utah State in exchange for other blocks of state public land.

The State of Utah then immediately transferred the land to the mining corporation for a tiny fraction of its real worth. Neither land administration agency gave any serious consideration to the overall environmental impact of converting such a large area of desertland adjacent to a major river into a vast chemical processing operation, although such "environmental impact studies" were required by federal law even then.

The chemical company then built the vast solar evaporation ponds in the winter, during minimal visitation to Dead Horse Point and Anticline Overlook, thus avoiding any public hue and cry until it was too late. The following spring, however, there was, indeed, a public outcry over the secretive land-swap, and the conversion of such a large area of public land in full view of developed scenic overlooks into a experimental industrial operation. The entire proceeding encountered strong criticism from environmental organizations and from a sizable segment of the public, beginning with the first spring visitors to popular Dead Horse Point State Park, from which the immense plastic-lined solar evaporation ponds were glaringly visible.

The experimental new operation worked by flooding the deep conventional mine. The water, warmed in the mine, soon became saturated with soluble Paradox Formation salts. It was then pumped to the surface and into the ponds for concentration by solar evaporation. The concentrated brine and salt slurry was then transported to the refining mill beside the river at the end of Utah 279. After the more valuable potash mineral was taken from this concentrate, the remaining waste, largely sodium chloride -- commonly known as table salt -- was discarded into the same reservoir behind the potash mill.

This reservoir alone posed a potential environmental hazard of major magnitude, as pointed out at the time by a local environmental organization. Should flash-flooding ever overflow and breach that earth-fill dam, thousands of tons of salts would pour into the Colorado River, severely damaging its aquatic and shoreline ecosystems all the way to the Gulf of California, and drastically reducing the agricultural and culinary uses of the river's water for several states and part of Mexico, perhaps for years. Concern over this potential hazard was ignored, even though such a disaster would strongly impact Canyonlands National Park, Glen Canyon National Recreation Area, Grand Canyon National Park, Lake Mead National Recreation

Area and other bodies of water farther down the Colorado River drainage.

After two ownership changes, one involving a foreign corporation, the "solution-mining" operation was taken over by the current owner, which also markets the sodium chloride. The potential environmental hazard of the earth-fill dam and its contents remains, however, as does the massive environmental destruction that created the entire "potash" operation, and the solar evaporation ponds are steadily leaking raw salts.

Thus, this entire mineral-extraction operation is an excellent example of how all-too-common human activities cause immense environmental damage. It is a prime illustration of the principles outlined in the sidebar, HUMANITY AS A GEOLOGIC FORCE.

While the various industrial firms involved doubtless profited from this operation, that profit cost the public dearly in several ways, both directly and indirectly. The public lost directly to the extent that the operation was subsidized by the taxes used to construct the road downriver to the mill, a road whose primary purpose was to serve that mill. Any recreational value was incidental, and that value has only very recently been recognized. Unusually high maintenance costs for this road continue to be a tax burden, largely because construction of the road and parallel railroad disturbed the river cliffline's natural erosional surfaces, thus accelerating exfoliation -- the massive rockfalls that periodically block and severely damage both the road and the tracks.

The public also lost vast expanses of federal land that should have been included in Canyonlands National Park when it was established, land previously unspoiled. The total environmental damage to the area was immense. First, a paved road was constructed down the rivergorge, with its blasting, bulldozing and destruction of prehistoric sites. The railroad spur that was built to the mine, from the east-west rail route that parallels Interstate 70, was even more destructive. Great canyons and a long tunnel were blasted from solid rock in order to maintain the required gradient for the railroad. Countless thousands of tons of rock were moved and used as fill along the entire route, blighting the outstanding views from U.S. 191 north of Moab, and from the visitor center and entrance station of Arches National Park.

Then, once the mining operation was shifted to the solar-evaporation process, an immense additional area of public land was devastated, turning what was the most beautiful redrock desert in the Moab area into an industrial slum along a popular route into Canyonlands National Park, and damaging forever the panoramic views from two developed scenic overlooks, Dead Horse Point and Anticline Overlook.

Still worse, none of this enormous loss and destruction of public land was preceded by an environmental impact study. There was no public participation or input. No mitigating provisions or alternate plans were even considered. The legality of the rapid and secretive land exchange was not challenged.

Beyond doubt, the various owners of this massive industrial operation that is so glaringly visible from a state park and a federal recreation area have profited, but that profit has been at public expense, and the public has come up considerably short in the deal, especially the many miners who died in the operation and their families. But big mining corporations are traditionally as calloused to the loss and ruination of lives as they are to the despoliation of the land and the destruction of natural values. To them, profit is what counts. Nothing else matters. And the damage continues to grow worse, with raw minerals now leaking many places from the supposedly leak-proof evaporation ponds into nearby drainages, aquifers and the Colorado River.

Unfortunately, this all-too-common human and environmental tragedy is not addressed in the bland interpretive information at Dead Horse Point and Anticline Overlook. It is easy to understand why not.

GEOLOGIC STRATA VISIBLE FROM THIS ROAD

Formation	Member	Page
Morrison Formation	Brushy Basin	Page 18
	Salt Wash	
	Tidwell	
Entrada Sandstone	Moab	Page 16
	Slickrock	
	Dewey Bridge	
Navajo Sandstone		Page 13
Kayenta Sandstone		Page 13
Wingate Sandstone		Page 12
Chinle Formation		Page 12
Moenkopi Formation		Page 12
Cutler Group	White Rim	Page 11
	Organ Rock	
	Cedar Mesa	
	Lower Cutler	

ROAD NAME: **UTAH 313**

SUMMARY: This state road leaves U.S. 191, the main highway through southeastern Utah, to enter a picturesque canyon system, then ascend through three geologic strata to the elevated terrain of the immense and complex Island-in-the-Sky peninsula. There, it travels across high-desert meadows surrounded by spectacular geologic erosional landforms toward Dead Horse Point State Park, which is located on one slender arm of the Island-in-the-Sky peninsula. The broad panoramic view from this lofty viewpoint includes three island mountain ranges, the outstanding beauty and geologic variety of Canyonlands National Park and Canyon Rims Recreation Area, the maze-like Colorado river gorge, and other unique erosional features.

ROAD SURFACE: paved for its entire length.

VEHICLE: any vehicle.

MAP: ***Canyon Country* OFF-ROAD VEHICLE TRAIL MAP - Island Area.**

GUIDEBOOK: ***Canyon Country* OFF-ROAD VEHICLE TRAILS - Island Area**

Sevenmile Canyon

MILEAGES:

TO START: 11.3 miles, from Main and Center, Moab.
ROAD: 23.5 miles, one-way to end.
RETURN: 34.8 miles, from end of described road.
TOTAL: 69.6 miles, from Main and Center, round trip.

TIME: allow at least a half-day for this trip, or a long day if the Island Road is added to the trip or if the return trip is made via the Long Canyon ORV trail by those with suitable vehicles.

ROUTE: Main Street in Moab, then U.S. 191 north to where Utah 313 heads west toward Dead Horse Point State Park and the Island-in-the-Sky District of Canyonlands National Park, staying on Utah 313 all the way to the main state park overlook at the southern tip of Dead Horse Point.

NAVIGATION: drive north from Moab on U.S. 191, then west on Utah 313 into mid-Sevenmile Canyon. At the junction where the Island Road forks from Utah 313, turn left on the state highway and continue into the state park to the road end. If a side trip into Canyonlands National Park is added, refer to the log for the Island Road, elsewhere in this book.

GEOLOGY ROAD LOG:

0.0 Stop at the visitor information kiosk at the pullout a few yards west of the junction of U.S. 191 and Utah 313. When facing the direction Utah 313 is traveling, the high cliffs ahead to the left and right are Wingate Sandstone above sloping terraces of Chinle Formation on both sides. Below the Chinle, visible on the left portal, are exposures of Moenkopi Formation and Cutler Sandstone that are largely obscured by rubble from the cliffs above. Near the ground level on the left portal, and on south toward Moab Valley, the red sandstone abutments that protrude from the lower rubble slopes are also upper Cutler strata. Beyond this pullout, the Moenkopi and Cutler appear at the base of the righthand portal as well, but from here are obscured by the rock ledge adjacent to the pullout.

The white rock slabs beside the rest stop are harder sandstone layers in the lower part of the Salt Wash Member of the Morrison Formation. The stratigraphic displacement that puts this formation at road level, rather than in its proper sequence more than 1,000 feet above the nearby cliffline, is due to the Moab Fault that approximately parallels the railroad tracks just ahead. In the vicinity of the tracks, note the abandoned uranium mining activity in the cliffline to the right that has produced slopes of tailings in the gray Chinle deposits. The open, brushy meadows surrounding this road junction are largely dune sand and alluvium held in place by cryptogams, grasses and shrubs. See the side bar, CRYPTOGAMIC SOILS. There are several thin layers of volcanic ash in this alluvium that date from less than one million years ago, indicating volcano activity to the west of this region at that time.

The road beyond the pullout enters mid-Sevenmile Canyon. This drainage goes east under U.S. 191, travels for a few hundred yards into the open desert there, then plunges abruptly down into its deep, narrow and twisting lower canyon. On west, where this road ascends from Sevenmile Canyon, the canyon forks into two major branches. The highway then travels for several miles on the higher ground between these two canyon branches.

0.5 Here, reddish Cutler sandstone appears just above road level in the righthand cliff. Note the beautiful, intricately carved surface of this rock produced by erosion, and the white color-banding that is typical of this marine tideland deposit. The cause of this color-banding is unknown, but the author of this book believes it to be related to the decay of vegetation deposited with the original sediments. See the sidebar, ROCK COLORS.

1.5 At the gate to an ORV trail going left, note the immense amphitheater in the cliffline to the left, and the sloping strata in the promontory on its left side. These strata were deformed by the fault noted earlier. The uranium-bearing Chinle Formation exposed in the walls of this amphitheater was heavily prospected and mined during past uranium boom days. The ORV trail here once provided access to these mines. The Sevenmile Canyon Rim ORV trail, described in the referenced book and map, skirts the southernmost cliffline of this amphitheater, offering good views of past mining activities and the surrounding scenic Wingate Sandstone cliffs. Beyond this junction, the descending geologic strata in the clifflines bring the Chinle formation to road level on both sides, with the looming Wingate cliffs just above that.

2.3 Stop at a wide pullout on the right. At this point, the cliffs on both sides of the canyon are Wingate Sandstone at the road level. The deposits at the bases of the cliffs are partly Wingate, partly remnants from the horizontal layering of Kayenta Sandstone just above the vertical cliffs, and partly dune sand held in place by vegetation. Note that the vertical cliffs here are heavily patinaed with black and brown desert varnish. The vertical streaking indicates that rain runoff and snowmelt have contributed to the varnish formation. See the sidebar, DESERT VARNISH.

Beside the pullout, in the righthand cliff, note the large arc of rock. This is a typical indication of what geologists call "stress-relief exfoliation." See the side bar, ROCK EROSION, for an explanation of this process. All massive (thick) and relatively uniform sandstones such as Wingate, Navajo, and the Slickrock Member of Entrada Sandstone, are subject to this kind of exfoliation. The small cave at road level beside the pullout has been formed by seeping water. The honeycombing in the cliff is produced by rain runoff, where flowing surface water erodes away softer portions of the rock. As the road continues, the canyon walls are Wingate Sandstone topped by the horizontal layering of Kayenta Sandstone.

Just ahead, a spur trail to the left descends into the drainage line where subsurface water has created a verdant oasis. Such springs are typical of long, gentle-gradient canyons such as this, where water flows continually beneath the sandy surface but is occasionally forced to or near the surface by underground rock strata, thus providing favorable conditions for water-loving plants such as cottonwoods, tamarisk, river willow and reeds.

2.7 At a wide pullout on the right, there are prehistoric petroglyphs on the cliff base beyond the patch of boulders just ahead. These petroglyphs were carved into the dark desert varnish by prehistoric Indians who used this canyon for hunting and gathering. There are more such petroglyphs on the cliff wall at left-front that separates the two major branches of upper Sevenmile Canyon. Both of these branches provide excellent hiking, with many views of the variety of erosional features that occur in Wingate, Kayenta and Navajo sandstones.

The Sevenmile Canyon system drains an enormous area of terrain. During the rare heavy rainstorms, floods of raging water descend these canyons, creating an immense waterfall at the lip of the lower canyon just east of U.S. 191. Such flash flooding is rare on the human time scale, but has occurred frequently enough during geologic time to carve deep canyons through solid rock. Such sporadic flooding has formed all the canyons in this region.

3.3 Here, the road begins to ascend from the Sevenmile drainage, climbing the slopes above the North Fork of Sevenmile Canyon. The rock on both sides of the road is at the interface between the Wingate and Kayenta sandstones. As the road climbs, it enters the Kayenta and then, near the top, the more rounded cliffs of Navajo Sandstone. Note how the rock cliffs have been drilled and blasted to create this road. Within the historic past, before this road and its earlier version were constructed, stockmen running cattle in the high country ahead received supplies via a cableway constructed near the head of the North Fork of Sevenmile Canyon. This cableway, its supports, and a two-section wooden ladder nearby, can still be seen by hiking up this branch of the canyon.

4.7 At the top of the grade, drive into a scenic pullout to the right. Here, the La Sal Mountains dominate the skyline to the southeast. To the east, the red-hued sandstone buttes looming above the layered Kayenta are decomposing remnants of Navajo Sandstone. The entire skyline to the south is dominated by terraced Kayenta Sandstone, with occasional remnants of Navajo. To the north, the two detached monoliths are known as the Monitor and Merrimac buttes. They were so-named because of their fancied resemblance to the two Civil War ironclad ships that bore these names. The uppermost vertical cliffs of the easternmost butte, the Monitor, is the Slickrock Member of Entrada Sandstone. The wrinkled layering between the vertical cliffs and lower white sandstone is the Dewey Bridge Member of the Entrada. The white sandstone upon which both monoliths are based is Navajo Sandstone. On Merrimac Butte, and Big Mesa on to the west, the white layer just above the vertical cliffs of Slickrock-Entrada is the Moab Member of Entrada Sandstone. The rubble slopes above that on Big Mesa are the Salt Wash Member of the Morrison Formation, just above a thin layer of red Tidwell Member of the Morrison. A special type of agate forms in the Dewey Bridge-Entrada and, as the matrix shale erodes away, the much harder "Dewey Bridge agate" accumulates on the exposed upper surface of the white Navajo Sandstone.

The rounded, light-hued sandstone adjacent to the pullout area and beside the road on ahead, is also Navajo Sandstone. Note the seep-line in the low cliff adjacent to the parking area. In this general vicinity, the author of this book discovered large deposits of petrified wood in Navajo Sandstone. Such fossil remnants were virtually unknown in this desert-aeolian deposit until a series of such discoveries were made in the late 1980s by the author and a local amateur geologist. Since these discoveries were reported, a University of Utah geologist has generated a scientific paper officially announcing this startling occurrence. See the side bar, PETRIFIED OASES.

4.8 Back on the main road, for the next mile or so note the predominance of dune sand on both sides of the road, partly held in place by vegetation. The desert-dune sand that originally formed the Navajo Sandstone easily weathers on its surface, as the minerals binding the original sand are dissolved and carried away by rain and snowmelt, thus releasing the sand granules to the wind once again. Such living dunes among the petrified dunes of Navajo Sandstone are fairly common in this region, and dominate this stretch of road on both sides.

As the road continues, the two main branches of upper Sevenmile Canyon are to the left and right. To the north, the looming cliffs of Big Mesa dominate the skyline. The cliff of that immense mesa continues to be, from top to bottom, remnants of Salt Wash-Morrison, a thin layer of reddish Tidwell-Morrison, and the Moab, Slickrock and Dewey Bridge members of Entrada Sandstone, all standing on a base of white Navajo Sandstone.

A pullout at Mile 6.6 permits detailed study of this spectacular mesa. Note how seep-caves have formed at the interface of the Slickrock and Dewey Bridge members. Rockhounds who hike along the base of this cliff will find a plentiful assortment of colorful Dewey Bridge agate.

7.8 At a large pullout on the right, stop and climb onto the nearby Navajo Sandstone dome for an outstanding view of the Monitor and Merrimac, Big Mesa and the upper drainage of the North Fork of Sevenmile Canyon. In Big Mesa, note the several large and very active seep-caves forming at the interface of the Slickrock and Dewey Bridge members of the Entrada, with vegetation growing in and below the seep-lines. This viewpoint affords an excellent look at the wrinkled nature of Dewey Bridge-Entrada. It is believed by geologists that this wrinkling of strata that were deposited flat by water was caused by some form of lateral movement of the dune-sand deposited on top of the Dewey Bridge sediments before they had hardened. This phenomenon is rare, but observed almost anywhere the Dewey Bridge-Entrada is exposed. In the projecting fin of this dark red rock, behind the first and second columns, is a slit-eyed natural window through which the Monitor and Merrimac can be seen and photographed from certain angles. This opening was named "M & M Window" by the author many years ago. A picture of this window appears in the book ***Canyon Country* ARCHES & BRIDGES.** Beyond this pullout for several miles, the road travels at or near the top of the Navajo Sandstone, with rounded domes of this white rock exposed here and there.

8.3 An ORV trail goes to the right here. Branches of this trail reach the base of Big Mesa, thus offering rockhounds with off-road vehicles easy access to its plentiful Dewey Bridge agate.

8.8 Turn into the pullout on the left here, which offers a beautiful full-circle panorama. On the skyline to the west, the distant slopes of the San Rafael Swell dominate the horizon, with exposures of Navajo and Entrada sandstones showing in the broad meadows in the foreground. The sheer-walled butte and adjacent rounded ridge of dark red sandstone are Slickrock-Entrada. This ridgeline is known as "Rainbow Rocks" because of its broad color-banding that is visible even from this distance. To the north, the sheer walls of Big Mesa and the more distant similar walls of Bartlett Wash, dominate most of the horizon.

To the east, the Monitor and Merrimac buttes loom above broad expanses of Navajo Sandstone. On the extreme eastern skyline, the northern slopes of the Uncompahgre Uplift loom above the Entrada Sandstone spires and fins of the Windows Section of Arches National Park. The white rock below those spires is Navajo Sandstone. Toward the La Sals, the more distant sheer-walled cliffs are Wingate Sandstone. Below the La Sals, the Entrada Sandstone cliffs of Wilson and South Mesas loom above the sloping strata and sandstone fins of Moab Valley and Behind-the-Rocks. The maze of fins and domes in Behind-the-Rocks is Navajo Sandstone. The entire southern horizon is dominated by layered and heavily-vegetated Kayenta Sandstone.

At certain times of the day when the lighting is just right, the huge opening of Skywalk Arch is visible in the rimland of Big Mesa, in the uppermost levels of the Slickrock-Entrada, just above a broad horizontal band of white. The first man known to have reached this lofty span was a California computer engineer. His story of that adventure appears in the book, ***Canyon Country* SLICKROCK HIKING & BIKING.** The slickrock adjacent to this pullout is Navajo Sandstone.

9.0 Back at the main road turn left.

9.1 Dubinky Well Road goes right here. Before Utah 313 was constructed, the only vehicle route into this higher country for access to Dead Horse Point and other Island-in-the-Sky sites was via Dubinky Well Road and the Blue Hills Road that goes west just south of the present Grand County airport. Reaching Dead Horse Point State Park via that route required a long, tedious day of driving over very rough roads.

12.9 At a road junction here, the road to the right goes out onto Horse Thief Point, then descends into the Green River gorge. The mountain range visible on the skyline to the southwest is the Henry Mountains, near Hanksville, Utah. To the right of the Henrys, the more distant slopes of the Aquarius Plateau and Thousand Lake Mountain are visible on clear days.

As the main road continues south from this junction toward Dead Horse Point, it travels out onto the immense, complex, sheer-walled peninsula called Island-in-the-Sky that lies between the Colorado and Green River drainages as they are converging to the south. Dead Horse Point State Park, an early development, occupies one arm of this peninsula. The Island-in-the-Sky District of the more-recently established Canyonlands National Park occupies a much larger part of it. As the road travels out onto this immense peninsula, it goes through broad, open meadows of alluvium and dune sand held in place by vegetation. The occasional outcrops of rock are either ledgy Kayenta Sandstone or remnants of Navajo Sandstone.

13.8 Arths Pasture Road goes left here toward the Gemini Bridges. This road and the two branching ORV trails that go to these spectacular twin natural bridges are described elsewhere in this book. From this junction, the high ridgeline visible below the three separate units of the Henry Mountains is the Orange Cliffs-Big Ridge highland that defines the western perimeter of the Maze District of Canyonlands National Park. This feature is described more fully in the geology log for the Island Road, elsewhere in this book.

15.3 At the junction here, the Island Road continues straight ahead, to enter the Island-in-the-Sky District of Canyonlands National Park. Utah 313 goes left toward Dead Horse Point State Park. The dome of sandstone just east of this junction is called The Knoll. It is a remnant of Navajo Sandstone capped with the harder rock formed from the mineralized sediments of a petrified oasis. See the side bar, PETRIFIED OASES. That harder, more weather-resistant rock has preserved this dome of sandstone, while similar surrounding rock has eroded away. What was once a desert spring, and probably the lowest point in an immense area of dune sand, is now the high central point on a large elevated peninsula.

Major drainages develop in all directions from The Knoll like spokes from the hub of a wheel. Some of these canyons drain into the Green River gorge, others into the Colorado River gorge. Beginning in the north and going clockwise, these canyons are the Sevenmile and Bull canyon complexes, Day Canyon, Long Canyon, Taylor Canyon, Mineral Canyon and Hell Roaring Canyon.

Beyond this junction, the road continues to travel across open meadows of alluvium and dune sand held in place by native and exotic vegetation, with more and more glimpses of the lower terrain to the east as the road travels out onto the slender arm of the peninsula known as Dead Horse Point.

17.0 The Long Canyon ORV trail, described elsewhere in this book, angles left here.

20.8 A spur road goes left here to the group campsite for Dead Horse Point State Park. The main campground, visitor center and overlooks are just ahead. From here on to the southern tip of Dead Horse Point, the cliffs of the Island-in-the-Sky District of Canyonlands National Park, below the peaks of the more distant Henry Mountains, dominate the southwestern horizon. The sheer cliffs are Wingate Sandstone, topped with thin layers of Kayenta Sandstone. The rounded domes glimpsed occasionally above these are remnants of Navajo Sandstone. To the south, the Abajo Mountains and the lofty plateau of Elk Ridge and associated highlands dominate the horizon. To the east, below the sharp peaks of the La Sal Mountains, the wildly distorted terrain between Dead Horse Point and the mountains is Moab Valley, Behind-the-Rocks, the vast Bull Canyon system, and the major erosional features described in other road and trail logs in this book.

22.0 Turn left into the Dead Horse Point State Park visitor center. After paying the required fee at the entrance station or visitor center and studying its exhibits, take the cliff-rim walk to the east. From the middle of this short loop walk, the panorama to the east and below to the left is dominated by Wingate Sandstone cliffs topped with horizontal strata of Kayenta Sandstone, with a few rounded slopes of white Navajo Sandstone above that. Immediately below the sheer cliffs, the rubble-covered slopes are Chinle Formation. Note the old prospecting trail in the uranium-bearing gray layer in this formation. The dark red strata below the Chinle are Moenkopi and Cutler deposits.

The elevated mesa at the same level to the southeast is Hatch Point in Canyon Rims Recreation Area. The strata in its cliffs are the same as those of Dead Horse Point. The Colorado River wends its circuitous route through the deep gorge between the two plateaus. The rimlands of the gorge are a gray-hued limestone that marks the interface between the upper and lower Cutler beds. Several existing books and maps on the geology of this region indicate that these lower beds were a separate geologic deposit called "Elephant Canyon Formation." The existence of such a separate deposit has since been disproven. See the sidebar, YES -- GEOLOGISTS ARE HUMAN.

The northern tip of Hatch Point is Anticline Overlook. The dark sandstone layers below and to the left of Anticline Overlook have been arched upward by the Cane Creek Anticline. Most of these strata visible from here are Cutler Sandstone topped with Moenkopi sediments, with a remnant dome just below the overlook capped with Chinle deposits. Below the clifflines, the broad sloping meadows are almost entirely decomposed rock and alluvium from the cliffs, plus dune sand, all held in place by sparse desert vegetation. Even so, sporadic rain runoff has cut mini-canyons into these alluvium slopes. The canyon walls beyond the nearby arm of Dead Horse Point on the left, and Anticline Overlook on Hatch Point, are the eastern cliffs of Cane Creek Canyon. For more details about that vicinity, see the geology log on Cane Creek Road, elsewhere in this book.

Beyond Cane Creek Canyon, the vast sloping maze of Navajo Sandstone fins and domes is the Behind-the-Rocks wilderness area. Just beyond that maze, but out of sight, is Moab Valley. Still farther toward the mountains, steep-walled Wilson and South mesas jut out from the La Sal foothills. The rims and slopes below these mesas are Entrada Sandstone. To the south, the panorama ends at the Abajo Mountains, with parts of the Lowlands District of Canyon Rims Recreation Area in the lower terrain. For descriptions of these, refer to the geology logs for Utah 211 and Needles Overlook Road.

In the red-desert terrain below this vantage point, the blue-tinted "lakes" are solar evaporation ponds. The salts that are concentrated there are produced from nearby mines that penetrate downward into Paradox

Formation salt beds. For more information about this operation, see the sidebar, POTASH.

In the flat terrain below this viewpoint, the faint vehicle trail is the Potash ORV trail described in the referenced guidebook and map, and elsewhere in this book. The faintly visible spur of that trail that comes into the canyon on the left once provided access to the uranium prospecting trail noted earlier. Between the Colorado River and Hatch Point, the large conical structure is called Pyramid Butte. It is largely Cutler Sandstone topped with a remnant of Moenkopi. The white rock around this canyon-rim trail and the nearby visitor center is Kayenta Sandstone.

22.1 Back on the road, turn left toward the main viewpoint on Dead Horse Point.

22.2 The main campground road goes right here.

23.1 A wide pullout here provides a view toward the west of the Island-in-the-Sky, with the tips of the Henry Mountains on the distant horizon. For more complete information about the terrain below Dead Horse Point, refer to the descriptions at the visitor center and at the end of this road. Below the cliffline beneath the Henry Mountains to the southwest, note the vehicle trail climbing into the canyon from the rimlands of the Colorado River's inner gorge. This is part of the historic Shafer Trail. The White Rim ORV trail continues left just above the level of the white sandstone there that rims the inner gorge, while the Potash Trail goes right.

Just beyond this pullout, the road crosses a narrow isthmus in Dead Horse Point. To the left of the road, sheer cliffs make careful driving advisable. The tangled deadwood fence at this narrow point is a reminder that there once was such a "ripgut" fence here that was used to confine livestock on the sheer-walled point beyond.

23.5 Here, this road ends at a developed overlook. From the parking area, walk the rock-walled rimlands around the end of the point for spectacular panoramic views to the east, south and west. This viewpoint overlooks most of Canyonlands National Park and even larger Canyon Rims Recreation Area. The surface rock here is Kayenta Sandstone.

From the southern tip of Dead Horse Point, the panorama to the east was described at the visitor center, mile

22.0. To the southeast, the lofty peninsula of Hatch Point continues to dominate the horizon. The projecting point just below the east end of the Abajo Mountains is the undeveloped Canyonlands Overlook. The more distant projection on Hatch Point is the popular Needles Overlook. Both overlooks are part of Canyon Rims Recreation Area, as is most of the lower terrain to the east of the Colorado River all the way to the Abajo foothills. As described earlier, the Abajo Mountains and the highlands of Elk Ridge and associated plateaus mark the southern skyline.

To the west, the soaring cliffs that are other projections of the Island-in-the-Sky peninsula are largely in Canyonlands National Park. As with Dead Horse Point and Hatch Point, the vertical cliffs of Island-in-the-Sky are Wingate Sandstone, with layered Kayenta Sandstone just above that. To the west, the uppermost cliffs are Navajo Sandstone.

The panorama from this point is equal in spectacular beauty and geological diversity to that of the famous Grand Canyon, although its total depth is not as great. From this viewpoint, several large serpentine curves of the Colorado River can be seen almost two thousand feet below. The loop in the river just below this point to the southwest is called The Gooseneck. See the sidebar, ENTRENCHED MEANDERS. Note the white sandstone rimland above the river toward the southwest. This is the White Rim member of the Cutler Group. On the whole, this member of the Cutler does not extend east of the Colorado River, but there is one anomalous exposure of it in Castle Valley. See the road log for Castle Valley Road. The sloping red-hued strata just above the White Rim are Moenkopi deposits. Above that, the gray-hued Chinle Formation extends upward to the base of the vertical Wingate Sandstone cliffs. The lower Cutler beds continue below the White Rim sandstone to the river level, although in some locations, traces of the upper Hermosa Group are exposed at river level, especially on downriver in the central part of Canyonlands National Park.

To the east of the river, the gray-hued flat expanse below the red Cutler sandstone cliffs is the interface between the lower and upper Cutler beds. This gray limestone layer is rich in shallow-marine fossils. The horizontal strata below it are a mixture of marine and tideland deposits, some of which also contain sealife fossils. In the river itself, note that sand bars are exposed when the water level is low. These sand bars shift and change constantly as the river flow varies, but those along the river shoreline are now stabilized by the exotic water-loving shrub, tamarisk. Photographs of the Colorado River gorge taken from the same vantage point as those taken by the Powell Expeditions of 1869 and 1870-71 show that the river bank has changed very little since then, beyond the growth of the riverbank tamarisk. Despite this slow rate of change, the Colorado River, in the ten million years of its existence, has managed to cut downward through solid rock for more than a mile, and while doing so has removed from this region about five thousand vertical feet of rock and sediments.

The winding road on the flat terrace below Dead Horse Point is still more of the Potash ORV trail noted earlier. It continues at this level, roughly parallel to the river gorge, until it enters Canyonlands National Park. There it branches, with the White Rim Trail continuing left at the level of the White Rim Sandstone. The righthand trail climbs steeply to the top of Island-in-the-Sky via the historic Shafer Trail. This route was first developed as a cattle drive trail, then improved for four-wheel-drive vehicle use during the uranium boom days. Even though the Shafer Trail is now in Canyonlands National Park, it is still too steep and rough for most conventional vehicles. For details, see the geology log of the Potash-Shafer ORV trail elsewhere in this book.

GEOLOGIC STRATA VISIBLE FROM THIS ROAD

Formation	Member	Page
Navajo Sandstone		Page 13
Kayenta Sandstone		Page 13
Wingate Sandstone		Page 12
Chinle Formation		Page 12
Moenkopi Formation		Page 12
Cutler Group	White Rim	Page 11
	Organ Rock	
	Cedar Mesa	
	Lower Cutler	
Hermosa Group	Upper Hermosa	Page 9
	Paradox Salts	

ROAD NAME: **CANE CREEK ROAD**

SUMMARY: Cane Creek Road offers exposures of a wide variety of geologic strata and erosional forms that are spectacularly colorful and scenic. It closely parallels the Colorado River gorge for a short distance then travels the deep lower narrows of a long stream-watered canyon to end at a ford of that stream. Part of the arching strata of the Cane Creek Anticline are visible in that more open part of the canyon. There are delicately balanced rocks, prehistoric rock art sites and fascinating sidecanyons to hike along the way. Those driving off-road vehicles can combine this road with the described Hurrah Pass and Chicken Corners off-road vehicle trails.

ROAD SURFACE: paved to the mouth of Cane Creek Canyon, poorly-maintained gravel beyond there, with rough surface and one short steep grade with switchbacks.

VEHICLE: any vehicle, for the described trip, in dry weather.

MAPS: ***Canyon Country*** **OFF-ROAD VEHICLE TRAIL MAP - Canyon Rims & Needles Areas** or ***Canyon Country*** **OFF-ROAD VEHICLE TRAIL MAP - Canyon Rims Recreation Area.**

GUIDEBOOKS: ***Canyon Country*** **OFF-ROAD VEHICLE TRAILS - Canyon Rims & Needles Areas** or ***Canyon Country*** **OFF-ROAD VEHICLE TRAILS - Canyon Rims Recreation Area.**

MILEAGES:

TO START: 0.7 miles, from Main and Center, Moab.
ROAD: 11.2 miles, one-way to end.
RETURN: 11.9 miles, from end of described road.
TOTAL: 23.8 miles, from Main and Center, round trip.

TIME: about 1/2 day, plus time taken for stops along the way. For a leisurely trip taking photos along the way, a lunch stop and time for some hiking, allow a full day. Allow for a long, full day if the Hurrah Pass and Chicken Corners ORV trails are to be traveled beyond this road.

ROUTE: Moab city streets to Cane Creek Road to where it ends, as described, where the road fords Cane Creek in mid-canyon. Those with off-road vehicles can continue across the creek ford and travel the Hurrah Pass and Chicken Corners ORV trails.

NAVIGATION: drive south on Main Street (U.S. 191) then west on Cane Creek Road. Return via the same route unless exploring the ORV trails beyond this road is planned.

GEOLOGY ROAD LOG:

0.0 The road first travels through the outskirts of Moab parallel to the western cliffs of Moab Valley. The rounded domes of sandstone visible ahead in the vicinity of the downriver Colorado River portal are Navajo Sandstone. The terraced layers that dominate the cliffs are Kayenta Sandstone, with some small exposures of fractured and decomposing Wingate Sandstone below that. Rubble from these strata cover the lower slopes. Little or none of the Chinle Formation below the Wingate is visible beneath the rubble. The low, rounded hills ahead near the portal are an upthrust of the Paradox Formation salts that formed this and other salt anticlines. Here, erosion has exposed a ridge of the Moab Valley salt anticline that extruded upward to near the surface. Note that the western cliffs beyond the river portal ascend steeply. This resulted from the displacement in the north end of Moab Valley caused by both the salt anticline and the Moab Fault. The rounded domes on the skyline south of this junction are more Navajo Sandstone, with the same formations below as noted earlier. See the sidebar, SALT ANTICLINES.

0.6 The road is traveling along the base of the exposed Paradox salts noted earlier. The mineral has been eroded and partially covered by soils and vegetation, but the gray-white salts are exposed at intervals on the surface.

0.8 Cane Creek Road angles left here.

1.1 As the road approaches the downriver portal of Moab Valley, the Wingate Sandstone becomes apparent in the vertical cliffs in the near side of the portal, with the same rubble-covered Chinle Formation below that. The strata in the far portal are the rounded Navajo Sandstone at the top, the horizontal layering of Kayenta, with some near-vertical slopes of Wingate below that, plus some traces of Chinle Formation covered by rubble on the lower slopes. On up the valley to the north, there are traces of reddish Moenkopi and Cutler deposits exposed within the rubble slopes. See the sidebar, PARADOXICAL VALLEYS, for more information about valleys crossed by rivers in this region. The road-cut on the right just ahead is an exposure of Paradox salts.

1.4 At a pullout here, there is a good exposure of Paradox salts to the left of the road, as rounded mounds lightly covered by vegetation, with the raw minerals exposed in several places. Within and above the Paradox salt mounds, the large, red boulders are Wingate Sandstone fallen from the cliffs above. The Paradox salt exposures just ahead have been eroded by surface water and melting snow, leaving small caves in the salts.

1.8 Beyond a pullout here the road enters the Colorado River gorge at the downriver portal, leaving Moab Valley behind. Note that the geologic strata on both sides of the river are steeply sloped. This tilt was given the strata by the Moab Valley salt anticline. Just ahead, the Wingate Sandstone and the ledgy Kayenta Sandstone just above it, dive below ground, leaving the round-topped vertical cliffs of the Navajo Sandstone at river level. This road, the one on the opposite side of the river, and the road that goes upriver from Moab Valley, are the only paved roads that closely parallel the Colorado River within its deep gorge along its entire length through four states and part of Mexico. Interstate 70 closely parallels the river in more open terrain in western Colorado, and other roads in Nevada and California travel near the river, or reservoirs built on it, after the river has emerged from its deepest gorges, the Grand Canyon in Arizona and Black Canyon in Nevada.

2.7 The Moab Rim ORV trail goes left here. The rock at road level here is a mixture of Navajo and Kayenta sandstones. The upper cliffs are all entirely Navajo Sandstone. The gravel banks ahead are remnants of river-gravel deposits left behind on terraces as the river has cut deeper. Such gravel deposits occur on ledges and benches above the road until it enters Cane Creek Canyon, and a short distance into that canyon. Note that along this stretch of river gorge the exotic shrub tamarisk has stabilized the riverbanks, even though during the high-water season in June, the river water sometimes still invades the lowlands along this stretch of the river gorge. From this pullout, note Little Arch high in the cliff on the opposite side of the river. Despite its name, Little Arch is far from "little." It can be reached by a short hike beyond a spur trail of the Poison Spider Mesa ORV trail, which is described in the guidebook, ***Canyon Country*** **OFF-ROAD VEHICLE TRAILS - Island Area.**

2.8 Note the large spring-seep cave to the left at the base of the Navajo Sandstone cliff. The water seep-lines are where the mosses and lichens are growing and mineral streaks are descending. In the winter the seeping water in this cave builds up an immense cascade of ice. Other seep-caves with beautiful desert varnish streaking around them can be seen still higher in the cliffs above.

3.0 Here, the Kayenta Sandstone has finally disappeared entirely underground, leaving the cliffs on both sides of the river Navajo Sandstone, with traces of rubble and ancient river gravel on its lower slopes and terraces. Note the several seep-caves in the cliffs on the opposite side of the river and the beautiful patterns built up by rain streaking and snow-melt.

3.1 Beautiful Moonflower Canyon cuts into the cliffs to the left. This short canyon was cut by drainage from the Behind-the-Rocks wilderness area high above. Note the ancient but still healthy cottonwood trees in the mouth of the canyon. These indicate the presence of permanent subsurface water. It is possible to hike from the Moab Rim ORV trail almost to the beautiful pouroff that drains into Moonflower Canyon. There are prehistoric petroglyphs along the cliffs adjacent to the road, beginning on the far corner of this canyon. Most of these are on the private land adjacent to the road, but some can be photographed from the road right-of-way with a telephoto lens. The land between here and Cane Creek Canyon on both sides of the road is private. Travelers should stay on the road right-of-way. At some points the near-vertical cliffs on the far side of the river actually overhang the road. Climbers who have rappelled down from one point on the clifftop have landed on the road's centerline. That same stretch of cliff for the next mile is popular with climbers practicing their techniques.

3.3 Beginning here, note the several caves that have been blasted into the cliff-base for various purposes along this stretch of private land. Such caves have been used for decades in canyon country by stockmen for the storage of animal feed and equipment and, in some locations, even for seasonal occupation by stockhands.

3.6 The road angling left is private. It climbs onto the higher terrace to the left. In the past, the great masses of abandoned river gravel left on this broad terrace were bulldozed over the cliff to the road level, then used for various construction purposes. Beginning here, note on the other side of the canyon the several deep cuts in the cliffline caused by jointing and water erosion. The bottomlands here of rich river-deposited alluvium were once used for agricultural purposes, with melons being one very tasty product. The cliffs on both sides of the river continue to be Navajo Sandstone, with deposits of river gravel on near-horizontal surfaces to the left.

4.6 Pritchett Canyon and the ORV trail that travels it go left from the road here. The caves in the left canyon wall are not natural but were blasted by the private landowner.

4.7 The pavement ends here, as the road leaves the Colorado River gorge and enters Cane Creek Canyon. From here to the end of the road as described it is graded and graveled and, although rough in stretches, is travelable with care by all kinds of vehicles during good weather. Note that the strata on both sides of the river begin to ascend downriver, exposing the layered Kayenta Sandstone below the rounded domes of Navajo Sandstone. The narrow lower gorge of Cane Creek Canyon is dominated by the Glen Canyon Group, i.e., Navajo, Kayenta and Wingate sandstones.

4.8 Beyond the pullout here, the road enters Cane Creek Canyon. Note the dense vegetation on the slope on the opposite side of the canyon near the powerline poles. A seeping spring there creates immense masses of ice during the colder months of the year. As the road enters the canyon, the upper cliffs are Navajo Sandstone, with some exposures of terraced Kayenta below that. On the far side at the mouth of the canyon, the rounded brush-covered slopes are remnants of Colorado River gravel abandoned there millennia ago. As the road enters Cane Creek Canyon, note the small arch in the upper cliffline straight ahead.

Cane Creek Canyon is immense. Its drainage begins in the lower slopes of the La Sal Mountains, crosses U.S. 191 at the Hole-'n-the-Rock tourist development, then enters a stretch of deeply-cut narrows. After several miles, the canyon widens and is joined by equally large Hatch Wash Canyon to create the broad, open sheer-walled canyon where this road as described ends. Below this open part of the canyon, the drainage enters the narrows through which it travels for several more miles. It then emerges from that to join the Colorado River, for a total length of approximately 25 miles. From its confluence with Hatch Wash Canyon, the Cane Creek Canyon drainage runs roughly parallel to the Colorado River gorge, but flowing in the opposite direction. Beyond the end of this road, off-road vehicle trails continue up the canyon or leave the canyon via Hurrah Pass to descend into the Colorado River gorge. Refer to the log for the Hurrah Pass - Chicken Corners trails for the geology of this last route.

5.2 At a narrow pullout, note how lower Cane Creek Canyon turns and twists below the Navajo Sandstone upper cliffs and lower terraces of Kayenta. For an explanation of this phenomenon, see the sidebar, ENTRENCHED MEANDERS. Up the canyon from here, note that the geologic strata are rising. The uppermost vertical cliffs and domes are Navajo, the terracing below is Kayenta Sandstone, and the vertical cliffs that wall the deep inner gorge are Wingate Sandstone. Cane Creek itself is fed by numerous springs along its length, in particular within its upper and lower narrows. It is also fed by subsurface water and occasional flood-water from its major tributary, Hatch Wash Canyon. At times during the year, Cane Creek flows continuously from about U.S. 191 to the Colorado River, but the rest of the year it is intermittent, except for even longer periods of time when it is continuous within the upper and lower narrows. During the colder winter months, Cane Creek, within its lower narrows, builds up and freezes over to form a continuous ribbon of ice that is sometimes smooth enough for ice-skating, and several miles long.

5.5 A road goes right from here, descending into the canyon bottom to a popular primitive camping area. The canyon going left from here has remnants of small prehistoric ruins and some rock art at the base of the vertical cliffs. A hiking route goes up this canyon into the spectacular saddle between Cane Creek Canyon and Pritchett Canyon. It is possible, although not easy, to descend into Pritchett Canyon via this route for a loop hike.

5.6 Note that the cliff on the left has been drilled and blasted to construct this road. There is an interesting arc-topped seep-cave beginning to form in the cliff above and ahead. The slopes and terraces here are covered by gravel carried here from the upper drainage by flooding in past times, then abandoned as the drainage cut deeper. The rock was brought all the way from the La Sal Mountains. The upper Wingate Sandstone is beginning to appear at the creek level here.

5.8 At a pullout here, there are two small but interesting arches about twenty feet above the road level straight ahead. This kind of arch forms readily in the horizontal layering of Kayenta Sandstone. Note an ORV trail on the terraced slopes lower in the canyon to the right-front. This is the beginning of the Amasa Back ORV trail that can be seen ahead to the right, ascending the rocky terraces on the far side of the inner drainage. This very steep and difficult trail eventually skirts around the upper rim of Jackson Hole, an immense abandoned meander of the Colorado River gorge. The terraced rock strata on both sides of the canyon here continue to be Kayenta Sandstone. The vertical cliffs and domes above that are Navajo Sandstone. Note on the opposite side of the canyon a large arc-topped alcove forming in the base of the Navajo Sandstone cliff. With good binoculars, prehistoric pictographs are visible in that alcove. There are petroglyphs on some of the rocks below.

6.0 The Amasa Back ORV trail drops steeply from the road here. This trail should be attempted only by experienced drivers with competent off-road vehicles.

6.1 Note the vertical Wingate Sandstone cliffs in the bottom of the gorge. The rising Kayenta Sandstone strata ahead are leaving only remnants of Navajo Sandstone at the uppermost levels on the right, although the Navajo cliffs continue high above that level on the left of the road. The large, light-hued boulders to the right of the road for the next half mile are Navajo Sandstone fallen from the cliffs above.

6.2 At a narrow pullout here, and another pullout just ahead, the large boulder below the road level has prehistoric petroglyphs on all four sides. On the nearest corner is a petroglyph depicting a woman giving birth. This has been dubbed "the oldest known birth certificate." There are a few petroglyphs on a boulder nearer the road 0.1 mile ahead.

6.4 At a pullout here, a short ORV trail goes right, out onto a lower terrace in the canyon. This route provides good hiking to views of the narrow inner gorge. From the terrace rim, the curving opposite canyon wall offers a good echo. Just ahead, note that the road drops into the lower gorge via a set of switchbacks. The road from here to the bottom of the canyon has been cut through Kayenta Sandstone. Park at this pullout for the hike up the drainage line just ahead to the left, to spectacular Funnel Arch. The first part of this hike is steep and difficult and should be attempted only by hikers in good physical condition. Funnel Arch is out of sight in the Navajo Sandstone above to the left. The hike begins right at the interface between the Navajo and Kayenta sandstones.

Funnel Arch

6.5 Another pullout to the right here can be used for parking for the hike to Funnel Arch. The route begins by ascending the slickrock slopes to the left of the road, then continues up the drainage narrows that appear in the cliffline here. The road-gravel here covers a long stretch of dinosaur tracks that were discovered more than two decades ago by the author of this book. The tracks were documented shortly afterward by Professor W. Lee Stokes of the University of Utah. Drive carefully between here and the bottom of the canyon, because there are no guardrails or shoulders and the narrow road is flanked by a sheer drop. This stretch of the road should not be attempted when covered by ice or snow, or during rain.

6.9 From a narrow pullout here, note the appearance of the Wingate Sandstone cliffs in the canyon-bottom opposite the road.

7.0 Stop and enjoy the spring in the cliff above the road. The main spring flows continuously at about the same temperature and volume all year long, although the surrounding seeps freeze during the winter. Ferns, mosses, cane grass and other moisture-loving plants thrive in this spring area. One species of tiny, delicate orchid thrives in the seep line just above the road, and watercress, algae and grasses flourish in the tiny stream below the spring. Unlike most water sources in the region that are thoroughly contaminated by cattle droppings, the water from this spring is pure and potable as it comes from the cliff.

The spring is fed by water absorbed by the great masses of relatively porous Navajo Sandstone high above it. The water seeps slowly downward through fractured Kayenta layers, then flows laterally along an impervious layer in the Wingate Sandstone to the cliff from which it emerges. This spring, and others in the vicinity, contribute water to Cane Creek. A hike downcanyon from here, to a convenient egress at the campsite area noted earlier, gives an excellent view of the inner gorge narrows. There is easy access from this road to the creek-bottom a short distance beyond the spring. The Wingate Sandstone cliffs along this stretch of narrows have been beautifully sculpted by the flowing stream as it has cut ever deeper, and by the surface runoff of rain and snowmelt. From here on through the narrows, the cliffs on both sides are Wingate Sandstone.

7.1 At a pullout here, note the small arch opening in the top of the promontory on the skyline ahead. Wingate Sandstone is exposed at road level, with layered Kayenta Sandstone looming above, with occasional domes of Navajo Sandstone visible at the uppermost levels.

7.9 Park at a large pullout on the left for the interesting hike up well-watered Hunter Canyon, with its many large spring-fed pools, cascades and lush creek-bottom vegetation. The hike ends at a very large pool below a sheer pouroff from the upper drainage of Hunter Canyon in the Behind-the-Rocks wilderness area. Hunter Arch is a large natural span in the right wall of this canyon about 1/2 mile from the road. Primitive camping at the mouth of Hunter Canyon is popular. The lower cliffs above the road beyond here continue to be Wingate Sandstone covered with patinas of lichens, plus streaks of dark brown and black desert varnish. See the sidebars, DESERT VARNISH and LICHENS.

Just ahead, the road dips down to cross the Hunter Canyon drainage. Standard highway vehicles should be driven very carefully here because the road is a little rocky and often cut by Hunter Canyon flooding. Drivers of low clearance vehicles should check this crossing before attempting it, especially during the off-seasons when road maintenance is very infrequent. During the winter, the springs and tiny stream of Hunter Canyon freeze over and build up immense masses of ice in the drainage bottom.

8.1 Note the cave straight ahead above the primitive campsites beyond the small meadow. The cave has been formed by seeping water that appears to have dried up in recent years.

8.2 Note the bulldozer cut in the opposite slope that was left by uranium prospectors decades ago. The large boulders beside the road have fallen from the Wingate cliffs above. Note the slit-type arch in the cliffline to the right-front. This opening is also visible on the return trip, from the opposite direction.

8.5 Beyond the narrow pullout here, the road emerges from the canyon narrows into the wider central valley of Cane Creek. Note that the rock strata on the right of the valley are ascending, with the gray deposits of the Chinle Formation appearing just across the canyon at road level. The cliffs above the Chinle slopes are Wingate Sandstone. Note the bulldozed vehicle trail ascending that same slope in the mid-levels of the Chinle Formation, and the banks of gray mine tailings about midway. See the sidebar, URANIUM. The crenelated reddish cliffs below the mine trail are Cutler sandstone. The upward arching of the strata here is a part of the Cane Creek Anticline. Both Cane Creek Canyon, and the Colorado River gorge to the west of the cliffs here, cut into this major salt anticline. The giant boulders on each side of the road for the next half mile or so have fallen from the vertical Wingate Sandstone cliffs high above. The surfaces of some of these boulders have been eroded by rainfall into exotic shapes and forms.

9.0 The mine road noted earlier begins here. The high point on the horizon at left-front is Anticline Overlook, on Hatch Point in Canyon Rims Recreation Area. The low saddle on the horizon straight ahead is Hurrah Pass. The Hurrah Pass ORV trail ascends the sloping strata to the left of the pass beyond the end of this road as described. From here, a beautifully eroded cliffline of red Cutler sandstone topped with layered Moenkopi deposits stretches along the lower canyon wall to the right. The sloping, terraced deposits above the red cliffs are uranium-bearing Chinle Formation. The canyon bottom here consists of alluvium and debris from the decomposition of all the formations that wall the canyon, plus some harder rock that has been carried all the way down from the La Sal Mountains.

9.4 From a pullout here, note to the left of this stretch of the road how harder caprocks fallen from

continued on page 121

URANIUM

by M. M. Barnes

The Utes and Navajos who inhabited the Colorado Plateau during historic times dug up colorful red vanadium ore and bright yellow uranium bearing rocks. They pounded these into dust, mixed that with animal fat, then slathered their bodies with this frightening war paint.

Actual mining of uranium began incidentally in the mid-1800s when gold diggers came upon pitchblende, a dark, radium-bearing ore with which uranium and vanadium ores are often associated. By the turn of the century, radium had been isolated and was soon hailed as a "miracle" treatment for cancer. Thus, at first, the radium was retained and the rest was discarded as tailings. Later, a market for vanadium developed from its use as a steel-hardening agent. Later still, the waste dumps themselves were mined when commercial uses were found for uranium oxides as pigments in dyes, inks, stained glass and ceramics.

Then, market demands for both radium and vanadium accelerated. World War I production increased the need for vanadium as a steel hardening agent, and radium was used in illuminating watch faces, compasses, gunsights, and airplane dials.

World War II brought some secretive use of uranium in government-sponsored research into the use of certain of its isotopes and derivatives as nuclear explosives. The end of that war, however, caught America without a developed local supply of uranium. The element had been used to create the atomic bomb that ended the hostilities, and held promise for other important uses. Thus, the uranium discoveries of the 1950s in southeastern Utah were fortuitous for both the country and the region.

Uranium was suddenly the most critical material the United States had ever known. Not only was it needed for atomic weapons, it also promised plentiful, inexpensive electrical power, which could be used for gas-free operation of cars, planes, and locomotives; the preservation of meat through irradiation; and even the cheap distillation of sea water. A domestic supply of uranium was vital to maintaining a nuclear edge in the cold war that was developing with Russia.

The Atomic Energy Commission thus decided that national security demanded immediate stimulation of the United States' almost nonexistent uranium production. Glittering inducements were offered. In earlier days, ore could not be profitably handled unless it contained at least two per cent uranium. Now the AEC agreed to buy rock assaying as little as one tenth of one per cent. Fat bonuses were given for new claims brought into production. The AEC also helped finance access roads through the broken Four Corners terrain, and granted haulage allowances for distances of up to a hundred miles.

On July 6, 1952, Charles A. Steen, graduate of the Texas School of Mines and Metallurgy, and former geologist for Standard Oil Company of Indiana, struck the biggest deposit of high-grade uranium ore in the country. Within the first six months his Mi Vida mine, southeast of Moab in Lisbon Valley, shipped a million dollars worth of ore that was so pure it assayed up to 87 per cent uranium. Ore prices were guaranteed by the AEC, the only purchaser until 1962, and Steen's strike triggered a mineral rush that rocked the entire Colorado Plateau, including the canyon country in the Moab vicinity.

Suddenly, snorting bulldozers were plowing their way toward every conceivable source of this treasured mineral. Mitch Williams, a long-time Moab resident, remembers this time well. As a young retired Air Force pilot he flew planes around the area with geologists and prospectors. ***"Suddenly air strips were everywhere that a 'dozer could reach to scrape some ground,"*** Mitch recalls.

By the mid-1950s, almost six hundred producers on the Colorado Plateau were shipping uranium ore. A bonanza in penny uranium stocks established Salt Lake City as "The Wall Street of Uranium." Southeastern Utah was booming, especially the general Moab area, as the backcountry swarmed with human gophers, burrowing into every exposure of the few uranium-bearing formations, all seeking instant fortunes.

But by 1964, after purchasing almost 9 million tons of ore valued at $250 million, the AEC announced that "it is no longer in the interest of the Government to expand production of uranium concentrate." The market was saturated and reserves were sufficient for the foreseeable future. Federal buying ceased, and for some time commercial purchases remained illegal. This brought uranium mining to a complete halt. Eventually, however, commercial purchase for nuclear power plant use was permitted, and limited uranium mining in the general Four Corners region resumed. It ceased again in recent years, except to complete existing long-range purchase contracts, when far richer uranium-vanadium ores were found in Canada, Australia and South Africa.

For the curious tourist, remnants of these past mineral searches are still quite visible throughout the region, wherever there are exposures of uranium-bearing geologic deposits, primarily the Chinle and Morrison formations. Wherever these ore-bearing formations are visible, there are signs of past prospecting activities -- the predictable traces of eroding bulldozer trails up impossibly steep inclines, alluvial fans of "tailings" dredged from tiny black-hole tunnels as if by busy ants, and sometimes collapsing external scaffolding and other supporting structures.

The anthills are quiet now and the uranium frenzy is hushed -- at least for a time. Still, the remains are there for the viewing, a part of the history of the Colorado Plateau and general Moab area. Such remnants in the backcountry should be viewed from afar, however, for even the powdery tailings of this radioactive ore are virulent, and all

continued ----

too often deadly radon gas accumulates in such mines.

The town of Moab itself also bears the scars of past uranium-mining boom times. Each boom brought with it a surge of growth in the town, as well as an influx of supporting industries. Each following bust left its legacy of empty buildings and sprawling acres of industrial junk, with these unsightly remnants still visible many places in Moab Valley. The most conspicuous remnant, of course, is the abandoned uranium-vanadium processing mill that mars the scenic splendor of the valley adjacent to Arches National Park. This stark vestige of the past has long since been decommissioned by the government and is supposed to be torn down and its conspicuous tailings pile buried, to reduce the health hazards both represent, but the mining company that owns the property is delaying this expensive task as long as it can.

Those interested in reading more about uranium mining in the general Four Corners region and its various effects will enjoy the following books:

URANIUM FRENZY, Boom and Bust on the **Colorado Plateau** by Raye C. Ringholtz, W. W. Norton & Company, 1989. A thrilling story of the uranium boom days featuring the rags-to-riches tale of Charlie Steen, the impact on communities and individuals, and the tremors of the aftermath. Available through book stores and libraries.

SAN JUAN COUNTY, UTAH, People, Resources, and History by Allan Kent Powell, Utah State Historical Society, 1983, the chapter entitled Uranium Mining on the Colorado Plateau by Gary L. Shumway. A 34-page comprehensive history from the viewpoint of a San Juan resident whose family was involved. Available through libraries.

GUIDEBOOK TO THE GEOLOGY OF UTAH, Number 9: Uranium Deposits and General Geology of Southeastern Utah by Utah Geological Society, Salt Lake City, Utah, 1954. Includes history, stratigraphy, and mineralogy as well as maps and other illustrations of the "boom" area. This book may be available through some library systems.

Charlie Steen's MI VIDA by Maxine Newell, 1992. The story of the world's richest uranium strike, the man who discovered it, and the effect of both on a small Utah town. Available through Moab-area book stores and mail-order outlets.

above have protected columns of softer Moenkopi sediments.

9.8 From another narrow pullout opposite a large boulder, note that just below the uppermost cliffs of Wingate Sandstone, the gray horizontal layering of the Chinle Formation has eroded into vertical columns. Note in the steep slopes ahead another bulldozed vehicle trail ascending to a uranium mine.

9.9 The mottled red sandstone that appears through the rubble to the left of the road is the first nearby exposure of Cutler sandstone.

10.0 Note the mine scaffolding to the left, below the sheer Wingate cliffs and to the left of a promontory of gray-colored Chinle Formation. This is the mine that the steep trail to the left-front once reached.

10.2 From the large pullout on the right, the promontories ahead that the road skirts around are Cutler sandstone. Below the gray Chinle Formation deposits to the left are thin layers of Moenkopi just above the more prominent red exposures of Cutler sandstone. The white sandstone ledges just above the road to the left are also Cutler. In the branching mine trail to the left-front in the steep slopes, a large balanced rock appears below where the trail forks. From one angle, this balanced rock resembles the profile of a smiling man wearing a large turban. The rock is thus called "The Happy Turk." A short, rough ORV trail that leaves Cane Creek Road just ahead goes to the base of the Happy Turk balanced rock. This column of Cutler Sandstone is so

Happy Turk

delicately balanced and undercut that it could collapse at any time. Any winter or rainstorm could spell its doom.

To the right, across the canyon, is the low saddle of Hurrah Pass, with columns of beautifully eroded Cutler sandstones below the pass and continuing down to the canyon floor. The massive tower to the left of Hurrah Pass is Chinle deposits on a base of Moenkopi and Cutler. The high promontory on the skyline still farther left is Anticline Overlook, which can be reached by driving paved and graveled roads in the Hatch Point District of Canyon Rims Recreation Area. See the books ***Canyon Country* OFF-ROAD VEHICLE TRAILS - Canyon Rims Recreation Area** and ***Canyon Country's* CANYON RIMS RECREATION AREA,** and the map, ***Canyon Country* OFF-ROAD VEHICLE TRAIL MAP - Canyon Rims Recreation Area,** for guidance to that and many other overlooks in this spectacular BLM recreation area. The roads to Needles Overlook and Anticline Overlook are described elsewhere in this book.

Beyond here, Cane Creek Canyon lies in the Lowlands District of Canyon Rims Recreation Area. The rounded gravel on the low mound just ahead and to the right of the road here was deposited by Cane Creek flooding during earlier millennia.

10.5 Note the white-turbaned Happy Turk to the left of the road, just above road level. A few yards ahead, the short ORV trail to the huge balanced rock angles left.

11.1 An ORV trail going left here leads to the long-abandoned Monte Cristo uranium mine complex. Note the several tailings slopes to the left in the lower cliff-line. To help trace the veins of ore, a crude road was constructed to the top of the cliffs here from U.S. 191, then a series of vertical drill cores were taken. The traces of these drilling operations are still visible from along the final mile or so of the Cane Creek Canyon Rim ORV trail, which is described elsewhere in this book.

From pullouts on the right or left, just before this road as defined ends, the geologic formations in the canyon cliffs are as follows. The uppermost cliffs on both sides of the canyon are Wingate Sandstone, with remnants of the horizontally layered Kayenta on the cliff rims. The slopes below the vertical Wingate cliffs are partially covered by collapsed debris but are, in descending order, Chinle Formation and Moenkopi Formation, with Cutler sandstone at the canyon-floor level. The vertical columns to the left of the road here, just above road level, are Cutler sandstone. The nearby promontories to the right of the road, and the lighter-hued sloping strata to the right just above road level, are also Cutler. The low hills in the canyon bottom are water-rounded rock carried from the La Sal Mountains, mixed with decomposing local rock, sand and alluvial deposits, all held in place by native vegetation.

The brush-obscured drainage line ahead and to the right is also sand and alluvium deposits. In the actual streambed, these recent deposits are continually being cut and eroded further by the flash-flooding that has been worsened by the denudation of native vegetation in this canyon by overgrazing. To the right, lofty Hatch Point and Anticline Overlook dominate the skyline. The isolated prominences between that jutting point and the high Wingate Sandstone cliffs farther to the right are mixtures of Chinle, Moenkopi and Cutler, in descending order. Hurrah Pass is the low, wide saddle in the cliff-line to the right-rear. The cliffline to the rear and left rear is as described earlier.

This part of the canyon is still under grazing lease and is used as a cattle-wintering area despite the pitifully sparse remnants of native and exotic grasses. Within the historic past, cattle that were summered in the Island-in-the-Sky vicinity, were driven in the autumn down the Shafer Trail into the Colorado River gorge. There, they were driven across the river at a ford, then herded up and across Hurrah Pass and down into Cane Creek Canyon. Once in Cane Creek Canyon, the herds were divided. Some were driven or trucked through Moab to the Thompson Springs railhead. The remaining cattle were wintered in Cane Creek Canyon before being returned to higher pasturage the following spring. While this open stretch of Cane Creek Canyon is still used for wintering cattle, Hurrah Pass and the Shafer Trail are no longer used as a cattle drive route.

11.2 Cane Creek Road, as defined and described in this book and others in the *Canyon Country* series, ends here where the vehicle route fords Cane Creek. The main road beyond this ford is defined as the Hurrah Pass ORV trail. About 1/2 mile beyond this ford, the ORV trail that continues up Cane Creek Canyon goes left from the Hurrah Pass ORV trail. This difficult trail should be attempted only by experienced drivers with competent vehicles. If this ford is in good condition, with little or no water flowing, standard highway vehicles can safety continue on the main road for several miles toward Hurrah Pass before encountering difficulties. High-clearance, low-geared highway vehicles can reach the summit of Hurrah Pass if carefully driven, but only off-road vehicles should continue beyond the summit.

LICHENS

There are many types and forms of lichens, even in arid, high-desert canyon country. All are combinations of various species of algae and fungi living together in symbiotic form, with each benefiting from the arrangement.

Some researchers have claimed that the relationship is parasitic, with the fungi "preying on" the algae, but experiments have shown that, if separated, each of the two components of lichens can survive alone, and that each benefits from the association, which more closely fits the definition of "symbiosis."

Both blue-green and green algae combine with various species of fungi to become lichens. The algal component contains chlorophyll that can photosynthesize plant nutrients, using solar energy and atmospheric carbon dioxide, plus minerals provided by the fungi. It then contributes these nutrients to the symbiotic partnership. The fungal component does not contain chlorophyll, but it provides water from the environment and its acidic secretions dissolve rock and thus provide minerals for use by both. It also provides physical protection for the alga from the more harsh environments.

Lichens grow in a variety of forms, but the type commonest in canyon country, the type referred to in some of this book's road and trail logs, grows in the form of thin, dry crusts on rock surfaces and on the surfaces of cryptogamic soils. See the sidebar, CRYPTOGAMIC SOIL.

Rock lichens display a variety of colors, from black and shades of gray, to pale blues and greens, and shades of orange and yellow. The brighter colored types seem to require more moisture, so tend to occur at higher elevations. Most grow as small, rounded leaf-like clusters that cling tightly to the rock and slowly spread outward, sometimes abandoning their centers and thus leaving behind a botanical form of "urban blight." Others grow as small dots or clumps that spread very slowly, if at all.

As with most life forms, lichens affect geologic processes. By coating sandstone surfaces, rock lichens slow surface weathering, even though they do dissolve and absorb tiny amounts of minerals from the rock. Some types of canyon country rock seem to provide more of such necessary minerals, in particular Navajo Sandstone, and are often so thickly covered by lichens that little of the actual rock surface is visible. Some other geologic strata host few if any lichens.

In some locations, lichens have left evidence of this region's slowly changing climate. Along and near many cliff rims of the Slickrock Member of Entrada Sandstone, the white mineral "skeletons" of dead lichens show that the past climate was moist enough to permit lichen growth, even on rock now largely inhospitable to these strange plant forms. These lichen-shaped crusts of white minerals occur almost anywhere that upper surfaces of Slickrock-Entrada are exposed, minerals that were extracted slowly from the rock by the fungal component of lichens, over many decades, then left behind when climate conditions became too harsh for survival in those marginal locations. Similar lichen remains are found to a lesser extent on other sandstones.

Although the growth rates of various kinds of lichens are largely a matter for speculation and controversy among botanists, it is generally accepted that the crusty types that live on rock surfaces in arid terrain grow very slowly, that it takes decades and even centuries for visible spreading, depending on the type of lichen and local conditions. While this very slow but unmeasurable growth cannot be used for precise dating, it can help provide some indications of the age of historic remnants.

While this author was searching for a long-lost historic paleontological site that was reported in historic accounts as being just above a sandstone cliffline, he found ascending toe-holds cut into the steep slopes of the cliff. That these steps were old enough to have been cut during historic times was indicated by the fact that most of the notched steps had sizable clusters of lichens growing in them.

From this clue, the lost paleontological site was eventually found and reported to interested scientists and historians. That long search, in which the slow growth of lichens played a critical part, is described in the book, CANYONLANDS NATIONAL PARK - *Early History & First Descriptions.*

GEOLOGIC STRATA VISIBLE FROM THIS ROAD

Mancos Shale		Page 20
	Ferron Sandstone	
Dakota Sandstone		Page 20
Burro Canyon/Cedar Mountain Formations		Page 19
Morrison Formation	Brushy Basin	Page 18
	Salt Wash	
	Tidwell	
Entrada Sandstone	Moab	Page 16
	Slickrock	
	Dewey Bridge	
Navajo Sandstone		Page 13
Kayenta Sandstone		Page 13
Wingate Sandstone		Page 12
Chinle Formation		Page 12
Moenkopi Formation		Page 12
Cutler Group	White Rim	Page 11
	Organ Rock	
	Cedar Mesa	
	Lower Cutler	

ROAD NAME: **LA SAL MOUNTAIN LOOP ROAD**

SUMMARY: This popular and highly scenic loop road is unusual in several ways. It travels the general interface between two large salt valleys and a range of intrusive igneous mountains; it presents numerous outstanding panoramic views of major erosional features in the lower terrain near Moab; and it travels for most of its length through the same geologic formation, even though it climbs and descends drastically over its length. The road also offers access to several outstanding spur off-road vehicle trails, including two that end at small but scenic mountain lakes, and it provides a good look at how salt anticlines and intrusive igneous mountains elevate, fault and distort otherwise horizontal sedimentary geologic strata. Although this loop route can be taken in either direction, it is described here in a clockwise direction so as to combine this geology road log with all or parts of the logs for Utah 128 and Castle Valley Road, both described elsewhere in this book.

ROAD SURFACE: mostly paved. A couple of short stretches that are difficult to maintain because of severe winter weather or seeping water are graded and graveled.

VEHICLE: any vehicle.

MAP: ***Canyon Country*** **OFF-ROAD VEHICLE TRAIL MAP - Arches & La Sals Areas.**

GUIDEBOOK: ***Canyon Country*** **OFF-ROAD VEHICLE TRAILS - Arches & La Sals Areas.**

MILEAGES:

TO START: 29.5 miles, from Main and Center, Moab.
ROAD: 21.0 miles, one-way to end.
RETURN: 13.5 miles, from end of described road.
TOTAL: 64.0 miles, from Main and Center, round trip.

TIME: allow a short day for this loop trip, which will include ample time for many scenic stops and a leisurely picnic lunch along the way. If the road logs for all or part of Utah 128 and for Castle Valley Road are read along the way to the northern end of the La Sal Mountain Loop Road, allow for a long day.

ROUTE: Main Street (U.S.191) to Utah 128, upriver on that road to Castle Valley Road, up that road to its junction with the northern end of the La Sal Mountain Loop Road, then return to Moab via Spanish Valley Road and U.S. 191.

NAVIGATION: Drive north from Moab on U.S. 191 for 2.6 miles, then turn upriver on Utah 128 just before the main highway crosses the Colorado River. After another 15.9 miles, turn right on Castle Valley Road. The La Sal Mountain Loop Road spurs right from this road in another 11.0 miles. At the southern end of the road, where it ends by connecting with Spanish Valley Road and Pack Creek Road, follow the directions given at the end of this log for the quickest return to Moab.

GEOLOGY ROAD LOG:

0.0 As this road spurs right off of the Castle Valley Road, it immediately begins to ascend steeply through terrain that consists largely of broken rubble and eroding deposits that are a mixture of Morrison Formation, intrusive igneous material, and remnants of older deposits. The conical red sandstone peak adjacent to this junction is one end of a ridge of Wingate Sandstone that has been severely tilted by the formation of the La Sal Mountains. See the sidebar, LACCOLITHIC MOUNTAINS.

As the road climbs to higher elevations through the heavily vegetated slopes of the La Sal foothills, note on the mountain flanks the exposed areas bare of vegetation. All of these slopes, whether bare or overgrown, are a fractured rock that somewhat resembles granite but is actually *"porphyritic diorite,"* or intrusive igneous material that is crystalline in form. In some areas ahead, this granite-like rock has been moved from its natural erosional slopes by glacial activity. See the sidebar, GLACIERS OF THE LA SAL MOUNTAINS.

1.1 Stop at a pullout on the right and study the scene to the north and northeast. To the northeast, in Castle Valley, take note of the severely distorted strata where the Castle Valley salt anticline and the intruding La Sal Mountains have created a complex pattern of faulting and strata deformation. The geology of Castle Valley is explained in the log for the Castle Valley Road, elsewhere in this book. To summarize briefly here, the sheer walls and spires in the upper rims of the valley are Wingate Sandstone. The smooth, red-hued slopes below that are the Chinle and Moenkopi formations, in descending order. To the right of the valley, below and to the left of the spire of Castle Tower, the angled dark red strata that form a low ridge are Cutler Group sandstone. The red and white near-vertical cliffs in the lower end of the valley are an anomalous remainder of White Rim Sandstone, which is a member of the Cutler Group.

The conical peak in the valley is Round Mountain, which is the eroding remnants of an upthrust laccolith. The nearest cliffline on the right contains deposits younger than Wingate Sandstone, such as layered Kayenta Sandstone. Some slanting walls of Navajo Sandstone, with a few vestiges of still younger Entrada deposits, remain above the Kayenta, although most of these are largely covered by glacial deposits and remnants of Morrison Formation.

The extreme left end of the valley's western cliffline also has an exposure of red-hued Entrada Sandstone. The nearby heavily vegetated slopes are glacial debris covering distorted remnants of Morrison Formation deposits.

1.9 Note above the road on the left, the massive quantity of broken igneous rock that has been carried downward by landslides and past glacial activity. The vegetation on the mountain slopes has difficulty establishing a foothold because the cascading rock has formed "critical slopes," that is, slopes so steep that landslides occur easily and frequently. This intermittent movement of the rock tends to discourage the establishment of vegetation. It also makes hiking across such slopes quite hazardous, because even the added weight of a human can trigger a dangerous slide.

Note, to the right-front, a barren ridge of exposed igneous material. Beyond here, the road climbs a steep series of switchbacks.

3.0 From a pullout on the right at the summit of the switchbacks, note the stark ridge of igneous material protruding from the slopes of Pinhook Valley, which is an extension of Castle Valley. In 1881, a battle was fought in this valley between pioneer cattlemen and a band of Ute Indians. The clash resulted in several deaths, which have been memorialized by a marker in the valley that can be reached by a rough spur road, which is visible in the valley floor below. The slopes on the far side of Pinhook Valley are almost entirely glacial debris completely covering older deposits. Note how the Wingate and Entrada cliffs to the right disappear into this relatively recent deposit. Faulting parallel to the cliffline has hidden most of the strata that are normally between these two cliff-forming sandstones. For the next mile or so, exercise great caution while driving, because the road has little or no shoulder and no protection from very steep drops.

4.7 The ORV trail to Miners Basin goes left here. This spur trail and others noted later in this log are described in the referenced guidebook and map.

4.8 The graded road to the Pinhook Battleground goes right. Beyond here, the road again climbs to higher terrain, cutting through steep slopes of heavily vegetated glacial debris, accumulated soils and narrow strips of several formations. Note that the road has reached the populus plant community, with its dense stands of aspen trees wherever there is enough subsurface water.

5.7 A narrow pullout on the right here provides another overview of Castle Valley and the Colorado River gorge beyond. The geology of the Book Cliffs, on the distant skyline to the north, is described in the log for U.S. 191 North, elsewhere in this book.

6.1 The road that angles sharp left at this point goes to the U.S. Forest Service Cold Springs picnic area. Unless it has been recently trampled and defiled by grazing cattle, as often happens despite its recreational use, this springs area is lovely and is bedecked with a variety of wildflowers in the spring and early summer, among them wild iris.

At this point, the road has reached the Morrison Formation deposits through which it travels for the rest of its length. This prevalence of Morrison Formation on the flanks of the La Sals, high above its normal elevation, is a clear indication that this formation was deposited long before the La Sals thrust upward. When the range did form, some 24 million years ago, its central stocks further tilted sedimentary strata already distorted by adjacent salt anticlines, and its sills and laccoliths raised the relatively level strata between these anticlines to lofty elevations. For about the next ten miles, the road travels such an uplifted area of relatively level Morrison Formation deposits. Some of the broken rock in the soils near the road along this stretch was left there by the first three episodes of La Sal Mountains glaciation.

6.3 A pullout loop on the right here provides a parking area for another superb view to the north. Castle Valley and Richardson Amphitheater are immediately below, with the sheer cliffs of Dome Plateau just beyond the Colorado River. The long, elevated peninsula that walls the far side of upper Castle Valley is Adobe Mesa, and the one just beyond that is Fisher Mesa, which divides Professor and Fisher valleys. ORV trails that travel these lofty peninsulas are described in the referenced book and map.

The sloping strata and red Entrada Sandstone ridges of Arches National Park lie to the northwest. The Book Cliffs dominate the northern skyline. On clear days, the jagged ramparts of the San Rafael Swell are visible farther west in the distance. The immediate slopes above and below the road are the Brushy Basin Member of the Morrison Formation, although this deposit is largely obscured beneath dense vegetation and accumulated soils. The nearest rimlands of Castle Valley that are forested with the pinyon-juniper plant community are the Salt Wash Member of the Morrison Formation, with the slopes to the left of that ridge being heavily vegetated Tidwell Member of the Morrison Formation. Beyond this pullout, for a number of miles, the road travels through Brushy Basin-Morrison deposits that are glimpsed occasionally in road cuts and erosional slopes bare of vegetation.

8.8 The upper end of the Sand Flats ORV trail spurs right here. About a mile beyond this junction, the road drops into the Salt Wash-Morrison, although this interface is obscured by dense vegetation. After about another two miles, the road re-enters the Brushy Basin-Morrison. This change is also inconspicuous.

10.8 From a narrow pullout here, note the drainage to the right of the road cutting down through Salt Wash-Morrison deposits. Most of the natural water drainages along the slopes of the La Sal Mountains have long since been diverted into ditches and catchments for use in support of domestic livestock grazing and very limited agriculture. The road continues to travel through Morrison deposits that are obscured under heavy vegetation.

11.3 A steep graveled road goes left here to Warner Lake and its developed campground. While the road is not described in this log, and visiting it will require adjustment of mileages for the La Sal Mountain Loop Road, Warner Lake makes an excellent noontime picnic site, with its campground among tall aspen trees and the small lake set in green meadows that host colorful wildflowers in early summer.

Immediately beyond this junction the road drops steeply downward to cross the Mill Creek drainage. It descends through the Brushy Basin-Salt Wash interface before reaching Mill Creek. Beyond that drainage, the road climbs again through Morrison strata.

12.3 From a narrow pullout here, the broken vertical cliffs above the road to the left are the upper sandstone ledges of Salt Wash-Morrison. As it descends, the road continues traveling in this member.

12.7 The Loop Road crosses Mill Creek here, where a road spurs left toward Oowah Lake and its campground. This spur road also offers delightful picnic sites, but again, if this side trip is taken, subsequent mileages along the Loop Road must be adjusted.

Before leaving this junction, park at a safe place and walk to the bridge for a look at the inner gorge of Mill Creek, which cuts deeply into sandstone strata of the Salt Wash-Morrison.

Along the next half mile of road, note the deep, sheer-walled inner gorge of Mill Creek where it is visible through the trees, and the seeping springs in the slopes to the left of the road. This perennial seepage makes the road difficult to maintain, so it is frequently in disrepair. Most of the time, trickling streams flow from this slope in several places.

13.4 The road crosses the Horse Creek drainage. Just beyond the drainage, a faint foot trail left leads to a shaded, grassy slope above the creek. This makes a good picnic site that can be used while traveling the Loop Road without having to adjust mileages. The sloping sandstone terraces above this drainage are Morrison Formation deposits.

14.1 The Geyser Pass trail goes left here. From the vicinity of this junction, look down the drainage to the right. The sheer-edged mesa far below is South Mesa. The vertical rim of this jutting plateau is the Moab Member of Entrada Sandstone. The red sandstone slopes below that are the Slickrock Member of Entrada Sandstone. The more distant cliffline is the western wall of Moab Valley. Its jagged upper rim is Navajo Sandstone above terraced Kayenta deposits, with some Wingate Sandstone showing through the slopes of rubble. On clear days the Henry Mountains are visible on the skyline to the southwest. The road continues traveling through Morrison Formation deposits that are largely obscured by dense vegetation, with several peaks of the La Sal Mountains looming high to the southeast.

16.1 Along the next stretch of road, note the "painted-desert" Brushy Basin-Morrison deposits in the cuts to the right of the road. The gray, almost treeless ridge that the road traverses up ahead is the same deposit. The panorama to the south includes the Abajo Mountains and Elk Ridge highlands. The tree-covered slopes nearby and in the near distance are all Morrison Formation deposits. The westernmost of the ridges in the near distance is Black Ridge. Black Ridge Road, described elsewhere in this book, traverses that narrow, elevated peninsula. On clear days, the Henry Mountains are visible to the southwest, while the vast expanses of Canyon Rims Recreation Area, Canyonlands National Park and Glen Canyon National Recreation Area stretch across the intervening terrain.

18.4 Here, the road is traveling the gray ridge mentioned earlier. Note the smooth, painted-desert slopes of the Brushy Basin-Morrison. This ridge offers a good panoramic view of the La Sals. The ledgy deposits below this ridge on both sides continue to be Morrison Formation, largely the Salt Wash Member. As the road continues and begins to descend again, note how the Morrison strata have been distorted by the interaction between the Moab Valley salt anticline and the La Sal Mountains uplifting.

19.5 Stop at a wide pullout on the right here, where the road swings sharply left, and study the nearby rimlands of Moab Valley. Because the salt anticline that ultimately became Moab Valley thrust upward gradually and in stages, beginning almost three hundred million years ago, the deposits along its flanks are mixed by faulting or are thinner than in adjacent terrain. Further, relatively recent glaciers flowing downward from the La Sals have contributed their rocky debris in many places, further complicating the sequence.

In the rugged canyon that lies between this viewpoint and Moab Valley, the high rubbly slopes to the right are Salt Wash-Morrison. The light-hued broken sandstone cliffs below the Morrison rubble on both sides of the canyon are Moab-Entrada capped with thin remnants of reddish Tidwell Member of the Morrison Formation. The barren reddish sandstone slopes, ledges and spires on down the canyon are Slickrock-Entrada. Farther northwest, the lighter-hued rounded cliffs are Navajo Sandstone. The ridges and peaks to the left of this narrow drainage are topped with remnants of Salt Wash-Morrison and possibly some glacial debris. Farther left, on down in upper Moab Valley, the sloping hill is Brushy Basin-Morrison topped with a thin layer of undivided Dakota-Burro Canyon sandstone. For more information about Moab Valley geology, see the log for U.S. 191 South, elsewhere in this book.

To the south of this viewpoint, the low hills in the immediate vicinity are also Morrison Formation. The converging drainages are Pack Creek and Brumley Creek. The green cultivated area beyond the ridge that separates the two drainages is part of historic Pack Creek Ranch. During the third of the nine La Sal Mountains glacial episodes, a glacier descending the Pack Creek drainage stopped just 1/2 mile above where the ranch is now.

The uppermost part of the high ridge beyond the Pack Creek drainage is Mancos Shale topped with glacial gravel and a mixture of other decomposed deposits. Below the summit of the ridge, the protruding masses of light brown rock are undivided Dakota-Burro Canyon sandstone. The lower part of the drainage is entirely geologically recent glacial debris, gravel, decomposed rock, alluvium, soil and sand held in place by native vegetation.

Beyond the nearby sloping ridge, the horizontal, elongated plateau of Black Ridge is silhouetted against the distant Elk Ridge highlands. The microwave tower near the west end of Black Ridge marks the end of the Black Ridge Road, described elsewhere in this book.

Beyond this switchback overlook, the road drops steeply into the Brumley Creek drainage. The distorted Morrison strata also dive downward toward the ridge between the drainages, so the road continues to travel within the

Morrison until its junction with Pack Creek and Spanish Valley roads.

20.0 From the vicinity of a curve here, as the road straightens out for its final descent, note to the left several exposures of Moab-Entrada sandstone in the walls of the drainage.

21.0 This geology road log ends here at the junction with Pack Creek Road and Spanish Valley Road, but navigational instructions are given for returning to Moab via U.S. 191.

Pack Creek Road leads to a Forest Service picnic ground beyond the Pack Creek Ranch area. The La Sal Pass ORV trail continues beyond the picnic ground. That spectacular trail is described in the referenced book and map. To return to Moab, continue straight ahead at this junction.

25.9 Turn left here on a paved road to return to U.S. 191.

26.5 Turn right on U.S. 191 for the return trip to Moab.

34.5 Main and Center, Moab.

Warner Lake

LACCOLITHIC MOUNTAINS

According to one dictionary of geologic terms, a volcano is *"a mountain which has been built up by the materials ejected from the interior of the earth through a vent."*

But what if such molten rock rising through a weakness in the earth's crust were to meet so much resistance from massively thick layers of stratified rock that it could not break through the surface, yet still had pressure from below to relieve?

In that case, the molten rock would flow laterally between the rock strata, forming broad fans, or sills, somewhat like great spatulate leaves growing horizontally from a gigantic central stalk. A "sill" is defined as *"an intrusive body of igneous rock of approximately uniform thickness and relatively thin compared with its lateral extent, which has been emplaced parallel to the bedding of the intruded rock."*

Then, if the subterranean pressure were still not relieved, and further lateral extension were blocked by faulting or other barriers, these sills of molten rock would fatten, bulging upward, forming what geologists call "laccoliths," or "stone cisterns." Without this final upward bulging of their sills, such mountains would be called "intrusive igneous," but not "laccolithic."

Once a balance was reached between the pressures far below and those in the vent's central stock and branching laccoliths, the molten rock would slowly cool under this pressure, allowing the minerals in it to form crystalline structures not unlike those in granites, but not truly granite. Granite forms deep within the earth when sedimentary rock is subjected to heat and pressure over a long period of time.

Thus, in non-technical terms, laccolithic mountains are volcanoes that "died a-borning," they are mountains built by volcanic forces that were not great enough to break through the earth's surface, but that still raised great mountains of sedimentary strata above their central stocks and lateral intrusions, and then fattened those intrusions. They are mountains built by "intrusive" molten rock, rather than "extrusive" flows of lava.

As time passes after the formation of such mountains, erosion wears away the fractured rock on their tops and sides, gradually exposing the tops of their central stocks and even a few lateral structures. The igneous rock thus exposed has the appearance of gray rubble, which often forms great barren slopes so steep that trees and shrubs have trouble taking root on them, slopes where landslides can be, and have been, triggered even by the weight of a hiker.

Within the general Four Corners region there are seven, possibly eight, laccolithic mountain ranges. These are the Abajos and the La Sals, which are directly involved in Moab area geology, the Henrys, the Ricos, the Utes, the La Platas and the Carrizos. The laccolithic nature of Navajo Mountain is uncertain because not enough studies have been made to fully determine its structure. It may simply be intrusive, without the bulging sills that would define it as laccolithic.

The three laccolithic mountain ranges within southeastern Utah, those that are visible from the Moab area, formed at different times. According to the potassium-argon dating method sometimes used to determine the age of ancient rock, the Henry Mountains formed about 48 million years ago (MYA), the Abajos about 28 MYA and the La Sals about 24 MYA. While various geologists differ as to the exact ages of the ranges, these figures are good approximations.

The erosional cycle that exposed the cores of these laccolithic mountain ranges began about 10 million years ago, with most of this erosion occurring within the last 5 million years. During this period, almost 5,000 feet of the region's rock was removed, leaving the harder, more erosion resistant mountains looming high above the surrounding terrain.

Rain and snowmelt runoff from the La Sal and Abajo mountains, as well as from the Uncompahgre Uplift and the slanting rock strata on the northern nose of the Monument Uplift, has played a major part in the formation of the magnificent canyons, basins, plateaus and lowlands that are characteristic of the general Moab area.

In the Abajo Mountains, the laccolithic range that has contributed most to the shaping of the southern districts of Canyon Rims Recreation Area and Canyonlands National Park, there are four central igneous stocks, or columns through which molten rock rose from beneath the continental crust. The three most prominent are Shay Mountain, East Mountain (Abajo Peak) and West Mountain. Inconspicuous Johnson Creek Dome is the fourth, surrounded and obscured by the other three.

The parent stocks of East Mountain and West Mountain have been exposed by erosion, but only the laccoliths of the other two stocks are presently exposed. Geologists have estimated the total volume of the igneous rock intruded to form the Abajo Mountains to be 5.2 cubic miles.

The La Sal Mountains range has only three central stocks, but these are much larger, and its peaks are higher, than those of the Abajos. The three stocks are aligned approximately north and south, and each has a family of radiating dikes, sills and laccoliths. Each is also topped by a large dome of igneous material that spread laterally between sedimentary layers, leaving central structures something like gigantic mushrooms.

The northern dome of the La Sals is the largest of the three. Since it intruded between parallel faults that were part of the Castle Valley - Paradox Valley salt anticline, it was shaped by these faults into an oblong about 5 by 10 miles in size. Round Mountain, in Castle Valley, is one of the eroding laccoliths of the northern stock. The southern dome is smaller, at about 4 by 6 miles. Both of these stocks and domes are about the same age, and both elevated the terrain above them by about 6,000 feet.

The middle dome of the La Sals is smaller, at about 5 miles in diameter, and is believed to be somewhat younger. Although its central stock rose much less than those of the other two, its present peaks are the highest of the three, at nearly 13,000 feet. Its dome intruded into much younger deposits than did the other two.

The domes that spread from the north and south stocks penetrated Paradox Formation salt deposits that had been intruded upward in salt anticlines. The northern dome contacted salts associated with the Castle Valley - Paradox Valley salt anticline, while the southern dome met salts associated with the Moab Valley salt anticline. In each case, the intruding material thrust and squeezed the salts, which continued to flow and distort surrounding sedimentary strata for a long time, thus creating the jumble of tilted and faulted strata that are characteristic of the upper ends of these valleys.

The laccolithic mountains of the Four Corners region are unique within the United States. They have played, and are continuing to play, a vital part in shaping the magnificent topography of Canyonlands National Park, adjacent Canyon Rims Recreation Area and much of southeastern Utah's canyon country.

ROAD NAME: **SAND FLATS ROAD**

SUMMARY: The Sand Flats Road provides access to a popular sand-and-slickrock recreation area in the highlands between Moab and Castle valleys. While geologic variety in the Sand Flats area is largely limited to various erosional forms of the Glen Canyon Group, the road that ascends through the area toward the La Sal Mountains also provides outstanding views of more distant panoramas that are geologically diverse, as well as access to the more varied Sand Flats ORV trail that continues beyond the graded Sand Flats Road.

ROAD SURFACE: paved to the Slickrock Bike Trail, poorly-maintained gravel beyond there, with rough surface and one short, steep grade that is not always passable to standard highway vehicles.

VEHICLE: any vehicle, for the described trip, in dry weather.

MAP: ***Canyon Country* OFF-ROAD VEHICLE TRAIL MAP - Arches & La Sals Areas.**

GUIDEBOOK: ***Canyon Country* OFF-ROAD VEHICLE TRAILS - Arches & La Sals Areas**

MILEAGES:

TO START: 1.3 miles, from Main and Center, Moab.
ROAD: 8.9 miles, one-way to end.
RETURN: 10.2 miles, from end of described road.
TOTAL: 20.4 miles, from Main and Center, round trip.

TIME: about 1/2 day, plus time taken for view-stops and hikes. For a leisurely trip taking pictures along the way, a lunch stop and time for some hiking, allow a full day.

GEOLOGIC STRATA VISIBLE FROM THIS ROAD

Formation	Member	Page
Entrada Sandstone	Moab	Page 16
	Slickrock	
	Dewey Bridge	
Navajo Sandstone		Page 13
Kayenta Sandstone		Page 13
Wingate Sandstone		Page 12

ROUTE: East on Center Street, south on Fourth East, east on Mill Creek Road, then the Sand Flats Road up the hill past the cemetery.

NAVIGATION: Drive Moab city streets to the Sand Flats Road, then that road to where it ends, as described, at the water tanks on the ridge between Negro Bill and Rill canyons, then return via the same route. As an alternate, those with off-road vehicles can continue beyond the tanks on the Sand Flats ORV trail to the La Sal Mountain Loop Road and return to Moab via U.S. 191. The return on the La Sal Mountain Loop Road via Castle Valley is a longer option.

GEOLOGY ROAD LOG:

0.0 From the road start, where it branches from Mill Creek Road, the Sand Flats Road ascends steeply from Moab Valley past the cemetery, climbing through rubble of decomposing Kayenta and Wingate sandstones and traces of older strata that have been tilted, deformed and fractured by the Moab Valley salt anticline.

0.6 The road tops out of the grade at the turnoff to a firing range, in view of the La Sal Mountains to the east. The vertical sandstone cliffs below the mountains are Navajo Sandstone. The rubble slopes in the foreground are Kayenta Sandstone.

0.8 The road goes around a sanitary land-fill. The cliffs to the east are Navajo Sandstone. The foreground rock ledges are Kayenta Sandstone with dune sand in between.

1.4 The road starts up a short, steep grade at the base of the Navajo Sandstone cliffs. Note the accumulated dune sand at the base of the cliffs to the northwest. The dark broken rubble of a petrified playa is barely visible at the top of the cliff to the north. See the sidebar, PETRIFIED PLAYAS.

1.6 At a pullout on the right, pause for a view of Moab Valley from an intermediate level within the Navajo Sandstone, with the sanitary land-fill below. This site once won second place in a nation-wide contest for dumps in the most scenic settings. One in Alaska won first place.

The cliffs on the western skyline are the western monocline of the Moab Valley salt anticline. The uppermost rounded domes and fins in the cliffs are Navajo Sandstone. The horizontal layering immediately below is Kayenta Sandstone. The few vertical-walled abutments showing through the rubble are Wingate Sandstone. Below that, some Chinle Formation deposits are visible through the rubble. Beyond the drainage line to the left of the dump, note the rock strata warped by the Moab Valley salt anticline during its formation. See the sidebar, SALT ANTICLINES. Some of the slopes adjacent to that drainage have remnants of gravels deposited there by rapid stream runoff during earlier glacial periods. See the sidebar, GLACIERS OF THE LA SAL MOUNTAINS.

1.9 Here, the road is traveling through domes of Navajo Sandstone set among sand dunes that are partly immobilized by vegetation.

2.1 Note the cottonwoods growing to the right of the road, indicating permanent water due to subsurface springs, plus a drainage catchment created by road construction.

2.5 The road passes the beginning of the Slickrock Bike Trail, then angles right. The biking route can be seen, marked across the sandstone slopes and domes to the north and west, traveling on exposures of Navajo Sandstone. Beyond here, the road enters the "Sand Flats" proper, traveling through open meadows that are dune sand set with exposed weather-rounded domes of Navajo Sandstone. See the sidebar, ROCK EROSION.

3.0 Here, a distant panorama opens up. To the left-front on the skyline is Porcupine Rim. That rim constitutes one edge of Castle Valley, another salt anticline. The rugged drainage of Negro Bill Canyon lies between the road and the rim, but is not yet apparent. The Porcupine Rim area is largely Kayenta Sandstone with a few remaining domes of Navajo Sandstone. One dome on the skyline of Porcupine Rim is topped by reddish remnants of Entrada Sandstone. Within the Negro Bill Canyon drainage, the rimlands are Kayenta Sandstone, with a few exposures of upper Wingate Sandstone visible as vertical cliffs. Straight ahead, the same formations are on the skyline, but more heavily vegetated because of the higher elevation. The mesa below the La Sal Mountains is Wilson Mesa and its associated unnamed rimlands. The uppermost rims of these tablelands are the Moab Member of Entrada Sandstone. The bare sandstone slopes below that are the Slickrock Member of the Entrada. The lower Dewey Bridge Member of the Entrada is visible as darker red rock on top of the lower rounded domes of white Navajo Sandstone in the near-foreground.

4.3 Note the beautiful cross-bedding and vertical striations in the Navajo Sandstone cliffs to the left of the road and for the next half mile. Note also the large cave at the base of the cliff that was created by water seeping from potholes above. Note the flat, water-affected layer at the top of the highest fin. This is a remnant of a "petrified playa." Some playa slabs found in this vicinity, since collected by paleontologists for study, have had numerous prehistoric tracks in their petrified surfaces. Some of the dune sand in this vicinity has been held in place by vegetation long enough to begin changing it to soil, that is, dune sand held with humus from decayed vegetation. In many areas, this soil-building process is enhanced by microscopic cryptogams. See the sidebar, CRYPTOGAMIC SOILS.

4.8 Here, the road climbs steeply between Navajo Sandstone fins. This short stretch of road is rough and steep, but can be driven by highway vehicles except when too eroded, wet or icy. All vehicles should be driven cautiously here. The road beyond the grade is better, but quite rough.

5.2 The road tops out of the steep grade and enters another stretch of terrain dominated by Navajo Sandstone domes and fins exposed among brushy flats and meadows of dune sand and decomposed rock.

5.4 Note the lovely angled cross-bedding in the Navajo Sandstone domes to the left here and for the next quarter-mile.

5.8 From the crest of a low rise, the large, rounded domes to the south are Navajo Sandstone. The major drainage between the domes to the west is Sandfork Canyon, which drains from the Sand Flats into the

North Fork of Mill Creek Canyon. For the next mile or so, the mountains loom above the Entrada Sandstone cliffs of Wilson Mesa, South Mesa and smaller adjacent unnamed tablelands.

6.3 Here, the road crosses one upper drainage of Sandfork Canyon. An interesting hike down this drainage ends at a spectacular pour-off. A few yards ahead, an inconspicuous ORV trail goes right, toward the rim of the North Fork of Mill Creek.

6.6 An ORV trail angles right here to rimland views of North Fork Mill Creek and Rill Canyon. The road continues from here through terrain dominated by exposures of Navajo Sandstone set among dune sand and decomposing rock.

7.6 The sandstone mesa to the left is topped by a horizontal layer of gray-colored petrified playa sediments. Some such deposits in this area contain petrified wood and petrified algae.

8.0 From the summit of a grade, the drainage of upper Negro Bill Canyon comes into view on the left. The bare sandstone exposures on each side of the canyon continue to be Navajo Sandstone, with ledgy Kayenta Sandstone below that and rimming the canyon. The highest dome on the skyline to the left is topped by the horizontal layering of dark red Dewey Bridge Entrada, with a smaller remnant of Slickrock Entrada above that. The high cliffline toward the mountains ahead is Slickrock-Entrada topped with Moab-Entrada and thin remnants of Morrison Formation deposits.

8.8 The deep inner gorge of Negro Bill Canyon comes into full view here. The vertical inner-canyon walls are Wingate Sandstone, with the vegetated slopes above being Kayenta Sandstone. The off-road vehicle trail visible below is a spur of the Porcupine Rim Trail that travels near the rim of Negro Bill Canyon for several miles.

8.9 Here, the described road ends at a pullout. Note the nearby spring-fed water tanks with grassy slopes above, indicating underground water-flow that has reached the surface on these slopes. Navajo Sandstone is a relatively porous rock and can absorb and hold large amounts of water. Most such water comes from slowly melting snow, because much of the scant rainfall runs off from this terrain too quickly for it to be absorbed into the rock. In this location, the water seeping from Navajo Sandstone has reached the less permeable Kayenta-Navajo interface and has traveled horizontally to appear on the surface of this slope. Because of the slope of the strata in this area, the spring may actually drain several square miles of higher terrain.

From the pullout, the skyline panorama to the south begins on the left with a dome of Entrada Sandstone. Next are rimlands of Moab-Entrada above slopes of Slickrock-Entrada. The projecting point is South Mesa. In the distance are the Abajo Mountains near Monticello, with the cliffs to the distant-right being the western wall of Moab Valley. The rounded domes silhouetted against the Abajos are the Navajo Sandstone of "Behind-the-Rocks," the area beyond the cliffs that wall Moab Valley. To the left of that is Black Ridge. The microwave tower near its western end is visible. The deep canyon below the pullout to the south is upper Rill Canyon, whose inner gorge is Wingate Sandstone below ledgy slopes of Kayenta Sandstone.

The skyline to the north is the same Porcupine Rim described earlier. The deep canyon is upper Negro Bill Canyon with its sheer Wingate Sandstone walls. The road here is in the Kayenta Sandstone, just below its interface with the Navajo.

Negro Bill Canyon is fed along its length by numerous springs, and is either a flowing or intermittent stream most of the year in its lower several miles. Rill Canyon has fewer springs, most of which are in its extreme upper end. In the winter these seeping springs create immense cascades of ice, or "ice falls," that often last well into the warmer season.

Note here that, while the Sand Flats Road has been traveling within or near the base of the Navajo Sandstone from the scenic dump on, the road has gained about 1,400 feet in elevation. This illustrates how the strata in this area have been severely tilted by the upthrusting of the La Sal Mountains. See the sidebar, LACCOLITHIC MOUNTAINS.

Highway vehicles can continue beyond this pullout for about 3/4 mile before the trail gets rough and steep, but should not attempt to go beyond a safe turn-around point. For the geology beyond this pullout, refer to the log for the Sand Flats Trail, described elsewhere in this book.

ROCK EROSION

Rock erosion sounds like a simple subject. It is -- if only one kind of rock is involved, if that rock has not been affected by any kind of tectonic activity, and if it is exposed to mild climate and weather.

In canyon country, however, many kinds of rock and unlithified sediments are involved. Further, the region has been drastically affected by several kinds of distorting tectonic activity, and both the climate and weather are extreme. Thus, it would require an entire book to fully explain the wide variety of rock erosion modes encountered in the region.

Even so, since rock erosion has created the complex assortment of topographic forms that make the region so beautiful and geologically unique, it is necessary at least to summarize the variables that affect the process, and offer a few useful generalities.

The major variables in rock erosion are: 1) the nature of each deposit of rock itself, 2) the physical configuration of the rock exposure, 3) the amount and type of tectonic activity to which the rock has been subjected, 4) the nature of the rock above and below each deposit, and 5) the climate and weather to which the rock is exposed. With this many primary variables, plus many other lesser factors within these, the complexity is almost endless.

However, rock erosion is the "name of the game" in canyon country. It is the fundamental force that has shaped the region's spectacular topography over the past several million years. In order to gain a basic understanding of the region's multitude of erosional landforms, it is necessary to have at least a rudimentary grasp of the underlying principles of its development, through the interaction of various kinds of tectonically distorted geologic deposits, the pull of gravity, and the primary elements of weather -- water and, to a very small extent, wind -- over millions of years of time.

As noted above, there are a number of major variables that affect how a type of rock erodes:

Variables in the nature of the rock include: 1) its degree of lithification, 2) its particle-size assortment, 3) its hardness and chemical composition, 4) its manner of deposition (by wind or water), 5) how it was affected by various environmental factors after deposition, 6) the nature of the ground water that saturated it before and during lithification, 7) the amount of life in its depositional environment and, 8) its thickness and uniformity.

Variables in the physical configuration of the rock exposure include: 1) exposure of its upper surface to slow weathering, 2) the angle of the exposure, 3) exposure of a cross-section to undercutting by flowing water, 4) whether or not this includes continuous water flow (a river or stream), intermittent water flow (flash-flooding), or contact with standing water (a reservoir), 5) whether the exposure is due to human activities (such as highway cuts that create fractured, unnatural exposures), 6) the amount of freezing and thawing to which the surface is exposed, 7) the amount of exposure to water runoff and atmospheric humidity, 8) whether the surface is exposed to seeping water, 9) whether the deposit is undercut, and 10) how much of the rock is exposed to these configurations. In general, upper surfaces of rock strata erode more slowly than their cross-sections, because flowing water in a drainage is a more violent force than surface precipitation, especially if the flowing water can undercut rock walls, thus exposing them to the pull of gravity.

Variables in subsequent tectonic distortion of a geologic deposit include faulting, jointing and tilting, which are caused by the formation of anticlines, synclines, uplifts, mountain building, general regional tilting and, on rare occasions, by the impacts of meteors. In canyon country, geologic strata that remain as horizontal as they were deposited are rare, yet any kind of fracturing or tilting accelerates rock erosion.

Variables in the rock strata above and below include softer or harder overlying or underlying rock, and combinations of these two variables. Softer rock above a particular deposit exposes its upper surface to various kinds of erosion. Softer underlying rock undercuts a deposit, accelerating its collapse from the pull of gravity. Harder overlying rock protects a deposit and delays various kinds of surface erosion. Harder rock below usually delays undercutting, but when the strata are tilted, it promotes undercutting of a softer deposit.

Variables in the climate and weather to which the rock is exposed include: 1) the type and amounts of precipitation, 2) the general humidity, 3) the extremes of temperature during a day and a season, 4) the rate of water and snow-melt run-off, and 5) variations in the climate over a long period of time. These factors in turn affect the amounts and types of vegetation that thrive in the region, and these variables strongly affect erosion rates. Human activities that adversely affect natural vegetation communities thus increase erosion correspondingly.

All of the above factors, plus the nearly infinite ways they interact, endlessly complicate the subject of rock erosion, yet by understanding the nature of a particular rock deposit, the type of exposure it generally receives, and the elements to which it is exposed, then examining the various types of erosion that can affect it, a general understanding of the nature of its erosion is possible to grasp. Conversely, once it is known how a particular type of rock erodes, it is possible to use this knowledge to help identify a geologic formation.

There are six general types of rock erosion, plus combinations of these. The general types include: 1) surface weathering, 2) chemical exfoliation, 3) surface exfoliation, 4) collapse from the pull of gravity, 5) stress-relief exfoliation, and 6) stress-pattern exfoliation.

Surface weathering means erosion by the elements of weather and climate -- rain, snow, atmospheric moisture, heat, freezing, vegetation growth and decomposition, and transient water runoff. The rate of weathering in turn depends on the composition and hardness of the rock, its exposure, and the manner in which it has been fractured or tilted locally by tectonic activity. Surface weathering is generally a slow but steady process, with occasional exceptions.

Chemical exfoliation is similar to surface weathering. "Exfoliation" means the falling away of relatively thin slabs or "leaves" of rock. Chemical exfoliation is caused by the slow removal, by seeping water, of the minerals that bind the silica granules of sandstone together, where the rock surface is not exposed to surface water, such as in caves and overhangs. Evidence of this process is a crust of white minerals and the scaling away of thin layers of rock, both found on some rock surfaces sheltered from the elements. This type of erosion can also be observed above the waterlines of

continued ----

sandstones exposed to standing water, such as in reservoirs. High atmospheric moisture can produce the same effect, but this condition is rare in high-desert canyon country. Massive aeolian, or partly aeolian, sandstones such as Navajo, Wingate, Entrada, Cedar Mesa and White Rim are all subject to chemical exfoliation.

Surface exfoliation means the scaling away of rock surfaces in relatively thin slabs, usually from horizontal or gently sloping exposures of the massive sandstones listed above. This process has two different causes. 1) Water that has seeped into porous rock for a few inches can freeze. This expands a thin slab of surface rock, causing it to separate from the underlying rock in the form of a surface "blister." 2) The expansion of rock surfaces from extreme heat can also cause such blistering. Once such thin sections of rock have separated from the underlying rock, they easily fracture into smaller pieces and erode away. Hikers crossing exposed upper surfaces of sandstones prone to this form of exfoliation will notice such loosened surface slabs in all stages of collapse and erosion, including some that are still in the form of unbroken rock blisters, often several feet in diameter. These are indicated only by the hollow sound they give when crossed on foot.

Rock collapse from the pull of gravity is common and a relatively simple and uncomplicated process. It depends largely on the strength of the various kinds of undercut or tectonically fractured rock, and the elements or conditions that have produced these. Rivers, streams, and flash floods undercut rock in cliffs and slopes, especially in narrow entrenched meanders, with most of the undercutting taking place on the outsides of meander curves. When undercut, massive aeolian sandstones shear away on near-vertical planes. If the cliffline is fairly straight, the rock shears away almost vertically. If the cliff has a concave curve, the collapse tends to have an undercutting slope. If the cliff curve is convex, the shearing away tends to produce an outward slope. This kind of erosion causes the countless miles of sandstone clifflines in canyon country slowly to recede, eventually leaving temporary thin walls and spires of rock, then terraces or mounds of underlying softer deposits, or ledges of harder deposits. Tectonically fractured rock, or rock broken by freezing water or by human activities, tends to be pulled down by gravity far more quickly than rock that has retained its original integrity and configuration.

Stress-relief exfoliation is also a fairly simple erosional process that combines the action of stress-cracking, water erosion and gravity collapse. In canyon country, it is limited to massive sandstones, such as Navajo, Wingate, Entrada, Cedar Mesa and, to a minor extent, White Rim Sandstone. When such very thick sandstones lithify, deep beneath the surface, they do so under pressures created by the weight of thousands of feet of heavy overburden. These pressures are retained internally by the rock. When the rock is eventually exposed on the surface, these internal pressures tend to be relieved by jointing, with the planes of these cracks being parallel to the exposed rock surface, whatever its angle. These slabs of slightly detached rock vary in thickness from a few inches to as much as twenty feet or more, depending upon the nature of the rock at a particular locality. If the exposed surface of the rock is undercut, the loosened slab of rock tends to collapse immediately. If the surface is vertical or angled back from its base, the loosened slab tends to stay in place until it is broken into smaller pieces. After such stress-relief joint-planes form, they are readily penetrated by surface water, with subsequent expansion from freezing, which accelerates the further fracturing and collapse of the slabs that remain in place, usually in relatively small sections. This major erosional process can be readily observed wherever there are near-vertical exposures of massive sandstones. Some jughandle-type arches are slender remnants of such exfoliated slabs of sandstone.

Stress-pattern exfoliation is a different process entirely. Although it is also based on internal rock stress, the origin of the stress is different, and the result is more localized. When a limited stretch of one of the massive sandstones is undercut slowly, by stream flow or by chemical exfoliation caused by seeping water, the rock above the undercut stretch often has time to adapt to the changing internal stress without collapsing on a near-vertical plain. The weight of the rock above the undercut section is redistributed, with the bridging arc of rock transferring the weight of rock above the undercut stretch to the rock on each side, in much the same manner as the weight of an arching highway bridge is borne by its abutments. Subsequently, the rock tends to break away in relatively thin, curved slabs within the arc of the opening, and parallel to the arcing internal stress-pattern. This slow process gradually enlarges the undercut cave, and changes its ceiling configuration toward an ideal stress-arc that is close to being a half-circle. With such an arc, the rock layers below the stress-arc are no longer under tension and thus are not so subject to easy fracture. Once the ceiling of such a cave has reached this near-half-circle shape, stress-pattern exfoliation slows and virtually stops, although in most locations, chemical exfoliation continues. Natural arches, bridges, windows and tunnels that form in massive sandstones also follow this pattern of development whenever the shape of the surrounding rock permits.

Famous Rainbow Bridge is a prime example of this last type of erosion. Much of its soaring arc has reached the point of stability, although the magnificent span is now endangered by the encroachment of reservoir water beneath its abutments, another example of humanity as a geologic force. The fascinating history of this encroachment, and an explanation of the endangerment, are summarized in the book, ***Canyon Country* ARCHES & BRIDGES.**

SOME USEFUL GENERALITIES

The shape of an exposed rock formation is a useful indicator of its identity, because each formation erodes into a characteristic configuration:

When undercut, massive sandstones such as Navajo, Wingate, Entrada, Cedar Mesa and White Rim tend to shear away to form near-vertical cliffs, and when their upper surfaces are exposed, they tend to form rounded mounds. If jointed by tectonic activity, as in the vicinity of an anticline, they tend to form thin fins on their upper surfaces, and colonnades on vertical surfaces. These sandstones make good aquifers in that they are relatively porous and can hold considerable water. Because of this, they are prone to form seep-caves and various kinds of natural rock openings. These can and do form in most other rock formations, but with far less frequency. Aeolian sandstones have relatively coarse, wind-sorted sand grains, with the finer granules blown away or collected in such places as desert playas. This granular coarseness leaves microspaces in the rock in which soluble minerals collect during its lithification, and which subsequently hold water as an aquifer.

Layered water-deposited sandstones such as Hermosa, some Cutler, Moenkopi, Kayenta, Dewey Bridge-Entrada, Salt Wash-Morrison, Dakota, Mesa

continued ----

Verde, Wasatch and Green River, tend to erode into terraces, with their harder layers undercut by erosion of their intermittent softer layers to form jutting ledges. When their upper surfaces are exposed, these are relatively flat and weather slowly to form loose sand, rough surfaces and, with the help of vegetation, sandy soils. Water-deposited sandstones tend to have a wide variety of sand and sediment granules, and this helps create harder rock with less mineral content and diminished ability to hold water.

Limestone strata, such as desert playa deposits in Navajo Sandstone and the shallow-marine deposits that occur in Lower Cutler and upper Hermosa strata, are even harder and more resistant to surface erosion. When exposed, by the loss of less durable overlying deposits and with this process enhanced when the strata are tilted, these limestones tend to resist erosion and thus create broad terraces and benchlands, with the limestone cracking and weathering away very slowly. Desert playa limestone layers in Navajo Sandstone tend to protect underlying aeolian deposits, and help to undercut overlying sandstone, thus creating truncated domes topped with this type of limestone. Erosion-resistant, shallow-marine limestone has formed durable rimlands and broad terraces adjacent to the inner gorge of the Colorado River, downriver from Moab Valley, where the strata in the vicinity have been tilted by two anticlines. The Potash and Chicken Corners ORV trails, described in this book, travel these rimlands.

Softer, unlithified formations, such as Chinle, Tidwell-Morrison, Brushy Basin-Morrison, Cedar Mountain, Burro Canyon and Mancos Shale, tend to erode into smooth slopes that are interrupted by occasional thin sandstone ledges. If their upper surfaces are exposed, and not protected by a sandstone layer, they erode into rounded hills of soft, soil-like sediments. The gray or color-banded hills and slopes within sight of the northern 20 miles of U.S. 191, parts of Utah 128, and all of Interstate 70 between Grand Junction, Colorado, and Price, Utah, are examples of this.

The shapes and configurations of canyon walls are determined by the kind of rock deposits in them. Massive sandstones make near-vertical cliffs that are undercut in places by seeping water, while water-deposited formations make terraced, ledgy cliffs that taper outward from the canyon bottom. Most of the deeper canyons have combinations of deposits, so their cliffs are accordingly varied. This endless canyon-wall diversity lends equal variety to the aesthetic beauty of canyon country scenery.

All canyon country clifflines and their terraced bases, whether in canyons or defining lofty buttes and peninsulas, are slowly receding due to a combination of erosion modes. This inevitable retreat of clifflines has been, and still is, a major factor in the creation of regional topography, and the greatest component in this retreat is the alternating hard-and-soft nature of the region's rock deposits. Without this variation, canyon country would not exist in its present form, and would very likely have few if any of the spectacular canyons after which it is named.

Major changes that affect rock erosion have occurred during extensive alterations in the region's climate. These have taken place many times in the recent geologic past, and even several times within the last thousand years. During periods of cooler climate, the region receives more precipitation, including enough snow in the higher elevations to produce glaciation. The increased moisture accelerates surface weathering of rock but promotes the growth of lichens on some kinds of rock, thus slowing weathering. By enhancing the growth of vegetation which slows water runoff, the wetter climate also drastically reduces erosion from flowing water. This in turn allows the build-up of alluvium in canyons, which protects the bases of their cliffs from undercutting, thus slowing the main erosion mode that has created canyon country as it is now. In the La Sal Mountains, the cooler temperatures produced massive glaciers, which eroded rock by abrasion and transported great quantities of rock to lower elevations, where it was deposited. The upper ends of Moab Valley and Castle Valley have countless tons of such glacially transported rock that obscures the original rock formations and now protects them from normal erosion.

Wind plays a very small part in the process of rock erosion although, in the past, various geologists concluded that wind was a major factor in the shaping of canyon country sandstones. Historical accounts frequently mention the importance of wind in the formation of the region's rounded domes, fins, caves and numerous natural rock openings -- its arches, windows and tunnels. Subsequent research and study, however, has since discredited the myth that wind shapes rock, and has shown that the principal factor is water -- falling water, running water, seeping water, freezing water -- with wind playing the limited part of locally transporting some of the sand and finer rock granules loosened by water. Any major transporting of loosened material was also done by water. Within the last several million years, water has eroded and carried from the canyon country region about 5,000 vertical feet of decomposed rock. Although the idea of rock abrasion by wind-carried sand is attractive, tests have shown that rock particles that are heavy enough to do this rarely rise more than a few inches above the ground, even in the strongest winds, and that the lighter particles borne higher by wind have very little abrasive power. The myth of wind as an agent of rock erosion is explored further in the book, ***Canyon Country* ARCHES & BRIDGES.**

GEOLOGY TOURS - GRAVELED ROADS

ROAD NAME: **ANTICLINE OVERLOOK ROAD**

SUMMARY: Anticline Overlook Road goes to one of the most outstanding overlooks in immense Canyon Rims Recreation Area. Along the way, the road offers broad panoramas of high-desert meadowlands set with domes and walls of colorful sandstone, as well as a variety of erosional landforms. At several points, pullouts afford spectacular views into the geologically-complex lowlands of Canyonlands National Park and Canyon Rims Recreation Area that lie to the west of the immense Hatch Point peninsula. At the end of the road, developed Anticline Overlook is one of the most outstanding viewpoints in the entire area, with its elevated look at the colorful and intricate geology of the Colorado River gorge, mid-Cane Creek Canyon and the arched strata of the Cane Creek Anticline, with other major geological features visible in the distance.

ROAD SURFACE: graded and graveled.

VEHICLE: any vehicle.

MAPS: ***Canyon Country* OFF-ROAD VEHICLE TRAIL MAP - Canyon Rims Recreation Area,** or ***Canyon Country* OFF-ROAD VEHICLE TRAIL MAP - Canyon Rims & Needles Areas.**

GUIDEBOOKS: ***Canyon Country* OFF-ROAD VEHICLE TRAILS - Canyon Rims Recreation Area** or ***Canyon Country* OFF-ROAD VEHICLE TRAILS - Canyon Rims & Needles Areas.** The book, ***Canyon Country's* CANYON RIMS RECREATION AREA,** provides an overview of this public lands recreation area and its many prehistoric, historic, scientific and recreational resources.

GEOLOGIC STRATA VISIBLE FROM THIS ROAD

Entrada Sandstone	Moab	Page 16
	Slickrock	
	Dewey Bridge	
Navajo Sandstone		Page 13
Kayenta Sandstone		Page 13
Wingate Sandstone		Page 12
Chinle Formation		Page 12
Moenkopi Formation		Page 12
Cutler Group	White Rim	Page 11
	Organ Rock	
	Cedar Mesa	
	Lower Cutler	

MILEAGES:

TO START: 49.0 miles, from Main and Center, Moab.
ROAD: 18.3 miles, one-way to end.
RETURN: 67.3 miles, from end of described road.
TOTAL: 134.6 miles, from Main and Center, round trip.

TIME: Allow for a full day to drive to Anticline Overlook and back to Moab. This will leave time for visiting nearby Needles Overlook, too. The trip will be a little shorter if the Hatch Wash Road cutoff is taken, with the geology of the missed part of the Needles Overlook Road studied on the return trip.

ROUTE: Main Street (U.S. 191) south from Moab, Needles Overlook or Hatch Wash Road west, then Anticline Overlook Road north.

NAVIGATION: Drive south from Moab on U.S. 191. If taking the Hatch Wash Road shortcut, turn southwest on that road, west on Needles Overlook Road to its junction with Anticline Overlook Road, then north on that. If taking paved Needles Overlook Road, continue south on U.S. 191 to that junction, then west to Anticline Overlook Road. Follow the mileages and directions in each of these road logs for details.

GEOLOGY ROAD LOG:

0.0 This road is in the Hatch Point District of Canyon Rims Recreation Area. As it branches from paved Needles Overlook Road, this well-graded graveled road heads northward across the open rolling terrain of a northern extension of Hatch Point that is defined by Hatch Wash Canyon, Cane Creek Canyon and the lower terrain to the west. In the past, once the native grasses were destroyed, these monocultural sage flats were chained in an attempt to improve grazing for domestic livestock. This expensive "vegetation manipulation" failed. The native sage returned but the destroyed native grasses did not.

From this road for most of its length, the La Sal Mountains dominate the eastern horizon, with occasional glimpses of the Entrada Sandstone bluffs of the Cameo Cliffs recreation area below the mountains. The red sandstone outcrop to the right of the road as it starts is the Slickrock Member of Entrada Sandstone. It is called Eightmile Rock and has a cattle support development at its base. The road to Eightmile Rock goes right just ahead, but is not included in this description. In the distance ahead, the cliffs of Island-in-the-Sky dominate the horizon. The bald rock knob to the left-front at this junction is called Horsehead Rock. It is a remnant of Navajo Sandstone.

For the next several miles, the road travels through the upper layers of Navajo Sandstone and lower Entrada deposits, but neither deposit is visible much. The red hue to the soils where they are exposed by erosion derives from decomposed Dewey Bridge Member of Entrada Sandstone, which is dark red. The brushy meadows are sand and decomposed rock held in place by remnants of native vegetation. Watch for pronghorn antelope along this road, primarily toward the east.

0.1 Eightmile Road goes right here. This graded dirt road provides access to several Hatch Point ORV trails that are described in the referenced ORV books and map about Canyon Rims Recreation Area.

2.4 From a stretch of road in this vicinity, a lower terrace of Hatch Point is visible to the left. The light-hued rock domes there are Navajo Sandstone. Several ORV trails documented in the referenced literature descend onto this lower terrace.

3.3 From the crest of a low rise, note the massive amounts of Navajo Sandstone slickrock visible to the right-front. The nearer canyons and slopes are on Hatch Point, while the more distant fins and slopes are in the Behind-the-Rocks area that lies just west of the cliffs above Moab Valley. The Cane Creek Canyon Rim ORV trail, described elsewhere in this book, provides a more detailed description of the geology of Behind-the-Rocks. The deep and spectacular gorge of Hatch Wash Canyon lies between the near and far sandstone masses, but is not visible from here. To the left of the road for the next half mile, the low rounded hills are remnants of Navajo Sandstone largely covered with decomposed rock and native pinyon and juniper trees.

5.3 Park at a wide pullout on the left here and walk a few yards for an excellent view of the terrain to the west. This viewpoint provides a good look at the lower terrace that rims Hatch Point in this vicinity, and the ornately eroded cliffs and domes of Navajo Sandstone that loom above its brushy meadows. From here, note the darker red bands of sandstone in the Navajo that mark its uppermost levels. This color-banding represents an area where Navajo dune sands were interlayering with the deposit that was to become the Dewey Bridge-Entrada. Such interlayering is fairly common at the interface between distinctly different deposits. In this particular case, however, the Navajo Sandstone deposits may have eroded for a long period of time before being completely covered by Dewey Bridge-Entrada in this area, and the Carmel Formation farther west, but there was obviously some reworking of the upper level of Navajo as the first deposits of the Dewey Bridge-Entrada were encroaching.

From this rim viewpoint, Island-in-the-Sky with its sheer Wingate Sandstone cliffs is prominent in the west, and on clear days the Henry Mountains loom above the southern tip of that peninsula. The nearby prominence below the Henry Mountains that has the gap in its cliff-line is part of Hatch Point. That gap is closely approached by an ORV trail described in the referenced Canyon Rims Recreation Area books and map. The viewpoint at the tip is called Lockhart Overlook. For a more complete description of geologic strata visible in the lowlands from here, refer to the description from

Needles Overlook, at the end of Needles Overlook Road. Just beyond this pullout, the Trout Water Canyon ORV trail spurs right.

6.1 Another wide pullout on the left provides a second look at the lower terrain to the west.

8.1 The large mass of white slickrock to the left of the road is Navajo Sandstone topped by the relatively flat surface of a water-affected layer that probably marks its ancient erosional surface. To the left-front in the distance, terraced exposures of Wingate and Kayenta sandstones mark the rimlands of a developing canyon that cuts into Hatch Point, creating a jutting peninsula. The canyon was formed by erosion along a syncline that extends from here to the far side of the Colorado River gorge. Notice how all the terrain in the immediate vicinity is sloping toward the centerline of the syncline, thus channeling the water run-off that created the canyon. Undeveloped Canyonlands Overlook is at the end of an ORV trail that travels to the westernmost tip of that spur peninsula. That trail is described in the referenced literature but not in this book. Straight ahead, the cliffs of Island-in-the-Sky and Dead Horse Point dominate the skyline, with the Book Cliffs in the distance to the north.

8.7 The graveled road to the right here goes to the developed Hatch Point Campground.

9.5 As this stretch of the road crosses wide meadowlands, note the gentle slopes of Navajo Sandstone toward the mountains to the east. Note also the several large seep-caves at the base of the nearest sandstone slope. The Hatch Point Campground is located in the trees above those bare slickrock slopes. The spectacular fins and domes of the Behind-the-Rocks wilderness area continue to dominate the horizon to the northeast.

10.4 Notice to the left the upper rimlands of the canyon described earlier. At this point that canyon is several hundred feet deep.

11.4 From along this stretch of the road, the deep canyon described earlier becomes visible to the left. The layered rimlands of that canyon are Kayenta Sandstone, while the vertical cliffs and gently sloped slickrock rimlands are Wingate Sandstone. On this side of the canyon, the nearby wall of lighter sandstone is Navajo. Notice to the right how the broad open meadows walled with Navajo Sandstone are sloping downward toward the east, eventually to drain into Cane Creek Canyon via Trough Springs Canyon. Hiking these color-banded Navajo Sandstone slopes is described in the book, ***Canyon Country* SLICKROCK HIKING & BIKING.** Hiking provides an intimate look at such intricately eroded sandstone masses.

11.8 From a pullout loop to the left, the solar evaporation ponds of a mineral refinement operation beyond the Colorado River gorge detract from the magnificent geologic beauty of the scene. See the sidebar, POTASH. The sheer-walled point in the near-distance to the left of the ponds is the northernmost tip of the peninsula formed by the deep canyon noted earlier. The convoluted Wingate Sandstone cliffs of Island-in-the-Sky soar above the ponds. The crenelated dark red cliffs and terraces on this side of the river gorge are Cutler Group deposits. In ascending order, the slopes above the Cutler are Moenkopi, gray-tinged Chinle Formation, then the sheer Wingate cliffs topped with layered Kayenta Sandstone. Straight ahead and to the left-front, isolated domes and one low ridge of Navajo Sandstone loom above the meadows. The flat-topped dome there is known as The Flatiron. As the main road continues, it more closely approaches that dome.

12.0 Notice how the natural terrain to the left of the road has been severely blemished by an oil-drilling operation. This destructive use of Canyon Rims Recreation Area continues even though recreation planners have recommended that all destructive and conflicting practices be phased out in this outstandingly beautiful and geologic diverse area of public land.

13.3 More of the cliffs and lower terrain to the west are visible from this stretch of road. Notice in the distance to the left-front the immense barren patch below the Navajo Sandstone bluff. That scarred terrain was left by another oil-drilling operation decades ago, but the natural vegetation has barely begun to recover. Such disturbances illustrate the destructive results of human activities on such fragile high-desert terrain. See the sidebar, HUMANITY AS A GEOLOGIC FORCE. There are other such barren drill sites on this point ahead, to the right of the road. All but one are out of sight of this road, however.

14.4 Stop at the northern end of the wide pullout on the left here. In the Navajo Sandstone bluff to the left-rear, the opening of The Wine Glass natural arch is visible against the skyline. Cork Arch is in the same cliff-line this side of the Wine Glass but is not visible from here. The long ridge of Navajo Sandstone to the east against the mountains provides challenging slickrock hiking, which is described in the book noted earlier. To the northeast, the light-hued Navajo Sandstone fins of Behind-the-Rocks are on the skyline. Below them, the sheer Wingate Sandstone walls of mid-Cane Creek Canyon are barely visible beyond the nearby meadows. The Cane Creek Canyon Rim Trail, described elsewhere in this book, travels that spectacular rim on the opposite side of the canyon from Hatch Point. Straight ahead, another low butte of Navajo Sandstone looms above the rolling meadows.

15.1 Turn left on the road to Minor Overlook, which is on the far side of the low Navajo Sandstone butte that the road approaches.

15.7 Minor Overlook was named after well-loved Dave Minor, a recreation planner for the Bureau of Land Management who died in 1991. The viewpoint was dedicated by a special ceremony in the spring of 1992. The books ***Canyon Country's* CANYON RIMS RECREATION AREA** and **CAMEO CLIFFS - Hiking - Biking - Four-Wheeling** were also dedicated to this outstanding federal employee who inspired their authors.

The panorama below the overlook, from left to right, begins with the canyon described earlier that forms a separate peninsula, the cliffs and tip of that peninsula, then the low terrain of the Colorado River inner gorge. Beyond the meandering river, the Island-in-the-Sky and Dead Horse Point spread across the horizon. The conical promontory straight across the river is Pyramid Butte, which is undivided Cutler Group deposits topped with layered Moenkopi Formation. Below the Pyramid, the crenelated cliffline is Lower Cutler deposits clear down to the river level. The gray-hued inner rim of the river gorge is a layer of fossiliferous limestone that is part of the Cutler. On this side of the river, the open expanses of bare sandstone are also Cutler.

The low isolated ridge of dark red sandstone close to the river on this side has a large number of deep caves in it. For a further description of this lower terrain, refer to the log for the Hurrah Pass-Chicken Corners ORV trail elsewhere in this book. The slickrock terrace immediately below this overlook is Kayenta Sandstone. Note in the lower terrain a faint vehicle trail approaching the viewpoint. Spurs of this trail are visible in the gray Chinle slopes to the left. These trails were constructed in support of uranium prospecting during earlier mining boom days. See the sidebar, URANIUM.

16.4 Back at the main road, turn left.

18.3 Park at the end of the road and walk out onto Anticline Overlook. The scene beyond the overlook is described in a clockwise direction, from southwest to southeast, but to view this spectacular panorama it is necessary to walk to different vantage points on this extreme northern tip of Hatch Point. First, follow the gravel path to the shaded display pavilion for a description of Cane Creek Anticline, after which this overlook was named. The same display attempts to rationalize the immense blight on the colorful and unique terrain below and to the left created by the extraction of Paradox Formation minerals from deep beneath the surface, but fails to tell the entire tragic story. See the sidebar, POTASH. The refining mill visible beside the river is at the end of Utah 279, which is described elsewhere in this book. Most of the solar evaporation ponds associated with that mill are visible beyond the river. Above and to the left of the mill, and continuing to between the river and the evaporation ponds, the arching strata were warped by the Cane Creek Anticline. The strata to the north of this overlook slope downward into Cane Creek Canyon, indicating that the river cuts through the anticline close to its center. The structures along the far bank of the river below the salt ponds are an attempt to build a tourist development on leased Utah State land.

To the southwest, the Wingate cliffs of Hatch Point and Island-in-the-Sky have been described already. Below these cliffs, the fantasyland of erosional forms are largely Moenkopi and Cutler deposits clear down to river level. The upper rim of the river gorge here is the same gray fossiliferous limestone mentioned earlier. The drainage line rims on both sides of Pyramid Butte, to the left of the ponds, are especially rich in marine fossils. The ORV trail visible below to the north, and continuing along the western base of the overlook, is the Hurrah Pass-Chicken Corners Trail described elsewhere in this book.

Viewed from the northernmost tip of this overlook, mid-Cane Creek Canyon dominates the terrain below to the east. Notice Cane Creek Road coming up the canyon, and the Hurrah Pass ORV trail continuing beyond where that road ends at the ford of Cane Creek. The Hurrah Pass Trail twists up the slopes below toward the summit of Hurrah Pass, which is in the saddle in the ridge below and north of this overlook. Directly below this overlook to the north, the gray-topped massive

The view from Minor Overlook.

column is Chinle deposits on top, Moenkopi below that, then tapering abutments and cliffs of Cutler Group sandstones on down to the valley floor. Cutler deposits also dominate the wildly eroded terrain to the west of Hurrah Pass on down to the river.

The light tan sandstone visible in the sloping ridge to the north, northeast and east is all Navajo Sandstone, while the more distant white sandstone on the same ridge is Kayenta Sandstone. In that same rim, to the north and a little west, the jagged cliffline of Little Canyon breaks the continuity of Poison Spider Mesa, Gold Bar Rim and Arths Rim. The Book Cliffs dominate the northern horizon. The walls of mid-Cane Creek Canyon are Wingate Sandstone topped with remnants of Kayenta. The lower slopes are Chinle and Moenkopi, with crenelated bluffs of dark red Cutler sandstone below that on the eastern wall of the canyon. The free-standing spires in the center of the valley are also Cutler. For further description of Cane Creek Canyon, refer to the logs for Cane Creek Road and the Hurrah Pass-Chicken Corners Trail.

In the distance to the northeast, the downriver portal of the Colorado River as it leaves Moab Valley is visible between two large domes of Navajo Sandstone. Beyond that portal, the dark red cliffs and spires are Entrada Sandstone in the Windows Section of Arches National Park. The broken rocky terraces all around this overlook are Kayenta Sandstone. For further information about Cane Creek Anticline, refer to the sidebar, SALT ANTICLINES. Other sidebars in this book describe further other aspects of this outstanding panorama.

ROAD NAME: **BLACK RIDGE ROAD**

SUMMARY: This short road provides a good look at the way a salt anticline distorts and offsets geologic strata, leaving them apparently out of their usual sequence, plus deposits left by geologically recent glacial activity. The road also offers a spectacular panorama from its end on the lofty tip of Black Ridge. The sweeping view from there includes laccolithic mountains, a great salt anticline, a maze of sandstone fins, a complex of stream-watered canyons cut deeply into redrock strata, the broad arid meadows of Dry Valley, and a birdseye view of the colorful sandstone bluffs of Cameo Cliffs recreation area.

ROAD SURFACE: graded and graveled for its entire length, but rough in places.

VEHICLE: any vehicle when the road is dry.

MAP: **Map of CAMEO CLIFFS Biking - Hiking - Four-Wheeling**

GUIDEBOOK: **CAMEO CLIFFS Biking - Hiking - Four-Wheeling**

MILEAGES:

TO START: 13.3 miles, from Main and Center, Moab.
ROAD: 7.0 miles, one-way to end.
RETURN: 20.3 miles, from end of described road.
TOTAL: 40.6 miles, from Main and Center, round trip.

TIME: about 3 hours, if taken alone, but this road is best explored as a side-trip while traveling U.S. 191.

ROUTE: Main Street (U.S. 191) south, then east on Black Ridge Road.

NAVIGATION: Drive south on U.S. 191 for the indicated mileage, then turn left on Black Ridge Road, just beyond the summit of the steep grade out of Moab Valley.

GEOLOGIC STRATA VISIBLE FROM THIS ROAD

Formation	Member	Page
Dakota Sandstone		Page 20
Burro Canyon/Cedar Mountain Formations		Page 19
Morrison Formation	Brushy Basin	Page 18
	Salt Wash	
	Tidwell	
Entrada Sandstone	Moab	Page 16
	Slickrock	
	Dewey Bridge	
Navajo Sandstone		Page 13
Kayenta Sandstone		Page 13
Wingate Sandstone		Page 12
Chinle Formation		Page 12

GEOLOGY ROAD LOG:

0.0 As this graveled county road leaves U.S. 191, it approaches terraced bluffs of the Salt Wash Member of the Morrison Formation, then climbs through a cut into a rocky, faulted maze of this same formation for the next several miles. This immense area of Morrison deposits has been heavily prospected for uranium. Numerous ORV trails branch from this road along the way, going to

such "prospects." Most of these side trails have since been made impassable by erosion, but a few are still interesting to explore by those with suitable off-road vehicles. Any mine shafts found should not be entered. They are extremely hazardous because of possible collapse and accumulations of deadly radon gas. See the sidebar, URANIUM.

0.5 Here, and for several more miles, the rounded blue-gray slopes within sight of the road are the Brushy Basin Member of the Morrison Formation. From a narrow pull-out here, the ridge of brown-hued sandstone to the right is undivided Dakota Sandstone-Burro Canyon Formation that has been displaced from its higher stratigraphic position by faulting at this end of the Moab Valley salt anticline. See the sidebar, SALT ANTICLINES. A thin layer of the same deposit tops Black Ridge, which dominates the skyline to the south. Note how the lower slopes of Black Ridge have been cut by a maze of bulldozed mine-trails.

1.0 Here, the road begins a steep descent into the upper drainage of Cane Creek. The painted-desert sediments to the left of the road are Brushy Basin-Morrison. The fractured and distorted strata ahead on each side of the Cane Creek drainage, are also Morrison Formation deposits. The outcropping of reddish sandstone in the distance is the Slickrock Member of Entrada Sandstone, with the La Sal Mountains looming high above that.

This upper drainage of Cane Creek Canyon continues for several more miles toward the La Sals and is joined along the way by a number of tributary drainages. Thus, Cane Creek drains an enormous area of the La Sal foothills. During periods of heavy run-off, such as during and following periods of glaciation in these mountains, this now mostly dry creek has managed to cut a canyon that is about 25 miles long from beginning to end. It joins the Colorado River about 3 miles downriver from Moab Valley.

Just ahead, note the extensive mining in the Morrison Formation deposits in the opposite canyon wall. As in much of southeastern Utah, uranium mining in this area was not economically feasible. The relatively poor-grade ore could not possibly have been extracted from such terribly broken terrain, then shipped to mills sometimes hundreds of miles away, in a commercially economic manner. At that time, however, the United States Government was subsidizing the development of every trace of uranium ore, at enormous expense to the American public, so as to build stockpiles of this mineral for purposes of making nuclear bombs.

1.3 Note how the Cane Creek drainage here is flowing along the surface of a stretch of exposed slickrock, and how that surface has been shaped by flowing water.

1.4 Here, the road crosses the Cane Creek drainage on an old wooden bridge. Beyond the bridge, the road continues to travel through a chaotic jumble of Morrison Formation deposits. Occasionally, the blue-gray painted desert sediments of the Brushy Basin-Morrison show through the fallen rubble that dominates the slopes. For the next three miles, the road climbs above the Cane Creek drainage through Morrison deposits, toward the rimlands of Black Ridge.

2.0 The fractured light brown sandstone that walls the canyon to the left of the road here is Salt Wash-Morrison sandstone that has been displaced by the massive faulting in this vicinity. This stretch of road is traveling on or very near one such fault-line. Note the remnant of painted-desert Brushy Basin-Morrison deposits just beyond the cliff rim. The Morrison deposits in this area have been covered by glacial debris, with some of that removed by subsequent erosion. See the sidebar, GLACIERS OF THE LA SAL MOUNTAINS.

2.3 From this stretch of road, note the tilting of the Salt Wash-Morrison strata to the left, and the red Slickrock-Entrada sandstone cliff to the left-front. The Entrada is older than the Morrison yet, due to local strata tilting, it is physically higher than the Morrison deposits in the foreground. This tilting may have been caused by a sill or laccolith of intrusive igneous material during formation of the La Sal Mountains, or by subsidence due to local salt anticline faulting and slumping. Note that the Moab Member of Entrada Sandstone is missing in this area, although it tops Wilson and South mesas farther north around the base of the La Sals.

2.9 From the vicinity of an ORV trail spurring left, note the low hills of red-hued sediments to the left, lying on slickrock slopes. These are the highly mineralized, iron-rich deposits of the Tidwell Member of the Morrison Formation. Stratigraphically, this member lies just above the Slickrock-Entrada, which walls the wide canyon to the left-front. Note also, on the red Tidwell slopes, large irregular boulders of the quartz, or agate, that often forms in this deposit. This mineral is common throughout the rimlands of the Cameo Cliffs recreation area, which the Black Ridge Road penetrates and helps define. See the referenced guidebook and map. Note also, in the red-walled valley, the glacial deposits on top of a low butte of red Slickrock-Entrada in the center of that valley.

3.7 Beyond where an ORV trail goes left here, the Black Ridge Road climbs steeply toward the summit of the ridge, going through heavily vegetated slopes of upper Morrison strata, and offering magnificent panoramic views to the north and northwest.

4.5 The road tops out onto Black Ridge. The upper surface of this lofty promontory has been "chained," that is, most of its native pinyon-juniper forest has been destroyed in a futile attempt to enhance grazing for domestic livestock on this expanse of public land. The relatively flat upper surface of Black Ridge is covered with rocky glacial debris and thin soils barely held in place by the remaining native vegetation, but the rock just beneath the surface is undivided Dakota Sandstone-Burro Canyon Formation. Beyond the summit of the grade, the road winds across the mesa-top toward a microwave installation, ending there, but offering access to spectacular panoramic views of the lower canyon country to the south, west and north.

4.9 At the road junction here, the graded Highline Road goes left, while the Black Ridge Road swings right. Highline Road forms the eastern perimeter of the Cameo Cliffs recreation area noted earlier. It travels the lower slopes of the La Sals to the southeast, gradually ascending to meet paved Utah 46, but descending to cross deep Black Canyon and several lesser drainages along the way.

6.0 At an ORV trail junction here, the road reaches the western rim of Black Ridge and a spectacular view into lower terrain. This panorama is described more fully from the end of the road. For those with off-road vehicles or low-geared, high-clearance highway vehicles, an equally spectacular view to the south and southeast is offered from where this ORV trail reaches the southernmost tip of Black Ridge, one mile from this junction.

From that viewpoint, the Benchland ORV trail far below, as described in the referenced book, provides access to the rimlands of upper Muleshoe Canyon, as well as one stretch of the Black Canyon drainage. The road-slashed, wooded hill to the south has been heavily prospected and mined, leaving it an environmental disaster area. The panorama to the right of the distant Abajo Mountains is described later.

6.4 For a short stretch here, the road travels the crest of a narrow ridge that ends at the western tip of this lofty peninsula, with the La Sal Mountains and upper Moab Valley and Behind-the-Rocks to the right, and a complex of canyons stretching toward the west.

7.1 The road, as defined, ends here at the microwave installation, but can safely be traveled by most vehicles around the fenced building to a panoramic overlook just beyond.

From the plateau rim just beyond the microwave installation, the panorama from left to right is as follows. To the east of U.S. 191, the Slickrock-Entrada sandstone bluffs of the Cameo Cliffs recreation area loom above that highway. These cliffs are topped with lower Morrison Formation deposits. Just west of the highway, the scar of a gas pipeline route cuts across Dry Valley through the sparse native vegetation that remains after more than a century of overgrazing by domestic livestock. The installation near the highway is a pumping station for that pipeline. To the south, the Abajo Mountains dominate the skyline, ranging west into the highlands of Elk Ridge, Wild Cow Point and Beef Basin.

Directly below to the west, is the deep gorge of lower Cane Creek Canyon, which is joined by Muleshoe Canyon coming in from the left below the viewpoint. The vertical inner walls of Cane Creek Canyon, and the more distant Hatch Wash Canyon that joins it from the south, are Wingate Sandstone topped with layered Kayenta Sandstone. The sloping inner gorge is Chinle Formation deposits covered with collapse-debris. The light-hued sandstone in the vicinity of the Muleshoe-Cane Creek confluence is Navajo Sandstone.

The cliffs farther to the right are Slickrock-Entrada standing on a thin base of darker red Dewey Bridge-Entrada. The rubbly deposits on top of the sloping Entrada-walled mesa are remnants of the Tidwell and Salt Wash members of the Morrison Formation. Farther to the right, the sheer clifflines of Island-in-the-Sky and the middle stretches of Cane Creek Canyon dominate the skyline. In the distance, on clear days, the three separate sub-ranges of the Henry Mountains loom above detached Junction Butte.

For a better view of the panorama to the north, walk out onto the spur of Black Ridge that goes to the northwest. From that viewpoint, the fantastic Navajo Sandstone maze of Behind-the-Rocks partly obscures the more distant Book Cliffs, which lie to the north of Interstate 70. Beyond Behind-the-Rocks, Poison Spider Mesa, Gold Bar Rim and Arths Rim angle left into the distance. To the right of Behind-the-Rocks lie the upper reaches of Moab Valley, with the tablelands of Wilson and South mesas above that. The La Sal Mountains spread across the entire eastern horizon.

View west from Black Ridge.

ROAD NAME: **BROWNS HOLE ROAD**

SUMMARY: This short road does not have much geologic variety, but it does offer interesting views of the upper drainage of well-watered Muleshoe Canyon, plus access to several ORV trails in the northern part of Cameo Cliffs recreation area.

ROAD SURFACE: graded dirt.

VEHICLE: any vehicle in good weather.

MAP: **Map of CAMEO CLIFFS Biking - Hiking - Four-Wheeling.**

GUIDEBOOK: **CAMEO CLIFFS Biking - Hiking - Four-Wheeling.**

MILEAGES:

TO START: 21.3 miles, from Main and Center, Moab.
ROAD: 3.1 miles, one-way to end.
RETURN: 24.4 miles, from end of described road.
TOTAL: 48.8 miles, from Main and Center, round trip.

TIME: about two hours when traveled by itself, as a round trip from Moab. The road is best traveled as a side-trip from U.S. 191. This takes less than 1 hour, round trip.

ROUTE: Main Street (U.S. 191) south, then east on Browns Hole Road.

NAVIGATION: Drive the noted distance south from Moab then turn east just beyond the gas pumping station to the west of the highway.

GEOLOGY ROAD LOG:

0.0 At road start, the large sandstone mass to the left is known as The Nipples because of the conical shapes on its summit. The red rock there, and to the left of the road for the next half mile, is the Slickrock Member of Entrada Sandstone. The flat terrain on either side of the road for the next mile or so is alluvium consisting of sand and decomposed rock held in place by native vegetation. To the right-front, the low hill that is laced by old bulldozer cuts is entirely Morrison Formation, with the lower part being the Salt Wash Member and the upper part the Brushy Basin Member. The tailings of one mine are visible below the highest point of the hill. There are remnants of other active uranium mines several places on this hill, most of them not within sight of this road.

GEOLOGIC STRATA VISIBLE FROM THIS ROAD

Morrison Formation	Brushy Basin	Page 18
	Salt Wash	
	Tidwell	
Entrada Sandstone	Moab	Page 16
	Slickrock	
	Dewey Bridge	

The La Sal Mountains dominate the eastern horizon, with their intrusive-igneous central stocks exposed in the treeless peaks, and with decomposed rock of a wide variety on the lower slopes. See the sidebar, LACCOLITHIC MOUNTAINS.

Browns Hole Road enters the northern part of the area known as Cameo Cliffs. This public land recreation area is described in the noted guidebook and depicted on the matching map. The off-road vehicle trails that spur from this road are described in these two publications.

0.4 A short ORV trail called "Camp Trail" goes left here. It provides hiking access to The Nipples promontory plus a variety of good campsites for those with vehicles capable of off-road travel.

0.6 The Muleshoe Point Trail goes left here. This trail provides access to a large area of Cameo Cliffs North, including the rimlands of beautiful Muleshoe Canyon. It should be attempted only by off-road vehicles. Beginning just ahead, lofty Black Ridge dominates the northern skyline, with its microwave tower at one end. The history of this installation is described in the referenced guidebook.

Black Ridge is almost entirely composed of three members of the Morrison Formation, standing on a base of Entrada Sandstone that is largely obscured by rubble when viewed from this direction. The uppermost level of Black Ridge is a thin remnant of undivided Dakota Sandstone-Burro Canyon Formation. The Morrison Formation continues to dominate between Black Ridge and the base of the La Sal Mountains at about the same level. Many of the steep slopes just above the Morrison level in the mountains ahead consist of glacial debris deposited there during the several glacial periods in this region's geologically recent past. See the sidebar, GLACIERS OF THE LA SAL MOUNTAINS.

1.1 The road begins to descend into the immense drainage basin of Muleshoe and Black Canyons. The rock outcropping to the left of the road is the lower levels of Salt Wash-Morrison. The open terrain on both sides of the road continues to be largely sand and decomposed rock held in place by native vegetation, with occasional outcrops of reddish Entrada Sandstone to the north.

2.0 The northern wall of Muleshoe Canyon is visible to the left-front of the road. The drainage for this canyon system begins in the foothills of the La Sal Mountains. About 3 miles east of U.S. 191, it is joined by Black Canyon, the drainage just south of Black Ridge. Muleshoe Canyon then goes beneath U.S. 191 and twists through more than two miles of sandstone gorge before joining upper Cane Creek Canyon. Hiking this canyon complex is described in the referenced guidebook about Cameo Cliffs. Just ahead, the spur road to the right goes to an abandoned uranium mine. The road can be traveled by highway vehicles if carefully driven. The open mine tunnels there are very hazardous and should not be entered. See the sidebar, URANIUM.

2.8 From a curve in the road here, the rimlands of upper Muleshoe Canyon are visible to the left, with low bluffs of Salt Wash-Morrison above that. There is an old cattle corral to the right here. Some of the Morrison slopes below the La Sal peaks have been cut by bulldozer trails made in connection with uranium prospecting. The slopes to the right of the road continue to be Salt Wash-Morrison deposits.

2.9 Here, the road is closely paralleling the upper drainage of Muleshoe Canyon. The abutments of an old wooden bridge that once crossed the shallow drainage can be seen to the left of the road at one point. The walls of the inner drainage here are alluvium deposited during earlier periods of wetter climate, but now cut deeply by accelerated erosion. Most of this erosion is due to the grazing of domestic livestock, mineral prospecting and developments in nearby Browns Hole that have severely damaged the water-holding capabilities of the area's native soils and vegetation. This severe alluvium-cutting started when the first settlers in this region brought in great herds of cattle and sheep, which quickly destroyed the native grasses that were holding most precipitation for gradual absorption into the ground, rather than allowing rapid run-off. See the sidebar, HUMANITY AS A GEOLOGIC FORCE.

3.1 Here, the public road ends. Just ahead the road enters Browns Hole, a large tract of private land that was first settled in 1882. Tourists should not enter Browns Hole. The off-road vehicle trail that goes left here is the Cameo Cliffs Benchland Trail, which also provides access to several other ORV trails.

View north from The Nipples.

ROAD NAME: **DUBINKY WELL ROAD**

SUMMARY: This road is described primarily because it is used for access to the Spring Canyon Point ORV trail, but it does offer broad colorful panoramas and a variety of geological strata and erosional landforms along its described length to Dubinky Well, just north of the road's junction with the Spring Canyon Point Trail. Along this stretch, the road provides access to several other ORV trails as well.

ROAD SURFACE: graded dirt that is poorly maintained and sometimes questionable for standard highway vehicles, especially during and following heavy rain.

VEHICLE: any vehicle in dry weather and when the road is in good condition.

MAP: ***Canyon Country*** **OFF-ROAD VEHICLE TRAIL MAP - Island Area.**

GUIDEBOOK: ***Canyon Country*** **OFF-ROAD VEHICLE TRAILS - Island Area.**

MILEAGES:

TO START: 20.0 miles, from Main and Center, Moab.
ROAD: 6.8 miles, one-way to end.
RETURN: 26.8 miles, from end of described road.
TOTAL: 53.6 miles, from Main and Center, round trip.

TIME: allow about three hours if the road is to be traveled by itself, but a short day if it is to be driven in combination with the Spring Canyon Point Trail.

ROUTE: Main Street (U.S. 191) north, Utah 313 west, then this road north to Dubinky Well, then return via the same route or explore one or more of the branching ORV trails.

NAVIGATION: Drive north out of Moab on U.S. 191, then west on Utah 313 to its junction with the Dubinky Well Road, then return via the same route or explore one or more of the branching ORV trails. The Spring Canyon Point Trail, which spurs from Dubinky Well Road near where this description ends, is described elsewhere in this book.

GEOLOGY ROAD LOG:

0.0 As this road heads north from Utah 313, red-walled Big Mesa looms to the right, with the northern cliff of Bartlett Wash in the distance ahead. The thin red sediments on top of Big Mesa are the Tidwell Member of the Morrison Formation. The white band of sandstone below that is the Moab Member of Entrada Sandstone. The color-banded slopes and sheer walls are the Slickrock Member of the Entrada, while the layered dark red sandstone at the base of the cliffs is the Dewey Bridge Member of the Entrada. The entire mesa is standing on a base of white Navajo Sandstone, which is largely obscured from the beginning of the road but becomes visible farther on. The sharp-edged mesa to the right of the road just ahead is also Navajo, topped with the harder limestone remnants of an ancient desert drylake. See the sidebar, PETRIFIED PLAYAS. Beyond Big Mesa, the slopes of the Book Cliffs are apparent on clear days.

GEOLOGIC STRATA VISIBLE FROM THIS ROAD

Morrison Formation	Brushy Basin	Page 18
	Salt Wash	
	Tidwell	
Entrada Sandstone	Moab	Page 16
	Slickrock	
	Dewey Bridge	
Navajo Sandstone		Page 13

The geology of Bartlett Wash is described from a better viewpoint ahead. The rounded slickrock fins in the near-distance to the left-front are also Entrada Sandstone. These are described in detail in the log for the Spring Canyon Point Trail. The rolling open meadows through which this road travels are decomposed rock, sand and alluvium held in place by the native vegetation that remains after a century of severe overgrazing. The low, rounded outcrops of white rock on both sides of the road as it continues are the upper levels of Navajo Sandstone, several of which are topped by the flat remnants of harder desert playa sediments. In the extreme distance to the left-front, the jagged line of white rock is the eastern monocline of the San Rafael Swell.

0.5 The ornately eroded bluff to the right of the road is Navajo Sandstone topped with the horizontal layering of a desert playa. Some such highly mineralized rock strata have the petrified foot-tracks of the dinosaurs and other animals that roamed the desert environment at that time. There are a few indistinct dinosaur tracks in this playa deposit, on detached slabs of rock along the rim of the mesa. The rounded sandstone bluff to the left of the road is also Navajo Sandstone. As the road continues, look back toward that bluff and note the large seep-cave in its base. The domes of sandstone in the distance to the left-front are also Navajo Sandstone.

1.1 From the summit of a grade, pull off to the right and study the scene to the east. This broad V-shaped valley is Bartlett Wash. Note how the red cliffs that form this large drainage converge like an immense funnel. This forces the several tributary drainages in the wash into a single narrow gorge at the lower end of the valley. Although little surface water flows down these various drainages except immediately following heavy rain, the lower end of the valley is always wetted by subsurface water that is forced upward by rock strata. Both sheer walls of Bartlett Wash are Slickrock-Entrada standing on bases of Dewey Bridge-Entrada and topped with a white band of Moab-Entrada, with Morrison Formation slopes above that. The reddish Tidwell-Morrison just above the Moab-Entrada is largely obscured by the debris from the higher Salt Wash Member of the Morrison. This deposit is rich in fossil material, with known petrified wood sites above the northern cliffs of Bartlett Wash.

All of the white rock in the valley floor is Navajo Sandstone, which lies just below the Dewey Bridge-Entrada at the base of the cliffline. Note the immense sand dunes deposited against the distant cliff wall by prevailing winds. The mountainous terrain in the distance to the right is the Book Cliffs, with the much nearer western slopes of the Salt Valley anticline in Arches National Park just below. The open meadows of Bartlett Wash are largely dune sand and alluvium barely held in place by native vegetation, but are undergoing accelerated erosion brought on by excessive grazing of domestic livestock. Notice the gray desert playa limestone deposit on top of the low white Navajo Sandstone dome to the right-front.

1.5 At this junction, the road that continues straight ahead is the Spring Canyon ORV trail, which is not described in this book. Turn right to continue on the Dubinky Well Road.

3.4 From this stretch of the road, note the large arc-topped alcove in the sheer cliffline to the right. This alcove has an opening at its top, thus forming a natural arch. Although this opening is only a foot or so wide, it is much longer than that, and the weight of the heavy span is borne entirely on its two abutments. This is part of the technical definition of a natural arch or bridge. For more information about how natural arches and bridges form, refer to the book, ***Canyon Country* ARCHES & BRIDGES.** The book, ***Canyon Country* SLICKROCK HIKING & BIKING,** describes the hike to this big cliff-rim arch.

The terraced red and white sandstone to the right of the road is Dewey Bridge-Entrada at its interface with the lower white Navajo Sandstone. Notice in the road and on bare patches of Navajo the countless fragments of white agate. This harder rock forms in highly mineralized Dewey Bridge-Entrada, then remains on the upper surface of the Navajo as the softer Dewey Bridge sediments erode away. The red-hued soils in this vicinity are all decomposition products of the Dewey Bridge-Entrada and also contain great quantities of the agate. In the distance to the right, note the dark red sandstone spires and cliffs of Klondike Bluffs in Arches National Park.

3.7 Along the next stretch of road, note the three separate highlands of the Henry Mountains on the southwestern skyline. All of the white domes in the broad meadows are Navajo Sandstone. The distant red and white layer of rock below the middle peak of the Henrys is Entrada Sandstone above Navajo Sandstone.

4.3 The rounded, eroding bluffs and soils in this vicinity and for the next mile or so are all decomposing Dewey Bridge-Entrada.

5.6 As the road descends just ahead to cross shallow Dubinky Wash, watch to the left for remnants of the original Dubinky Ranch on each side of the wash. Dubinky Wash meanders across the open terrain to the southwest, then drops abruptly into a deep sheer-walled tributary of Hellroaring Canyon, which in turn joins the Green River gorge. The Spring Canyon Trail, which branched off earlier, crosses Dubinky Wash just above its spectacular pouroff. Beyond here, the low rounded hills continue to be decomposing Dewey Bridge-Entrada.

6.6 The Spring Canyon Point ORV trail goes left from this junction. That trail is described elsewhere in this book. Dubinky Well is just ahead.

6.8 This geology road log ends here at historic Dubinky Well, with its large stone water tank, metal tank, inactive windmill pump and livestock watering trough. From here, the road continues north across open terrain, eventually to join Blue Hills Road, as described in the referenced book and map. Several decades ago, before Utah 313 was constructed and before Canyonlands National Park was established, the only vehicle route from Moab to Dead Horse Point State Park and the rest of Island-in-the-Sky was via the Blue Hills and Dubinky Well roads, making a trip to the state park at that time long and rough. To explore the Spring Canyon Point Trail, return to its junction, then follow its description.

ROAD NAME: **HATCH WASH ROAD**

SUMMARY: This graded dirt road is generally not a destination in itself, but is traveled as a part of day-long routes on Needles Overlook Road, Anticline Overlook Road, or off-road vehicle trails that spur from these. It does, however, provide a look at several colorful geologic strata, some interesting erosional phenomena, and broad scenic panoramas that include the La Sal and Abajo mountains, the Cameo Cliffs area east of U.S. 191, and parts of the Hatch Point and Flat Iron Mesa districts of Canyon Rims Recreation Area.

ROAD SURFACE: graded dirt with occasional stretches of rock or sand. The road receives occasional maintenance.

VEHICLE: any vehicle, when the road is dry.

MAPS: ***Canyon Country*** **OFF-ROAD VEHICLE TRAIL MAP - Canyon Rims Recreation Area,** or ***Canyon Country*** **OFF-ROAD VEHICLE TRAIL MAP - Canyon Rims & Needles Areas.**

GUIDEBOOKS: ***Canyon Country*** **OFF-ROAD VEHICLE TRAILS - Canyon Rims Recreation Area** or ***Canyon Country*** **OFF-ROAD VEHICLE TRAILS - Canyon Rims & Needles Areas,** and ***Canyon Country's*** **CANYON RIMS RECREATION AREA.**

MILEAGES:

TO START: 23.6 miles, from Main and Center, Moab.
ROAD: 11.5 miles, one-way to end.
RETURN: 35.1 miles, from end of described road.
TOTAL: 70.2 miles, from Main and Center, round trip.

TIME: allow an hour for getting to and driving this road, in addition to the trip planned beyond the road's southern end.

ROUTE: Main Street (U.S. 191) south, then Hatch Wash Road southwest, plus whatever else is planned for the day.

NAVIGATION: Drive south from Moab on Main Street (U.S. 191) to a short distance beyond La Sal Junction, then turn right onto graded Hatch Wash Road.

GEOLOGY ROAD LOG:

0.0 As the road angles southwest from U.S. 191 into the Flat Iron Mesa District of Canyon Rims Recreation Area, the low hill to the right is a remnant of the Morrison Formation on top of a low dome of the Slickrock Member of Entrada Sandstone, with the entire hill largely covered with native vegetation. The rounded cliffs to the left-front are also Slickrock-Entrada topped with Morrison remnants.

GEOLOGIC STRATA VISIBLE FROM THIS ROAD

Formation	Member	Page
Morrison Formation	Brushy Basin	Page 18
	Salt Wash	
	Tidwell	
Entrada Sandstone	Moab	Page 16
	Slickrock	
	Dewey Bridge	
Navajo Sandstone		Page 13

The open meadows that dominate this trail for the next several miles are decomposed rock, sand, alluvium and soils held in place by remnants of native vegetation. The deep grasses that once filled this part of Dry Valley have long since been grazed to extinction, leaving just low desertland shrubs that domestic livestock will not eat. This drastic and prolonged overgrazing has immensely accelerated erosion throughout Dry Valley, causing deep cutting in drainage lines and the removal of countless tons of soils. Places where such cutting is visible are noted further ahead in this description. See the sidebar, HUMANITY AS A GEOLOGIC FORCE.

1.9 Turn left into the parking area for Looking Glass Rock. From the sandstone slopes adjacent to the parking area, study Looking Glass Rock which was created by an immense south-facing seep-cave breaking through the thin-walled dome of Slickrock-Entrada. This prominent feature was given its name because of its resemblance to the way a hand-held "looking-glass," or magnifying glass, tends to focus sunlight. When viewed from the south or east, the afternoon sunlight shining through this opening into the shaded cave beyond illustrates this resemblance. An early wagon road traveled just east of Looking Glass Rock, offering travelers such a view. Visitors who choose to climb up into the opening should exercise caution because of the steep drop beyond.

In the panoramic view from just above the parking area, the large dome of Slickrock-Entrada to the southwest is Hatch Rock, past which this road travels farther on. To the south, the Abajo Mountains loom above the convoluted cliffline of Rone Bailey Mesa. For a description of that immense elevated plateau, refer to the log for Needles Overlook Road, elsewhere in this book. All of the red-hued rock domes rising above the meadows to the southwest are Slickrock-Entrada sandstone. The low domes to the northwest, visible along this road from here on, are Navajo Sandstone set in open meadows of erosion products.

2.1 Back at the main road turn left.

4.6 Stop at the crest of a low hill and study the scene to the east and south. The shallow drainage line to the left is lower Joe Wilson Canyon, which originates in the sandstone bluffs to the east, all of which are Slickrock-Entrada. The eroded white slopes of this canyon are Navajo Sandstone. Note, just above the Navajo, a low ridge of reddish rock. This is the Dewey Bridge Member of Entrada Sandstone. To the left-front, the broad drainage of Hatch Wash is also walled with slopes of Navajo Sandstone topped with reddish remnants of Dewey Bridge-Entrada. Note in the distance how the alluvium in the center of the wash is being cut by flash-flooding. Straight ahead are Rone Bailey Mesa and the Abajo Mountains. See the sidebar, LACCOLITHIC MOUNTAINS. The small detached dome of red sandstone to the left of Rone Bailey Mesa is historic Jail Rock. Beyond this summit viewpoint, the road drops gradually toward the Hatch Wash drainage. As the road descends it passes intermittent exposures of eroding Navajo Sandstone.

6.1 Note the caves blasted into the Navajo Sandstone cliff to the right of the road. For a short distance here, the road is traveling through private land that was homesteaded in the early 1880s. At some time, these two caves and other structures in this vicinity were built to support local livestock ranching. The ranch dwelling was built in a nearby canyon. That canyon was rich with springs and grassy meadows at that time, but after more than a century of overgrazing in this area, the resultant flash-flooding has removed most of the soils from that canyon, leaving remnants of the ranch structures perched high on remaining shelves of soil. For more information about that ranch, refer to the book ***Canyon Country's* CANYON RIMS RECREATION AREA.** For more information about the historic 1859 Macomb Expedition that crossed that canyon and used its biggest spring as a campsite, refer to the book **CANYONLANDS NATIONAL PARK - Early History & First Descriptions.** Just ahead, beyond the drainage of Hatch Wash, the road enters the Hatch Point District of Canyon Rims Recreation Area.

6.3 Here, just before this road crosses the Hatch Wash drainage, a short spur road goes right, down into the wash below a sandstone pouroff toward a large pool of water. By studying the terrain surrounding this pouroff, it becomes readily apparent that, even during historic times, this pouroff and pool did not exist because the soils had not yet eroded below the current road level. The steep cutbanks of soil on both sides of the drainage below this ford have been created by the accelerated erosion caused by overgrazing throughout the immense Hatch Wash drainage area.

The sandstone cliffs and domes in this vicinity continue to be Navajo Sandstone topped with thin remnants of reddish Dewey Bridge-Entrada. Beyond the wash-crossing, the road travels at wash level for about a mile, then climbs gradually onto higher terrain, traveling at approximately the Navajo-Entrada interface.

7.3 Here, for a short distance, the road parallels Wind Whistle Draw. Note how the alluvium has also been drastically eroded in this drainage line. The low reddish hill to the left-front is Dewey Bridge-Entrada standing on a base of white Navajo Sandstone. The same combination of geologic strata dominates the western horizon here, while the upper tip of Hatch Rock looms on the skyline.

8.2 From this vicinity, note the developments in Hatch Rock to the right. This prominence was named after Alonzo Hatch, whose 1880s homestead was noted earlier. His ranch headquarters were in the lower end of a canyon out of sight just beyond Hatch Rock. This colorful monolith and surrounding terrain are currently leased by a private family. For more details about this, refer to the sidebar, HATCH ROCK DEVELOPMENT, in the book ***Canyon Country's* CANYON RIMS RECREATION AREA.** The colorful cliffs of Rone Bailey Mesa continue to dominate the southern skyline.

9.2 The road to the right here is private access to the Hatch Rock development.

9.6 At the junction here, the Hatch Wash Road goes left. The road to the right is Eightmile Road, which provides access to several ORV trails in this recreation area but is not described in this book.

11.0 The red sandstone domes to the left and straight ahead are Slickrock-Entrada. The trail spurring left at this junction goes to grazing developments on the other side of the rock. Note the large natural seep-caves in the Slickrock-Entrada dome ahead, which is called Soup Rock. There is a large, curiously-shaped pothole in the highest part of Soup Rock, which can be climbed from its eastern end.

11.5 This road ends here by joining the paved Needles Overlook Road, with lovely Soup Rock to the right-front. Turn left for access to the BLM's Wind Whistle Campground at the base of Rone Bailey Mesa and to return to U.S. 191. Turn right to go to Needles Overlook and to reach Anticline Overlook Road, both described elsewhere in this book. The description for the geology along Needles Overlook Road can be picked up at this point, because this junction is noted in that road log at mile 7.4.

TRAIL NAME: **ARTHS PASTURE ROAD - GEMINI BRIDGES TRAIL**

SUMMARY: this combination road-and-trail route goes to the popular Gemini Bridges, two huge parallel natural spans that bridge a deep canyon cut in Wingate Sandstone, and offers glimpses of several curious arches. The road and trail also provide excellent views of the myriad erosional landforms that the combination of Kayenta and Wingate sandstones can take when these strata are tilted by tectonic activity, then eroded for eons. The route also crosses Roberts Rift, with three exposures of this unusual miles-long rift visible from the Gemini Bridges Trail.

Gemini Bridges

GEOLOGIC STRATA VISIBLE FROM THIS TRAIL

Navajo Sandstone		Page 13
Kayenta Sandstone		Page 13
Wingate Sandstone		Page 12
Chinle Formation		Page 12
Moenkopi Formation		Page 12
Cutler Group	White Rim	Page 11
	Organ Rock	
	Cedar Mesa	
	Lower Cutler	

ROAD AND TRAIL SURFACES: eroded graveled road, rutted dirt, ledgy sandstone and rough slickrock.

VEHICLE: off-road vehicles only for the entire route, but carefully-driven high-clearance, low-geared highway vehicles can get to within walking distance of the Gemini Bridges.

MAP: ***Canyon Country* OFF-ROAD VEHICLE TRAIL MAP - Island Area.**

GUIDEBOOK: ***Canyon Country* OFF-ROAD VEHICLE TRAILS - Island Area.**

MILEAGES:

TO START: 24.7 miles, from Main and Center, Moab.
ROUTE: 14.1 miles, one-way to end.
RETURN: 10.0 miles, from end of described route.
TOTAL: 48.8 miles, from Main and Center, round trip.

TIME: allow a full day for this trip, a long day if either the Bull Canyon or the Sevenmile Canyon Rim ORV trail is added to the trip.

ROUTE: north on U.S. 191, west and south on Utah 313, then east on Arths Pasture Road to the Gemini Bridges Trail.

NAVIGATION: drive north from Moab on Main Street (U.S. 191) for 11.3 miles, west and south for 13.4 miles on Utah 313 toward Dead Horse Point State Park and the Island-in-the-Sky District of Canyonlands National Park, then east on Arths Pasture Road to the start of the Gemini Bridges Trail.

GEOLOGY ROAD & TRAIL LOG:

0.0 Arths Pasture Road penetrates an enormous sloping area that is essentially Kayenta Sandstone topped with soils held in place by vegetation of the pinyon-juniper plant community. As the road leaves Utah 313, remnant ridges of Navajo Sandstone are visible on both sides. The area the graded road enters is drained by the various arms of Sevenmile Canyon to the north of the road, and the Bull Canyon system to the south. These canyons can be glimpsed occasionally from the road before it reaches the Gemini Bridges Trail junction.

1.7 Note one upper arm of the Bull Canyon system to the right of the road.

2.1 Park at the base of a short downgrade here. To the north, Big Mesa and the Monitor & Merrimac buttes dominate the skyline, with glimpses of the Book Cliffs beyond. The geology of these prominent buttes is described under the Utah 313 log elsewhere in this book. Below the road in that direction is one upper arm of the south fork of Sevenmile Canyon. The inner gorge of Sevenmile there is Wingate Sandstone topped with whiter layers of Kayenta Sandstone. Ahead, the Kayenta Sandstone has been distorted by the various anticlines and synclines, and by major faulting in the vicinity of U.S. 191. The canyons visible there cut down into Wingate Sandstone. To the southeast, the La Sal Mountains loom above rounded domes and fins of Navajo Sandstone. Beyond this, the tablelands that jut out from the La Sal foothills are rimmed with the Moab Member of Entrada Sandstone, with the slopes below the rims being the Slickrock Member of Entrada.

3.3 The trail is still traveling through the lower strata of Kayenta Sandstone. Note to the right the upper walls of several branches of the complex Bull Canyon system. The sloping canyon walls are Wingate Sandstone that has not been undercut by erosion and therefore has not sheared away to form vertical cliffs. Rather, it erodes somewhat like Entrada-Slickrock and Navajo Sandstone do before they are undercut, with rounded sloping surfaces. The canyon walls are topped with the lower layers of Kayenta Sandstone. From this stretch of the road, the sandstone fin fantasyland of the Behind-the-Rocks wilderness area becomes more and more prominent.

3.8 The Sevenmile Canyon Rim ORV trail angles left here, while the Arths Pasture Road continues straight ahead.

4.0 At the junction here, go right on the Gemini Bridges Trail. Arths Pasture Road continues straight ahead. The Gemini Bridges Trail should not be attempted by standard highway vehicles, although high-clearance, low-geared highway vehicles can travel to the short spur that goes to the twin natural bridges if carefully-driven. From there, it is a short hike to the bridges.

4.8 The Four Arches ORV trail goes right here and descends into Crips Hole. Crips Hole, and the several tributary canyons that branch from it, are all upper tributaries of the Bull Canyon system. One spur of this ORV trail enters the narrows of Four Arches Canyon where four spectacular natural spans appear in the Wingate Sandstone walls of the canyon. This trail is described in the referenced map and guidebook. The Gemini Bridges Trail goes left at this junction, continuing to travel through rough and broken layers of Kayenta Sandstone.

5.3 At this point, look directly right at the far cliffline. About eighty feet below the white Kayenta Sandstone that rims the cliff, look for what appears at first glance to be a seep-cave in the reddish Wingate Sandstone, with a large rounded boulder sitting on the ledge above the cave. This is Boulder Bridge, which was first found, named and reported by the author in the early 1980s. An obscure spur of the Four Arches ORV trail closely approaches the span. Although some geologists have reported that Wingate Sandstone is not prone to form arches as it erodes, the Bull Canyon system has, at last count, more than three dozen natural rock openings of significant size, with the Gemini Bridges being the largest. There are also a number of arches in nearby Sevenmile and Little canyons, all in Wingate Sandstone.

5.6 Turn right here on the spur to the Gemini Bridges.

5.7 Continue straight ahead at this junction. The trail beyond this junction toward the Bridges is confusing. At the several trail forks, keeping to the right is the best approach.

5.8 Here, the spur trail crosses an expanse of slickrock, descending from it to the right, then swinging sharply left. Beyond this, continue to keep right at confusing trail forks.

5.9 At a trail junction here, keep right. In a few yards, another spur to the right climbs a few yards toward the rim of one tributary of Bull Canyon. A short hike beyond the end of the vehicle trail along the canyon rim provides spectacular views of the Bull Canyon system and various erosional forms in Wingate Sandstone. The trail left from this junction goes for a few more yards, then crosses one of the Gemini Bridges. For those who choose not to do this, there is room to turn around and go back the same way, either here or a little nearer the ends of the twin natural bridges. The mileages given for this spur include driving across the upper bridge.

6.0 Park just beyond the bridges for hiking in this vicinity. These magnificent twin natural bridges, at the interface between the Kayenta and Wingate sandstones, were first formally reported in late 1968 by Moab tour guide, Lin Ottinger. Ottinger, together with the author of this book and his wife, plus several friends, were the first to document the bridges and to reach their tops with a vehicle. A few days later, the first vehicle route into the canyon below the bridges was explored by Ottinger, this author and a friend.

Later, the bridges were measured and a chain ladder was dropped from the southern rim of the hole behind the bridges for the first access beneath them. The measurements in feet, taken at that time were, with the upcanyon (western) bridge noted first:

lengths, 89 and 70;

widths, 26 and 26;

minimum thicknesses, 40 and 18;

depths to the canyon below, 100 and 103;

spans beneath the bridges, 67 and 44;

maximum distance between them, 10.

These measurements, the first photographs of the bridges, and the story of their discovery were reported by this author in the November-December 1970 issue of OUTDOOR WORLD magazine. With Ottinger's approval, the author filed for official names of these natural bridges with the U.S. Board on Geographic Names. Their official names are "Gemini Bridges," per a decision dated March 7, 1969.

The canyon below the Gemini Bridges is a short spur of the main drainage line in this complex canyon system. The canyon below the bridges at this point is approximately 250 feet deep. The bridges are at about 5,000 feet elevation.

For the return trip to the main trail, follow your own fresh tracks, keeping to the left at each confusing spur.

6.3 Back at the main trail junction, turn right.

6.5 An alternate spur trail to Gemini Bridges goes right here. Continue straight ahead for the return to Moab or to reach the Bull Canyon ORV trail, described elsewhere in this book.

6.7 A slickrock pullout on the right provides a good view of the lower part of the Bull Canyon system. The canyon walls continue to be Wingate Sandstone capped with remnants of Kayenta Sandstone. Note that these normally horizontal strata are sloping, due to the general tectonic distortion of this area by faulting, anticlines and synclines. The Bull Canyon ORV trail drops into and explores the two main forks of lower Bull Canyon. A

View from near the Gemini Bridges.

spur of the Bull Canyon Trail enters the branch canyon below the Gemini Bridges. Note straight across the canyon the angled white streak in the red canyon wall. This is one exposure of a strange miles-long rift that cuts across this area. See the sidebar, ROBERTS RIFT.

6.9 The Gemini Bridges Trail angles left at a junction here. To the northeast of this junction, the white sandstone ridge is Gold Bar Rim, named after the canyon system that drains much of it, Gold Bar Canyon. This canyon drains directly into the Colorado River gorge. At the left end of the rim, where the reddish Wingate Sandstone is revealed beneath the white Kayenta Sandstone layers, the Little Canyon drainage goes to the right, following its own separate course to the Colorado River gorge. Farther ahead, this trail descends into Little Canyon on its route to U.S. 191.

7.4 This stretch of trail provides an excellent panoramic view to the south and east, beginning with the upper reaches of Little Canyon straight ahead, the lofty summit of Arths Rim to the left of that, and the summit of Gold Bar Rim and adjacent Poison Spider Mesa to the right. Note that most of the Gold Bar Rim area is bare of Navajo Sandstone, while that red-hued rock still dominates much of Poison Spider Mesa. The downriver portal, where the Colorado River leaves Moab Valley, is between two prominent sandstone domes toward the La Sal Mountains. The reddish sandstone to the left of the portal is Navajo Sandstone standing on a tilted base of whitish Kayenta Sandstone. To the right of the river portal, the Navajo Sandstone fin maze of Behind-the-Rocks dominates the rimlands above Moab Valley, which is not visible from here. The left spire of the two Navajo Sandstone prominences in the near-foreground is known as The Tortoise. The white rock in the immediate foreground is all Kayenta Sandstone.

7.8 At the junction here, the Arths Pasture Road goes left, to meet Utah 313 where this log began in 7.1 miles. The Gemini Bridges Trail continues straight ahead. Although a part of this ORV trail has been graded by oil exploration operations, it is still defined in various *Canyon Country* publications as the Gemini Bridges Trail, all the way to where it ends at U.S. 191. Those who prefer to return to Utah 313 should turn left here.

8.6 At this junction, the Bull Canyon Trail goes right while the Gemini Bridges Trail continues ahead. Note that The Tortoise monolith is prominent just below the southeastern skyline. Farther to the right from The Tortoise, the jagged cliffline on the skyline in the near distance is the eroded rimlands of Day Canyon and the Dry Fork of Bull Canyon. A geology log for the Bull Canyon Trail is found elsewhere in this book, but not covered in this description.

8.7 A short undescribed spur trail to the right here ends near a prominence that offers a wide panoramic view to the south. The main trail angles left beyond this junction, then begins its steep descent into Little Canyon.

9.0 At this point, about one hundred feet before the trail goes through a fenceline, there is a sizable natural arch in the slickrock rim to the left of the trail. A short spur from the trail, as it travels in the canyon bottom, goes directly to this arch. From this viewpoint, the spur of upper Little Canyon that the trail travels is directly to the left. Note the partially detached spire of Wingate Sandstone in the righthand side of this canyon that is known as "The Gooney Bird." The main drainage of this wash comes from another branch of the canyon that is out of sight to the left, then continues down the canyon that goes to the right. The Gold Bar Rim ORV trail goes down that drainage for about 1/2 mile before climbing up the tilted strata to the summit of Gold Bar Rim. Little Canyon offers good hiking, both upcanyon to the left, and downcanyon beyond where the Gold Bar Rim Trail leaves the drainage. This second hike is described in the book, ***Canyon Country* SLICKROCK HIKING & BIKING.**

9.1 As the trail reaches the floor of Little Canyon, the Gold Bar Rim ORV trail goes right, while the Gemini Bridges Trail goes left. The Gold Bar Rim ORV trail is described in the referenced book and map. At this trail junction, two exposures of Roberts Rift are visible in the canyon walls. An alcove in the cliff adjacent to the steep grade just descended was created by the erosion of the fractured rock on each side of this unique rift. It is interesting to hike to the top of that alcove, then follow this distinct straight seam in the rock until it disappears under layers of Kayenta Sandstone. From this same trail junction, an alcove in the opposite canyon wall also has an exposure of Roberts Rift as a vertical crevice about one inch wide filled with white mineral salts. For more details about this rifting, see the sidebars, ROBERTS RIFT and UPHEAVAL DOME.

Just beyond this junction, on the main trail, a short spur to the left goes to the arch noted earlier. The low canyon walls in this vicinity are the upper levels of Wingate Sandstone, topped with the lower horizontal strata of Kayenta Sandstone. Note the ornate way in which the exposed Wingate has been weathered by rain and snowmelt. As the trail continues up one major drainage of Little Canyon, the terrain rises. It has been tilted by the Moab Fault, thus exposing, in this branch of the canyon, older strata than appear in the other upper arm of Little Canyon.

9.4 Note the other upper arm of Little Canyon going to the left, and the large undercut alcove in its righthand wall. The Gooney Bird monolith becomes more prominent against the skyline straight ahead until the trail skirts around its base. Most of the Gooney Bird is

Wingate Sandstone, but the horizontal strata in its neck and head are at the Wingate-Kayenta interface.

9.8 From the base of the Gooney Bird, notice the large arch to the left-front, about midway up the sloping canyon wall. The arch opening is partially obscured by fallen rock and is obvious only when viewed by afternoon back-lighting. Note also, the cliff straight ahead that has been honeycombed by water runoff.

From the trail, just beyond the base of the Gooney Bird, note the large alcoves near the top of the cliffline ahead. The righthand alcove has an opening into still another one around the prow of the cliff, creating the vertical span of rock between the two alcoves called Owl Arch, so named because of its resemblance to an owl's eyes and beak when viewed from the canyon below.

9.9 The tall spire in the sidecanyon to the right is known as "The Bride." A spur trail just ahead goes into this canyon. From the end of that spur, when viewed head on, The Bride appears very slender. There is an interesting multiple arch hidden in the cliff spires behind The Bride, as well as another arch nearby in the same cliff that is difficult to spot from the canyon floor. One spur of the Gold Bar Rim Trail reaches the cliff top opposite The Bride, offering a good view of that spire, the nearby arches and the canyon below. All of the several arches mentioned in Little Canyon are also in Wingate Sandstone, as are all of those in the Bull Canyon system.

10.2 The massive span of Ojo Arch is just above the canyon floor to the left here. A hike across the canyon to beneath the span will reveal its long, narrow opening. It is possible to climb onto the span itself from a little farther back down the canyon.

Note that the rising strata have now exposed the upper deposits of the Chinle Formation on both sides of the canyon. These older deposits are indicated by their parallel stratification and darker color, which contrasts with the lighter Wingate Sandstone in the vertical cliff. Ojo Arch developed in the Wingate Sandstone at the interface between these two deposits.

Just a few yards ahead, watch for afternoon light slanting through the opening of Lin-Lynn Arch in the Wingate Sandstone, very close to the jagged upper skyline ahead of the trail. This oblique opening is a few dozen feet to the right of a short spire with a horizontal cap rock. It is possible to hike to this spectacular natural span from the end of the branch of the Sevenmile Canyon Rim Trail that ascends Arths Rim.

Note just ahead that the rising strata are bringing the gray uranium-bearing deposits of the Chinle Formation into view. In the cliffline to the left of the trail just ahead, there is a short cylindrical spire that stands out slightly from the upper cliff. Petrified dinosaur tracks have been found at the base of this spire and in a nearby drainage. An unmapped spur of the Little Canyon Rim ORV trail goes to this drainage. The small but numerous three-toed tracks at this site were made by relatively small carnivorous dinosaurs. Petrified tracks in the Wingate Sandstone are relatively rare. This is one of only two such sites known at this time. The scarcity of fossil life and animal tracks in the lower Wingate is largely because a major extinction of life took place between the Wingate and older Chinle. See the sidebar, GREAT EXTINCTIONS.

10.8 The ridge to the right along this stretch of trail is entirely Chinle Formation deposits. Note on the Chinle slopes to the left-front, the old uranium prospect-

The Gooney Bird

ing trail and mine tailings. The upper cliffs here are entirely Wingate Sandstone. Specimens of agate, petrified wood and other fossils erode out of the Chinle deposits in this area. The Little Canyon Rim ORV trail, described in the referenced book and map, closely approaches the upper rim of the large amphitheater to the left here, providing a spectacular view of this part of Little Canyon and the broken terrain on toward the La Sal Mountains.

11.0 A tilted hill appears to the right. The gray-hued deposits near its summit are Chinle Formation, but the strata just below these are Moenkopi Formation. The red bluff beyond and to the left of the hill is Cutler Sandstone. To the right, the upper cliffs are Wingate Sandstone with sloping Chinle Formation deposits continuing down to the canyon floor, although there are traces of Moenkopi Formation beneath the rubble at the extreme left of the slope beyond the nearby hill. Note again, the old mining trails in the Chinle Formation.

11.2 Note the abandoned uranium mine shafts in the gray Chinle deposits to the left-front. Do not enter such mines. They are made hazardous by falling rock and accumulations of radioactive radon gas. See the sidebar, URANIUM.

11.4 At a junction here, the main trail goes left up the slope. The other trail is now abandoned.

11.7 Stop at the wide pullout on the right here to view one of the most spectacular panoramas in this region. From this viewpoint, a wide variety of geologic strata, distorted by various tectonic activities, can be observed to the south and east. Beginning at the left, the Book Cliffs loom on the distant skyline. Below these are the Entrada Sandstone fins and spires of Klondike Bluffs and the Devils Garden, in Arches National Park. The other red sandstone on the horizon to the east is Slickrock-Entrada, with the broad flat expanses of white sandstone that are closer being the Moab Member of Entrada. The more rounded, light-hued sandstone to the east below the Entrada bluffs is Navajo Sandstone. To the east in the distance, the northern nose of the Uncompahgre Uplift in Colorado is visible on clear days, above the sheer Wingate Sandstone cliffs of Top-of-the-World and Fisher Mesa. The upper tip of Castle Tower and The Priest and Nuns, both in the eastern wall of Castle Valley, protrude above the general cliffline.

To the south and southeast, the La Sal Mountains and their foothills dominate the horizon, with slopes and fins of Navajo Sandstone in the near-distance, beyond the old highway. The present highway below this viewpoint is traveling through Cutler Sandstone. The old highway follows the approximate alignment of the Moab Fault. This massive fault has offset the geologic strata on both sides for almost two thousand feet in this vicinity. For more information about the Moab Fault, refer to the road log for U.S. 191 North, elsewhere in this book.

To the right of the main highway as it descends Moab Canyon, red Cutler Sandstone cliffs rim this branch of Little Canyon. Just above the red sandstone, the thin layering of Moenkopi sediments continues upward into the gray layering of the Chinle Formation. The cliffline beyond the Chinle-topped hill that angles down to the right into Little Canyon is Wingate Sandstone.

From this viewpoint, note that in this "hanging valley" branch of Little Canyon, the drainage goes to the right rather than plunging into the lower canyon to the left. This is because of the sloping strata adjacent to the Moab Fault.

Beyond this pullout, the trail cuts through the previously horizontal strata of Moenkopi Formation. Farther on, where the trail begins to descend, it cuts down into upper Cutler strata. Drive this stretch of trail carefully, especially when it is wet from rain or melting snow in the off-season. Note in the steep slope behind this viewpoint, an abandoned vehicle trail ascending steeply to an old uranium mine, with its gray mine-tailings.

12.1 From a safe pullout on the right here, the looming Entrada sandstone spires of Klondike Bluffs appear to the right-front. The break in the cliffline ahead is Sevenmile Canyon. For the geology of Sevenmile Canyon, refer to the log for Utah 313. At this point, the trail has descended into dark red Cutler Sandstones and remains in these deposits until it reaches its end at the main highway. To the left of the highway below, note how the railroad track was cut deeply into the Cutler Sandstone in order to retain the required maximum gradient. As this trail reaches that level, it travels for some distance on the embankment constructed to protect the railroad cut from flooding. For more information about this railroad, see the sidebar, POTASH.

13.2 Here, the trail reaches the embankment that protects the railroad cut. Note the Entrada Sandstone in the Windows Section of Arches National Park on the skyline to the east. To the northeast, the sloping Entrada Sandstone of Klondike Bluffs looms above painted desert slopes of the Brushy Basin Member of the Morrison Formation. For more details concerning the geology on both sides of U.S. 191, refer to the road log for that highway, elsewhere in this book.

14.1 The trail ends where it joins U.S. 191. Turn right for the 10-mile return trip to Moab.

TRAIL NAME: **BULL CANYON TRAIL**

SUMMARY: The stratigraphy along this relatively short trail and its two spurs does not vary much, but the trail does offer several unique geologic highlights. One of these is a spectacular view of the twin Gemini Bridges from below. Another is easy access to two exposures of Roberts Rift and a distant view of another. See the side bar, ROBERTS RIFT. Still another highlight is a close look at a breathtaking canyon confluence. Perhaps the greatest geologic novelty is a chance to study the myriad beautiful shapes into which Wingate Sandstone erodes when its base is not undercut.

TRAIL SURFACE: eroded graded dirt, rough slickrock, rocky wash-bottom and packed sand and soils.

VEHICLE: off-road vehicles only.

MAP: ***Canyon Country* OFF-ROAD VEHICLE TRAIL MAP - Island Area.**

GUIDEBOOK: ***Canyon Country* OFF-ROAD VEHICLE TRAILS - Island Area.**

GEOLOGIC STRATA VISIBLE FROM THIS TRAIL

Navajo Sandstone	Page 13
Kayenta Sandstone	Page 13
Wingate Sandstone	Page 12
Chinle Formation	Page 12

MILEAGES:

TO START: 32.6 miles, from Main and Center, Moab.
TRAIL: 8.4 miles, one-way to end.
RETURN: 36.2 miles, from end of described trail.
TOTAL: 77.2 miles, from Main and Center, round trip.

Bull Canyon in the winter.

TIME: allow at least a half day, plus time taken for stops and hikes along the way, if only this trail is driven. Normally, this trail would be taken as a continuation of the Arths Pasture Road-Gemini Bridges Trail. When these two described trails are taken together, allow for a long day.

ROUTE: U.S. 191, Utah 313, Arths Pasture Road to the trail start, then return either via Arths Pasture Road and the two paved highways or via the eastern leg of the Gemini Bridges Trail and U.S. 191.

NAVIGATION: if this short but unusual trail is to be taken as a continuation of the Arths Pasture Road-Gemini Bridges Trail, drive north on U.S. 191, west and south on Utah 313, then east on Arths Pasture Road and the Gemini Bridges Trail to the natural bridges, then continue east on the Gemini Bridges Trail to the start of the Bull Canyon Trail. If the Bull Canyon Trail is to be explored by itself, drive north on U.S. 191, west and south on Utah 313, then east on Arths Pasture Road to the start of the Bull Canyon Trail. Alternately, the Gemini Bridges Trail can be driven from U.S. 191 to the same junction, but if that approach has been made hazardous by recent precipitation, the longer route is suggested. The listed map and guidebook are strongly recommended for navigating this complex terrain and trail system.

GEOLOGY TRAIL LOG:

0.0 As the trail starts, it travels through a broad expanse of layered, water-deposited Kayenta Sandstone toward a series of canyons that have cut deeply into Wingate Sandstone and the older Chinle Formation. As the trail starts, the canyons seen ahead to the south are the two main branches of the complex Bull Canyon system, both of which this trail enters farther on. To the left of the Bull Canyon system, the upper rims of Day and Long canyons are barely visible on the southern horizon.

1.2 Here, the Little Canyon Trail goes left at a junction near the summit of a low hill. If time permits, park and walk to the top of the rise to the left for a spectacular panoramic view to the south and east. The slickrock rimlands of Arths Rim, Gold Bar Rim and Poison Spider Mesa dominate the eastern horizon, with the notch of Little Canyon separating Arths Rim and Gold Bar Rim. The La Sal Mountains appear to the southeast. Below the La Sals, the fantastic maze of Navajo sandstone fins and domes is the Behind-the-Rocks wilderness area. Moab Valley lies just beyond the jagged rim of that maze.

On each side of Little Canyon, the slickrock slopes of Arths Rim and Gold Bar Rim are covered with thin remnants of layered white Kayenta Sandstone on top of reddish Wingate Sandstone. Farther to the right, the tilted rimlands of Poison Spider Mesa are dominated by open areas and rounded domes of Navajo Sandstone. The upper rims of the canyon systems directly to the south are on the other side of the Colorado River gorge.

Beyond this summit, the trail continues through layered Kayenta Sandstone, descending toward its interface with the older Wingate Sandstone, which is not yet in sight. Note that all the strata in this area have been warped and tilted by various anticlines and synclines.

1.6 Turn right at the junction here, where the trail is dropping into the drainage line of the main fork of Bull Canyon. For a short distance, the trail travels up the rocky wash-bottom through the shallow narrows of the lower canyon. As the trail continues, notice how the tilted terrain causes the canyon walls to become higher and higher as the canyon is penetrated.

1.9 Note that the Kayenta Sandstone is rising above the level of the trail to the left and straight ahead, with Wingate Sandstone exposed below.

2.2 Along this stretch of trail, note how the Wingate Sandstone erodes into steep slopes, while the layered Kayenta Sandstone above it fractures and collapses into rocky terraces. The Kayenta was fresh-water deposited, while the Wingate is primarily desert-dune deposits. Even so, these dunes were at intervals inundated by water. These water-affected intervals are noticeable along this stretch of Bull Canyon as bands of darker rock within the light reddish-brown Wingate Sandstone.

One outstanding aspect of the Bull Canyon system is that it presents an unusual opportunity to study the myriad lovely shapes Wingate Sandstone takes as it erodes. Although most Wingate Sandstone in the region has long since formed its typical sheer cliffs, in places where this has not yet happened, as in much of the Bull Canyon system, the Wingate erodes into countless other shapes and forms. Only when erosion has cut down into the softer underlying Chinle Formation and undermined the Wingate, as it has in the upper ends of the several Bull Canyon tributaries, does undercutting of the Wingate produce the collapse that leaves near-vertical cliffs. See the sidebar, ROCK EROSION.

Notice in the canyon how seeping springs reaching the canyon walls have created numerous large, oval-shaped alcoves. Although some geologists have claimed that Wingate Sandstone does not readily form arches, the Bull Canyon system alone contains at least three dozen such novel erosional phenomena, including the magnificent twin Gemini Bridges. Several smaller natural rock openings are visible from this trail, one just ahead.

2.3 Note the large cave in the sandstone wall to the left of the trail near a large rockfall. At certain times of the day, a patch of sunlight in the cave reveals that it has an opening in its top, making part of the cave roof a natural arch.

2.7 Park here as the trail travels in the wash-bottom along the base of the cliff, where gnarled old cottonwood trees stand on each side of the wash. One exposure of Roberts Rift can be seen in the cliff above here as a pair of vertical cracks with discolorations in the sandstone on each side. A few yards ahead, where the trail climbs out of the wash, a similar line in the more distant cliff wall to the left-front indicates another exposure of Roberts Rift. The bleaching of the natural red coloration of the rock adjacent to the rift was caused by soluble minerals thrust upward into this giant crack in the earth from the marine deposits at least 2,500 feet below. In the wash-bottom that goes toward the more distant exposure of the rift, large boulders of hard conglomerate that have eroded from the rift are scattered along the wash. Some of these contain marine fossils that were forced upward by the meteor impact that created Roberts Rift. See the sidebar, UPHEAVAL DOME. This gigantic crack in the earth's crust continues southwestward and also crosses the Dry Fork of Bull Canyon, which is described later in this log.

3.0 Along this stretch of the trail, notice how seeping underground water has been forced to the surface by subsurface strata of rock. This moisture supports a small community of water-loving vegetation. Cottonwood trees require a constant supply of water for their roots. Their presence here indicates that even during the driest time of the year, there is subsurface water along this stretch of cliff-base.

3.3 Here, the trail swings right to leave the main canyon and enter a short tributary canyon. This vehicle trail into the spur canyon did not exist until the massive Gemini Bridges were first documented in 1968. For more details, refer to the log for the Arths Pasture Road-Gemini Bridges Trail elsewhere in this book.

3.6 Note a short spur canyon coming in from the left, and the massive, sheer cliff to the right of the trail with its rich patina of desert varnish. See the sidebar DESERT VARNISH. Here, erosion has long since undercut the Wingate cliff-base, causing it to collapse along near-vertical planes. In the spur canyon to the left, erosion has not yet completely undercut the Wingate but has started, as indicated by the rock-collapse to the left-front.

3.7 The trail ends in a large cave formed by moisture seeping at the cliff-base. High above this alcove is another seep-cave. The Gemini Bridges are high in the opposite cliff. For more information about how such natural spans form, refer to the book, ***Canyon Country ARCHES & BRIDGES.*** After studying the geology and spectacular beauty of this spur canyon, return to the trail junction at Mile 1.6.

5.8 Back at the trail junction, turn right. The trail shortly returns to the same wash. Continue driving down that drainage.

5.9 Here, the wash joins the drainage of the Dry Fork of Bull Canyon. The trail to be taken later goes right, crosses the wash and climbs into higher terrain. Now, however, turn left and continue down the drainage line, on the trail that is in or beside the wash.

6.2 Park where the trail ends at a large sand dune, then very carefully explore the vicinity on foot. Walk out onto the rocky point just ahead for a breathtaking view of the confluence of the Bull and Day canyon systems. Hike left across the big dune into the slickrock drainage, then down that to where it has cut a deep water-filled grotto. Skirt left around a large alcove below the grotto to see where the drainage plunges over a 400-foot pouroff into the short spur of Day Canyon. Note that the rim of this alcove is deeply undercut, and that a narrow crack back from the edge makes the rim a long natural bridge. Just a few feet above the final drop is a small natural bridge that formed when water finally breached the wall of a pothole.

Short hikes in this vicinity offer still other spectacular views of this very unusual canyon confluence, where the immense and complex Bull Canyon system pours over a sheer cliff into a deeper adjacent canyon. Notice how the prevailing winds that blow up the tributary of Day Canyon are forced vertically upward through a notch in the cliff rim, to form the enormous sand dune at the trail's end. Although many wheeled vehicles disturb this dune, it is renewed almost daily. Fresh sand from the canyon below constantly replaces that lost in the wash behind the dune.

After studying the spectacular erosional phenomena at the end of this spur trail, turn around and return to the last junction.

6.4 Note the gap in the rock wall beyond the drainage to the left. During the relatively rare periods of heavy flash-flooding, water overflows the shallow wash and plunges through that gap directly into Day Canyon.

6.5 Back at the wash junction, go left on the trail into the Dry Fork of Bull Canyon. The stratigraphy of this spur canyon is much like that of the main fork traveled earlier. It begins in remnant layers of Kayenta Sandstone but, since the strata are tilted more steeply than the wash-bottom and trail, the canyon walls grow higher as the trail goes up the canyon. This spur, however, continues until the canyon-bottom cuts into the upper Chinle Formation, which erodes more easily than the Wingate Sandstone and thus undercuts it. As noted earlier, this undercutting produces sheer walls, rather than the more gentle slopes created by erosion before the Wingate is undercut.

6.8 Note that the lower canyon walls are now Wingate Sandstone that is eroding into a stone fantasy-land. From here on up the canyon, study the endless beautiful shapes that have formed in the Wingate Sandstone where it has not been undermined. This canyon is an outstanding example of this relatively rare phenomenon. Most of the hundreds of miles of Wingate exposed in the region have long since been undercut and shaped into sheer walls. At this point, look for a small arch in the opposite canyon wall, a few feet above the trail level.

7.1 Note the walk-through natural tunnel in the cliff-base to the right of the trail.

7.7 The trail crosses a drainage line. Just beyond, notice the immense relatively fresh rock fall in the left cliffline. This collapse took place in the late 1960s and was first noticed during the original search for a way to the bottom of the Gemini Bridges. The Dry Fork of Bull Canyon was mistaken for the less conspicuous main fork. The present obvious trails and guiding signs did not exist then. From this vantage point, note that as the Chinle Formation appears at the base of the Wingate cliffs, they are undercut and collapsing.

7.8 Note that a spring in the base of the Wingate cliffs to the right of the trail has been piped down to water troughs for domestic livestock. The rubble-covered slopes at the base of the Wingate cliffs obscure the upper levels of the Chinle Formation. Here, as in both canyons, the canyon floor is covered by rounded masses of decomposed rock and sand held in place by the remnants of native vegetation left after a century of overgrazing.

8.0 From the summit of a low rise, note in the cliff to the left how the thin wall of remaining Wingate Sandstone has eroded into a series of columns and spires. Park here, walk toward the righthand cliffline and start ascending the exposed slopes of gray Chinle deposit toward a large alcove high in the cliff. From the vicinity of the large red boulders there, note another exposure of Roberts Rift in the cliff, which appears as a notch in the cliff summit, a vertical crack below that, with a band of light-colored rock on each side of the crack. From there, sight toward the southwest up a long narrow tributary canyon. That canyon was formed by erosion along the weakness created by the rift, with the rift-plane itself being close to the canyon's left wall.

For a look at a large natural arch that is not visible from the canyon floor, continue the steep climb into the elevated alcove to the left of the rift exposure. From this vicinity, it is also possible to hike up the rift-formed canyon by starting up its drainage line, then climbing up on the right side to higher levels, to avoid an impassable pouroff. The Chinle deposits along that route contain petrified wood, which is fairly common in this formation.

Roberts Rift exposure in the Dry Fork of Bull Canyon.

8.1 Note delicate Pop-Top Arch that is visible in the rim of the cliff to the right along this section of trail. The arch is so high that drivers of vehicles with tops will need to stop and step out to see this graceful arch, or watch for it on the return trip.

8.4 The trail forks here in a wash bottom. The left fork climbs a hill and ends shortly. The right fork goes up the rocky wash, climbs over an old reservoir that has long since been breached, then continues to a series of uranium mines in the Chinle Formation on up the canyon. The reservoir was built to catch water for use by the uranium miners. Note that at this point the trail is more than one hundred feet below the Wingate-Chinle interface, which begins at the base of the sheer Wingate cliffs. It is hazardous to drive beyond this point, so the trail log ends here. Those who continue hiking up the trail to the mining area should stay out of the mines, which are dangerous because of possible rock collapses and accumulations of deadly radon gas. See the sidebar, URANIUM.

In earlier decades, there was a large quantity of very dark petrified wood in this vicinity that had been exposed during construction of the trail and reservoir. Most of this ancient fossilized wood has since been collected by rockhounds, although some remains in the higher levels of the canyon and small pieces are sometimes found in the wash-bottom.

From this point, return to the confluence of the two Bull Canyon drainages, then leave by the same route to the trail beginning.

12.1 From here, where the trail began, it is possible to return to Moab via Arths Pasture Road, Utah 313 and U.S. 191, or via the shorter route that goes right here. This route returns to U.S. 191 via Little Canyon, with one narrow and hazardous stretch. The return to Moab via this shorter route is 15.4 miles.

Gemini Bridges. For scale, note human figures on ledge below bridges.

ROBERTS RIFT

In a professional paper dated 1975, geologist Robert J. Hite described a strange rift -- a special kind of fault -- that is exposed in several canyons to the northwest of Moab, Utah. He named this gigantic vertical crack in the area's geologic strata "Roberts Rift," after a deceased miner, J.E. Roberts, who had been developing uranium claims in the area of the rift exposures.

This miles-long rift is one of several northeast-southwest trending faults in the region, all of them above the ancient Paradox Basin, so was apparently related to the others in some general structural manner. Yet it was quite different in a very important way -- the rift contained material, including large rocks and datable fossils, from a marine tideland deposit 2,500 feet below then known as the "Honaker Trail Formation," but now known to be upper Hermosa Group deposits. See the sidebar, YES -- GEOLOGISTS ARE HUMAN.

Further, some of the original material injected from below into the rift had been fluids containing minerals capable of chemically affecting the iron oxide within the younger strata in which the rifting occurred. These included strata known as the Chinle, Wingate, Kayenta, Navajo and Entrada. Where the rift is exposed in the iron-rich Wingate Sandstone, the reddish color of the surrounding rock has been bleached to white or pale yellow.

Roberts Rift is visible in several canyon systems near Moab. As reported by Hite, it begins in Arches National Park, as a series of "calcite veinlets" in the vicinity of The Windows, but is clearly visible in the park -- if you know what to look for -- only in parts of Courthouse Wash. In that area, to the east of highway U.S. 191, the rift is exposed within the Slickrock Member of Entrada Sandstone. In one canyon wall, it is visible as a sloping crevice mostly covered with rubble, at the end of a short eastern spur canyon of Courthouse Wash. On the western side of the canyon, the rift has formed a much deeper sidecanyon, with the near-vertical plane of the rift becoming one wall of the sidecanyon.

To the west of U.S. 191, Roberts Rift is clearly visible on both sides of Little Canyon. On the eastern side, very near the confluence of the two major upper tributaries of the canyon, it consists of a near-vertical crack in the Wingate Sandstone wall about one inch wide and filled with white mineral salts. On the opposite side, erosion and rock collapse caused by the rift has largely covered the actual crack in the vicinity of the off-road vehicle trail that ascends the dugway out of the canyon at the confluence. However, the rift is clearly visible to hikers as an arrow-straight, southwest-trending joint in the surface rock on the relatively level terrain above and beyond the canyon, until it eventually disappears beneath higher, unaffected layers of the Kayenta Sandstone through which it cuts.

The rift, together with erosional forces, has also contributed to the formation of a dogleg in the upper end of the Little Canyon tributary that contains the rock spire known as The Bride.

Roberts Rift also cuts across the two main tributaries of the Bull Canyon system farther southwest. There, it is again visible in two walls as near-vertical cracks filled with mineral salts and smaller rocks, with the surrounding native rock bleached of its natural reddish color. In the Dry Fork of Bull Canyon, the rift has contributed to the erosional formation of a large tributary canyon. Beyond this tributary, the rift heads across Grays Pasture, in the Island-in-the-Sky, in a direction generally tangent to the roughly circular structure of Upheaval Dome. See the sidebar, UPHEAVAL DOME.

The author of this book has walked out all surface exposures of Roberts Rift, and has noted that to the west of U.S. 191 the rift apparently did not progress upward entirely through the Kayenta Sandstone strata when it first formed. The visible rift thus "goes underground" about midway in the Kayenta when it is being traced on the ground. The fact that the rift appears in the much younger Entrada Sandstone east of U.S. 191 and the Moab Fault is a good indication that the Moab Fault is older than Roberts Rift. The large vertical Moab Fault offset allowed Roberts Rift to cut into much younger strata east of the fault than west of it -- at the same approximate level.

The accompanying map shows the approximate alignment of Roberts Rift, as described by R.J. Hite and in a subsequent professional paper authored by geologists Peter W. Huntoon and Eugene M. Shoemaker.

The later report proposed a possible cause for Roberts Rift. As the authors noted --

"We concur with (Hite's) conclusion that altering fluids circulated upward from deep sources tapped by the rift. (However) we respectfully differ with the genesis of the rift as proposed by Hite .. "

Huntoon and Shoemaker then proposed a cause for the rift that would fully explain the origin of the enormous amount of energy that would have been needed to form the long rift and force up into it materials from 2,500 feet lower. Hite had indicated that the rift probably formed as a result of severe "wrench faulting," and hinted that Paradox salt flow had somehow thrust the older material upward into the resulting rift, but did not address the matter of the enormous amount of energy required for such a process, or the absence of any remnants of Paradox Basin evaporites within the rift.

The Huntoon-Shoemaker report did address these factors and concluded that the rifting, and the subsequent filling of the rift with older and much lower material, were directly related to the immense forces that formed Upheaval Dome. The report first analyzed the geologic setting at the time of the meteorite impact and rifting. It noted that --

"Deposition of the Mesa Verde Group was nearing completion .. (and) .. what is now the rim of Upheaval Dome was buried by almost a mile of younger sediments .. "

-- and that the surface deposits in the region then were either on a near-sea-level alluvial plain or in a shallow marine environment. It also pointed out that the much deeper marine tideland deposits were highly pressurized by still deeper Pennsylvanian organic deposits that were confined

continued ----

by the effective seals of higher and younger strata, thus producing conditions that could be triggered by some violent geologic event -- such as a meteorite strike. As noted by the Huntoon-Shoemaker report:

"Linking the origin of the Roberts rift to the impact of the Upheaval meteorite near the close of the Cretaceous time very conveniently accounts for the tremendous energy required to form the rift. What was needed was a sudden influx of energy. Such energy requirements implied an extraordinary event unlike any geologic process known to have occurred spatially near the rift. Extraordinary occurrences require extraordinary explanations. A meteorite impact satisfies both."

The report supported its conclusion with a stress analysis that showed a definite relationship between Roberts Rift and Upheaval Dome, and showed how the explosive impact of the meteorite could have provided the enormous energies required for the rift formation, and for the injection of solids and liquids into the momentarily-open rift from as much as 2,500 feet below.

SUMMARY AND CONCLUSION

When all of the presently available evidence is considered, little doubt remains that Upheaval Dome and Roberts Rift were formed at the same time by the impact of a meteorite or cometary body about 1,600 feet in diameter and probably composed largely of ice, as many such astronomical objects are. The shape of the disturbed geologic strata in the vicinity of the crater matches that of other known impact craters, and the immense energy required for the formation of Roberts Rift could easily have been provided by such a violent impact.

Since the time of that impact, at about the close of the Cretaceous, the impact site rose from about sea level to where its present remnant is still about 6,000 feet above sea level after more than a mile of rock has eroded from the original sedimentary surface. This erosion has left only remnants of the geologic strata deformed by the impact that were below even the lower tip of the original impact crater, and has exposed in some Moab-area canyons the upper zones of Roberts Rift. See the illustration accompanying the sidebar, UPHEAVAL DOME.

Since the meteorite impact took place at about the same point in geologic time as the much larger one that brought on the Great Extinction of dinosaurs and other forms of life at that time, it is probable that both of these meteors were part of a general swarm of such space debris that our solar system has encountered periodically over an immense span of time. Each such encounter has produced mass extinctions of life on Planet Earth, and has left deep scars on its surface, some of them obvious -- such as Upheaval Dome -- and others obscured by this planet's seas. Only recently have scientists found -- in the Gulf of Mexico -- the remains of the crater of the immense meteor that is believed to have caused the Great Extinction at the end of the Cretaceous.

Thus, visitors to the Moab area have a chance to marvel at two rare phenomena -- visible remnants of one of the visitors from outer space that have so devastated this planet in the past -- Upheaval Dome and Roberts Rift.

Roberts Rift exposures in Bull Canyon. The exposure in the photo above appears across the canyon in the photo below.

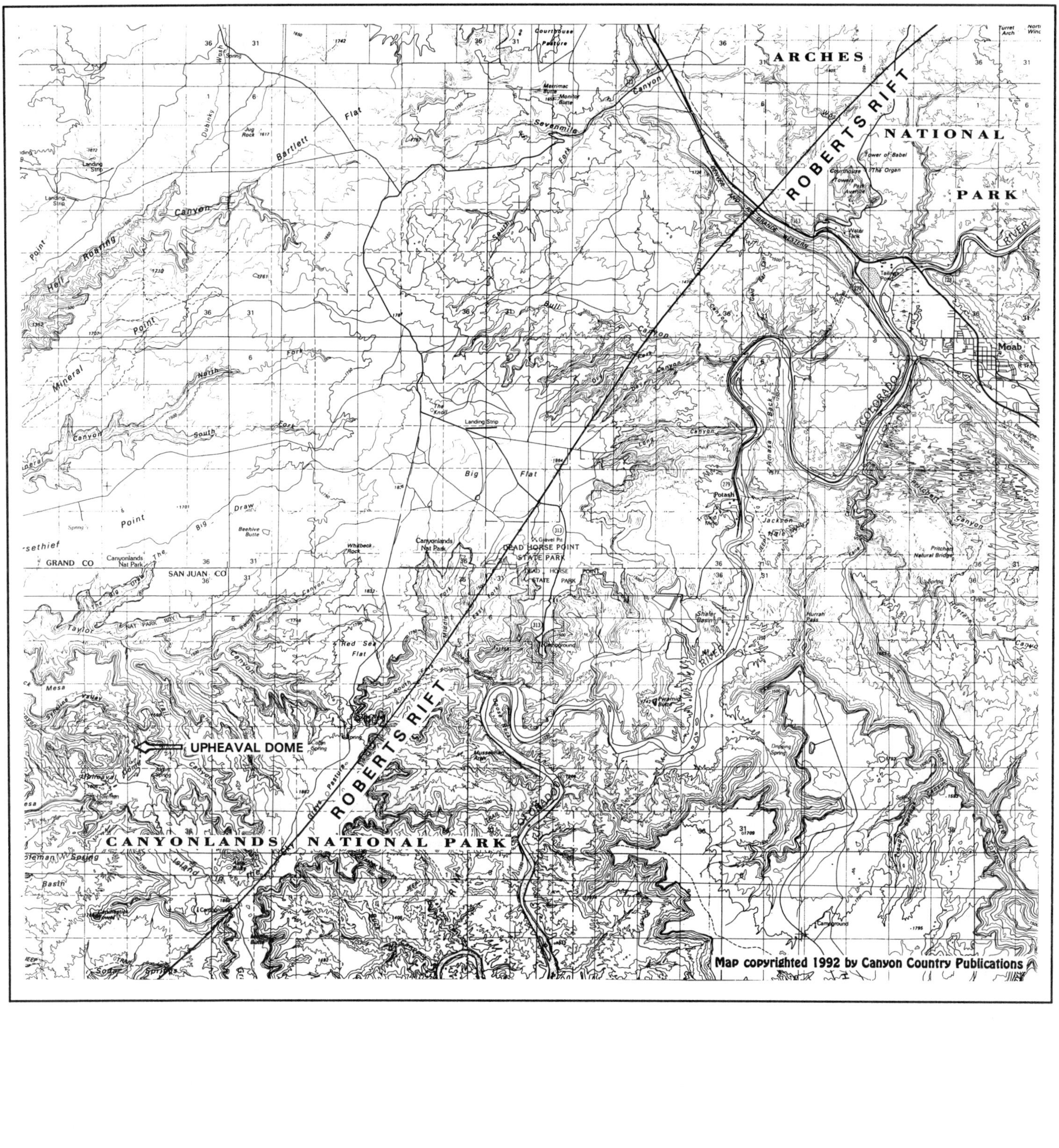
ARCHES NATIONAL PARK
ROBERTS RIFT
ROBERTS RIFT
UPHEAVAL DOME
CANYONLANDS NATIONAL PARK
DEAD HORSE POINT STATE PARK
Moab
COLORADO
RIVER
Big Flat
Landing Strip
Potash
GRAND CO
SAN JUAN CO
Map copyrighted 1992 by Canyon Country Publications

TRAIL NAME: **CANE CREEK CANYON RIM TRAIL**

SUMMARY: This fairly long trail travels across broad high-desert meadows, down rock-walled valleys and on the lofty rim of spectacular Cane Creek Canyon. Along the way, it offers outstanding views of a broad spectrum of geologic strata and erosional landforms. The last mile or so of the trail provides a chance to study a rarity -- the eroding upper surface of Wingate Sandstone, a deposit normally protected by the more durable overlying Kayenta Sandstone. The canyon-rim trail also provides views that include Hatch Point, mid-Cane Creek Canyon, Hurrah Pass, the Cane Creek and Moab Valley salt anticlines and Kings Bottom Syncline, the spectacular sandstone fins of the Behind-the-Rocks wilderness area, and the broad benchlands and soaring peaks of the laccolithic La Sal Mountains.

TRAIL SURFACE: packed sediments, broken rock, rough slickrock, steep slopes and grades, eroded cuts, rock ledges and sand.

VEHICLE: off-road vehicles only beyond the first five miles. Most standard highway vehicles can travel the first five miles during dry weather if carefully driven.

MAP: ***Canyon Country*** **OFF-ROAD VEHICLE TRAIL MAP - Canyon Rims & Needles Areas.**

GUIDEBOOK: ***Canyon Country*** **OFF-ROAD VEHICLE TRAILS - Canyon Rims & Needles Areas.**

MILEAGES:

TO START: 13.3 miles, from Main and Center, Moab.
TRAIL: 14.8 miles, one-way to end, as described.
RETURN: 27.5 miles, from end of described trail.
TOTAL: 55.6 miles, from Main and Center, round trip.

TIME: Allow for a long, full day to take this trail to its end. This will provide ample time for stops along the way, for lunch, and for a short hike into the spectacular slickrock wilderness beyond the end of the trail.

ROUTE: U.S. 191 south from Moab, then this trail to its end and back by the same route.

NAVIGATION: Drive south on Main Street (U.S. 191) from Moab for the indicated distance, then west on the Cane Creek Canyon Rim Trail, as described in these pages and the referenced map and book.

GEOLOGIC STRATA VISIBLE FROM THIS TRAIL

Morrison Formation	Brushy Basin	Page 18
	Salt Wash	
	Tidwell	
Entrada Sandstone	Moab	Page 16
	Slickrock	
	Dewey Bridge	
Navajo Sandstone		Page 13
Kayenta Sandstone		Page 13
Wingate Sandstone		Page 12
Chinle Formation		Page 12
Moenkopi Formation		Page 12
Cutler Group	White Rim	Page 11
	Organ Rock	
	Cedar Mesa	
	Lower Cutler	

GEOLOGY TRAIL LOG:

0.0 As this trail leaves U.S. 191 at the summit of the grade out of Moab Valley, it enters an area dominated by remnants of the Salt Wash Member of the Morrison Formation. Little else but this white, ledgy sandstone is visible for the first quarter-mile, as the trail travels through sparse pinyon-juniper woodland.

0.3 From the vicinity of a turn to the left and a spur trail to the right, the Navajo Sandstone fins of Beyond-the-Rocks come into view to the north. The upper end of Moab Valley can be glimpsed in the lowlands to the right, below the looming escarpment of South Mesa. The rimland of South Mesa is the Moab Member of Entrada Sandstone, with the slopes of the Slickrock Member of Entrada Sandstone below the rim largely covered by debris and vegetation. The light-hued sandstone still lower and in the far rim of Moab Valley is Navajo Sandstone. The layered white slickrock in the near-foreground at the junction is Salt Wash-Morrison. The La Sal Mountains sprawl across the eastern horizon.

The brushy slopes and tan-colored cliffs below the mountains are remnants of Mancos Shale. This relatively young deposit is out of its stratigraphic position here because this end of the Moab Valley salt anticline had formed, and most of the older deposits had eroded away, before the Mancos Shale was deposited. Such out-of-position deposits are found throughout Moab Valley. See the sidebar, SALT ANTICLINES.

In the distance to the left-front, the bare red-hued sandstone ridge is Slickrock-Entrada. The more open terrain is decomposed rock and sand from various formations held in place by native vegetation. Beyond this junction, the trail drops into that lower terrain.

0.4 At a trail junction here, stay left. The trail to the right, and some of the other spurs along this trail, are described in the referenced book and map.

1.3 From the vicinity of a spur trail to the right, the rounded sandstone cliffs to the left of the trail are Slickrock-Entrada topped with crumbling remnants of Morrison Formation deposits. Slickrock-Entrada also dominates the nearby terrain to the right of the trail. On the distant skyline straight ahead, the flat-topped peninsulas are Navajo Sandstone. Those ridges are on Hatch Point, beyond the deep gorge of mid-Cane Creek Canyon.

Coming into view in the lower terrain is an area of "living" sand dunes, which are rare in this region. Normally, sand is anchored against wind and water erosion by cryptogams, which grow on the surface of the sand. See the sidebar, CRYPTOGAMIC SOIL. In a few locations, however, active and shifting winds move the sand so frequently and vigorously that cryptogams are unable to bind the sand surface. Note that much of the rolling terrain surrounding the living dunes has been bound by cryptogams, and that this has permitted other forms of vegetation to take root and thrive, still further binding and enriching the sandy soil. This process clearly demonstrates one way in which biology plays a part in geology. Note beside the trail, here and elsewhere, how bulldozing has destroyed the binding effect of vegetation and cryptogams and is thus promoting erosion of the soils. This is an example of how human affairs often become geologic processes. See the sidebar, HUMANITY AS A GEOLOGIC FORCE.

2.3 Note in the cliffline to the left of the trail that lower Morrison deposits dominate for a stretch, between the smooth, red sandstone walls of Slickrock-Entrada, giving the appearance that the Morrison-Entrada interface sags downward there. This unusual effect was created by a large curving fault with several lateral branches that runs just behind the visible Slickrock-Entrada exposure. Note also here the remnants of bulldozed mineral exploration trails in the Morrison slopes, made in the search for the uranium ores that are sometimes found in this deposit.

3.0 The trail crosses a shallow sandy drainage. The slickrock adjacent to the drainage is the upper surface of Navajo Sandstone. Note how the very sandy soils in this vicinity are barely being held in place by vegetation. Part of this is because the cryptogamic surface has been disturbed by trail building and by vehicles that have recently left the trail many places as it travels through this broad valley. Part is also due to long periods of overgrazing, during which cattle have devastated the native grasses and their hooves have damaged the erosion-resisting cryptogamic surfaces. It is quite possible that this damage has caused, or at least contributed to, the creation of the patch of living dune sand in this valley.

3.1 This trail goes straight ahead at a junction here. The beautifully eroded sandstone domes and fins on the northern skyline are all in the Behind-the-Rocks area and all are Navajo Sandstone. Beyond here, for some distance, all of the slickrock exposed on both sides of the trail is also Navajo Sandstone.

4.0 At a junction here, this trail angles left.

5.0 Note how the soils on each side of the trail in this vicinity have a reddish tinge. This is because these soils consist of decomposed Entrada Sandstone, which has a high iron content. The immense slickrock ridge that the trail is approaching is Entrada Sandstone. The lower part that consists of wavy, layered deposits is the Dewey Bridge Member of Entrada Sandstone, while the upper part is Slickrock-Entrada. Note the several seep-caves in this ridge, most of them at the interface between the Entrada members, but two big ones at a water-affected layer within the Slickrock-Entrada. The largest of these, below the saddle in the ridge, is actually a pothole arch, which will be described just ahead in this log. Note to the right more of this reddish Entrada Sandstone ridge partly obscuring the Navajo Sandstone fins of the Behind-the-Rocks wilderness area. On the meadows in this vicinity, the grasses are not native but are exotic species that have been sown in an attempt to enhance grazing once the native grasses had been destroyed. Such non-native grasses do not thrive and do not hold the sandy soil as well.

5.1 Angle right on a spur trail that approaches the large slickrock ridge, for a short but interesting side-trip.

5.3 Where this spur trail joins another one that travels along the base of the slickrock ridge, stop and study massive Balcony Arch, which is the name of the pothole arch noted earlier. Like many other such natural spans, this one was created when an immense pothole that held water broke through into the seep-cave that the water had created in the cliff-wall below it, thus isolating a massive arc of rock. For more information on how natural arches and bridges form, see the book, ***Canyon Country* ARCHES & BRIDGES.**

As the trail continues near this ridge of sandstone, note the undulating interface between the two members of

Balcony Arch

the Entrada. For an explanation of this, refer to the earlier part of this book where Entrada Sandstone deposits are described.

5.6 A short spur trail to the left here approaches a large seep-cave that is commonly used for primitive camping. This cave may also have been used by prehistoric Indians, but modern use has long since destroyed any evidence of this. Beyond here, continue on the trail as it passes between the two slickrock masses and skirts around to the far side of the mass on the left. As this trail continues, stay left at each of the spur trail junctions.

5.7 Park here where this spur, as described, ends and study Picture Frame Arch, a large pothole span in the cliff to the left. It is possible to climb into and behind the arch from an access route to the left of the span.

Most of the white rounded sandstone to the north and northeast of here is Navajo Sandstone, but two red remnants of Entrada Sandstone top low domes to the

Picture Frame Arch

north. The larger of these is partly Slickrock-Entrada above darker Dewey Bridge-Entrada, while the smaller is entirely Dewey Bridge. On the skyline in the distance to the north, the Entrada Sandstone cliffs of Big Mesa are visible. To the left of Big Mesa, but nearer, a short stretch of cliffline north of the Colorado River appears above the white sprawl of Navajo Sandstone.

From Picture Frame Arch, turn around and return to the main trail by way of the trail spur that travels along the base of the red slickrock ridge.

6.1 At the trail junction below Balcony Arch, continue straight ahead on the shorter route to the main trail.

6.3 At the main trail, turn right.

6.7 From a narrow pullout on the right, as the trail begins to descend into lower terrain, study the Navajo Sandstone fins area to the right. Note that where this aeolian sandstone has been undercut at its base by drainage lines, it has sheered away into near vertical cliffs. The white rock-shards beside and near the trail in this vicinity are an agate-like mineral that forms in the Dewey Bridge-Entrada. As that water-deposited dark red sandstone erodes away, this harder mineral is left on upper surfaces of the underlying Navajo Sandstone.

Beyond here, the trail descends into lower terrain. The main drainage below eventually pours into a deep sheer-walled tributary of mid-Cane Creek Canyon. As in the open terrain this trail has already crossed, the meadows in this drainage area have been denuded of native grasses by grazing domestic livestock, leaving only sparse vegetation to hold the sandy soil in place. This has led to accelerated erosion, which can be seen several places as the trail nears or crosses the drainage.

7.7 From this stretch of trail as it is descending steeply into the drainage, note the gray-hued horizontal layering to the right and ahead within the Navajo Sandstone. This anomalous deposit, which is more than twenty feet thick, represents highly mineralized sediments that collected in a large and long-lasting drylake in the Sahara-like Navajo Sandstone desert. See the sidebar, PETRIFIED PLAYAS. Such lithified remnants of desert drylakes are fairly common in Navajo Sandstone, but few are as large and conspicuous as this one. As the trail continues on down the drainage, note that this exposure continues along the base of the sloping slickrock walls of Navajo Sandstone to the right of the trail. Note also in this area, that the rounded slopes on both sides of the trail are dune sand barely held in place by remnants of cryptograms and native vegetation. Just ahead, careless off-road vehicle drivers have further damaged the fragile sandy soils. Conscientious users of the public lands in this region stay strictly on designated and described trails.

7.9 Here, another trail joins this one and then leaves immediately. Stay left at both junctions.

8.0 The trail crosses the main drainage here. Note the cottonwood trees growing farther up the wash. These indicate that there is permanent subsurface water here. Note to the left how ATV users have damaged the sandy slopes with hill-climbing activities. This further promotes the erosion that has already been started by domestic livestock grazing.

9.2 Somewhere in this vicinity, park on an exposure of slickrock beside the trail and walk left down toward the main drainage line for a breathtaking view of its abrupt pouroff. As the trail continues, the low bluffs and domes of sandstone on both sides of the valley are still Navajo Sandstone. Watch the drainage to the left for glimpses of the upper rimlands of its deep inner gorge. Those rimlands are Kayenta Sandstone.

10.1 Stop at the summit of a low rise here and study the surrounding scene. Note in the nearby sandstone ridge to the right that the desert drylake deposits described earlier are still visible here as horizontal, water-affected layers within the cross-bedded aeolian sandstone. Ahead, the layered sandstone is Kayenta, while the lighter-hued canyon rimland is the upper surface of Wingate Sandstone. The deep canyon to the left of the trail is mid-Cane Creek Canyon. Its vertical walls are Wingate Sandstone topped with layered Kayenta. The sloping light-colored cliffs still higher are Navajo Sandstone. The higher terrain to the west beyond Cane Creek Canyon is part of immense Hatch Point. Both Hatch Point and Cane Creek Canyon are in Canyon Rims Recreation Area. Two of the roads on Hatch Point, Needles Overlook Road and Anticline Overlook Road, are described in geology logs elsewhere in this book.

In Cane Creek Canyon ahead, the upper part of the wide sheer-walled gray tower is Chinle Formation, while the lower part is Moenkopi Formation. The layered strata below the tower are largely Cutler Group deposits. The promontory above and to the left of that sheer-walled tower is Anticline Overlook. Note again that the sloping meadows on each side of the trail as it continues are largely dune sand barely held in place by remnants of native vegetation. To the left-rear, the tributary drainage from the valley that the trail has been traveling joins Cane Creek Canyon.

10.4 Note the seep-line in the cliff to the right of the trail. This occurs when water seeping downward through the relatively porous Navajo Sandstone reaches an impermeable water-affected layer and travels laterally, to form a seep-cave in the nearest cliff face. Such phenomena are common in massive sandstone such as Navajo, Wingate and Entrada. Note the young arch in the cliff to the right of the seep.

10.6 Beginning here, the trail ascends once more into Navajo Sandstone, where it reaches the level of the drylake deposits noted earlier before continuing to climb.

10.9 Here, just a few feet before the steeply ascending trail curves left across a shallow drainage, it reaches a fault. Note how the exposure of Navajo Sandstone to the right of the trail ends abruptly, interfacing with soils and rubble, with a layer of white minerals several inches thick subsequently built up within the juncture. This fault has formed the steep drainage to the left of the trail here and extends across Cane Creek Canyon into Hatch Point, where it has helped form a large tributary gorge called Trough Springs Canyon.

11.0 Pause near the summit of the grade for a good view to the south of Cane Creek Canyon and its rimlands.

11.1 At a trail junction here, go right. As the trail continues, it travels near low ridges of Navajo Sandstone.

12.0 As the trail begins to descend into somewhat lower terrain, note Anticline Overlook looming above the more distant solar evaporation ponds beyond the Colorado River, which cannot be seen from here. The entire scene in that direction is described in detail from the end of this trail. As the trail continues, it drops into and travels in various levels of layered Kayenta Sandstone, with only occasional remnants of Navajo Sandstone. In the distance, straight ahead, the gap to the left of the sloping ridge of white sandstone is Little Canyon. The Book Cliffs dominate the northern skyline.

12.6 At an obscure trail junction here, the Cane Creek Canyon Rim Trail goes left, to continue closely parallel to the rim of Cane Creek Canyon, while the trail to the right drops into lower terrain and goes to Pritchett Arch and Pritchett Canyon. Beyond this junction, the described trail continues to travel through layered Kayenta Sandstone.

13.2 Note to the right of the trail the steel pipe protruding from the ground that marks an old drill site. This particular drilling was probably done in association with petroleum exploration rather than uranium prospecting. The more distant scene to the east is described ahead from a better vantage point.

14.1 From another drill pad to the right of the trail, the view to the east is breathtaking! The full range of the La Sal Mountains stretches across the southeastern horizon. From left to right, the peaks and passes of the La Sals are: Grand View, Horse Mountain, Mount Waas, Manns Peak, Mount Tomasaki, Haystack Mountain, Geyser Pass, Mount Mellenthin, Mount Peale and Mount Tukuhnikivatz somewhat overlapping, La Sal Pass, and South Mountain.

The Navajo Sandstone domes and fins of Behind-the-Rocks dominate the scene below the mountains. Moab Valley lies just beyond this jagged rimline. Note that some of the sloping drainages that descend from the Behind-the-Rocks rimland have removed all of the Navajo Sandstone and cut deeply into layered Kayenta Sandstone.

All of the lower terrain below this viewpoint eventually drains into Hunter Canyon which, in turn, drains into lower Cane Creek Canyon. The sloping white sandstone ledge just below this drill-pad and its adjoining catch-basin is Kayenta Sandstone. The near vertical walls of the deeper canyon just beyond this ledge are Wingate Sandstone.

14.2 Park here at the summit of a short grade and walk a few yards for a breathtaking view of mid-Cane Creek Canyon. Exercise care when near the canyon rim.

14.4 Here the trail enters an expanse of barren slickrock. The trail angles right within a few feet, passing to the right of a group of pinyon trees. Beyond the trees, bear somewhat left and watch for signs of tire tracks in the loose sediments below the expanse of slickrock. From this lower level, the trail almost immediately climbs back onto another terrace of slickrock. Ahead, the trail angles right to skirt along the lower edge of a sloping expanse of slickrock, traveling mostly in the broken rock and sediments along that edge. As the trail continues, watch for tire tracks crossing patches of soils, and for crumbled sandstone where it crosses slickrock. The trail continues from here for another half-mile, but careful navigation is required in order to find the way without creating new tire tracks.

14.8 Here, the trail drops off of slickrock into the reddish sediments of a depression between the Cane Creek Canyon rim to the left and a red-and-white banded ridge of slickrock to the right. At the far end of this micro-valley, the trail angles right, then left, and enters an area that was extensively bulldozed by uranium miners. The mines were in the Chinle Formation deposits near the base of the canyon-wall cliff to the left. The drilling up here was done in an attempt to find additional veins of uranium ore. For practical purposes, this trail as described ends here, although remnants of earlier bulldozing continue ahead in several directions. Conscientious drivers park in this tiny valley to permit the slow-growing native vegetation beyond to recover and hide the bulldozing traces.

The spectacular 360-degree panorama visible from this vicinity is described from the summit of the low dome of white sandstone near the rim of the canyon just ahead.

The ledgy slickrock remnants in the nearby canyon rim are Kayenta Sandstone. On to the north, and in red-banded Trincline Ridge to the east, the rim is entirely Wingate Sandstone that has been exposed because the sloping terrain east of Cane Creek Canyon has permitted rapid water runoff to remove most of the Kayenta. This slope is part of the Cane Creek anticline. As described in the book, ***Canyon Country* SLICKROCK HIKING & BIKING,** there is excellent slickrock hiking beyond the end of this trail, with unique views of erosional landforms in the bare, upper surface of Wingate Sandstone, including a number of curious natural arches. Distinctive Trincline Ridge was given its name because it lies between the Cane Creek and Moab Valley anti*clines* and overlooks the Kings Bottom Syn*cline,* for a total of three *clines*.

Across Cane Creek Canyon, the high prominence is Anticline Overlook. With good binoculars, some of the developments on the overlook can be seen from here. Beyond that, in the distance, is the conical structure of Pyramid Butte, with the soaring cliffline of Island-in-the-Sky on the western horizon. The solar evaporation ponds so prominent along the base of that elevated plateau are beyond the Colorado River, which is out of sight because of the arching ridgeline of Hurrah Pass. For more information about the ponds, see the sidebar, POTASH. Note the slender ribbon of the Hurrah Pass Trail crossing the saddle in that ridgeline. Cane Creek Road and other parts of the Hurrah Pass Trail are visible on the canyon floor below. Both of these are described elsewhere in this book. The Cane Creek drainage winds sinuously along the canyon floor as a line of green vegetation in stark contrast to the surrounding red sediments.

The arcing strata in the vicinity of and beyond Hurrah Pass are part of the Cane Creek Anticline, where all the strata in this vicinity have been arched upward by pressures from below. The upper rimlands of Hatch Point directly across the canyon are Kayenta Sandstone, with sheer Wingate Sandstone walls below that. Rubble-covered slopes of Chinle, Moenkopi and Cutler deposits continue from there down to the canyon floor. All of the dark red strata in the vicinity of Hurrah Pass are Cutler Sandstone. Note the old uranium mine trail that travels within the gray Chinle deposits on the steep slopes to the right of Hurrah Pass.

The Navajo Sandstone fins and domes of the Behind-the-Rocks wilderness area dominate the eastern skyline. The low valley between Cane Creek Canyon rim and Behind-the-Rocks marks the Kings Bottom Syncline, where the geologic strata dropped while the Cane Creek and Moab Valley anticlines pushed them upward on each side. Short stretches of the Pritchett Canyon ORV trail can be seen in the lowest part of that rocky valley. The reddish cliffs in the distance to the northeast are Entrada Sandstone in the Windows Section of Arches National Park. The gap in the nearer cliffline to the left of the Windows Section is The Portal, where the Colorado River leaves Moab Valley and continues through its downriver gorge.

The continuous ridgeline to the left of The Portal is Poison Spider Mesa and Gold Bar Rim, respectively. Some Navajo Sandstone remains on Poison Spider Mesa but it is largely gone on Gold Bar Rim farther left, leaving the sloping surface there largely Kayenta Sandstone, although a few drainages there cut deeply into the underlying Wingate Sandstone. The break in that cliffline is Little Canyon, through which the Gemini Bridges Trail passes. The Book Cliffs dominate the extreme northern horizon, with sheer-walled Big Mesa taking over farther west. The geology of Big Mesa is described in the road log for Utah 313. The geology of the Book Cliffs is described in the road log for U.S. 191 North.

The high ridge in the distance to the southeast is topped by the microwave tower that marks the end of the Black Ridge Road, which is also described elsewhere in this book.

Above, Anticline Overlook looms above the arching strata of Cane Creek Anticline.

Below, color-banded Trincline Ridge frames the distant La Sal Mountains.

TRAIL NAME: **HIDDEN CANYON RIM TRAIL**

SUMMARY: This short ORV trail offers several unique geologic highlights, including a desert spring in an unlikely location, a major fault, two deep and lovely canyons cut into colorful Entrada Sandstone, dozens of strange "tripod rocks," a hanging garden in an immense seep-cave, a very hikable slickrock drainage line, and a close look at an anomalous deposit between two members of Entrada Sandstone. It also passes the collapsing remnants of an historic dugout cabin. This trail can be traveled by itself, or in combination with the Monitor & Merrimac Trail, described elsewhere in this book, depending on the time available.

ROAD SURFACE: graded dirt, rocky rubble, eroded slickrock and both wash and dune sand.

VEHICLE: off-road vehicles only beyond the first mile.

MAP: ***Canyon Country* OFF-ROAD VEHICLE TRAIL MAP - Island Area.**

GUIDEBOOK: ***Canyon Country* OFF-ROAD VEHICLE TRAILS - Island Area.**

MILEAGES:

TO START: 21.2 miles, from Main and Center, Moab, directly via the Blue Hills Road.

3.7 miles from the described loop of the Monitor & Merrimac Trail.

TRAIL: 4.3 miles, one-way to end.

RETURN: 25.5 miles, from end of trail via Blue Hills Road.

TOTAL: 51.0 miles, from Main and Center, round trip, if this trail is traveled by itself, rather than as an addition to the Monitor & Merrimac Trail.

GEOLOGIC STRATA VISIBLE FROM THIS TRAIL

Formation	Member	Page
Mancos Shale		Page 20
	Ferron Sandstone	
Dakota Sandstone		Page 20
Burro Canyon/Cedar Mountain Formations		Page 19
Morrison Formation	Brushy Basin	Page 18
	Salt Wash	
	Tidwell	
Entrada Sandstone	Moab	Page 16
	Slickrock	
	Dewey Bridge	

Hidden Canyon from its rim.

TIME: allow one-half day, plus time taken for hikes and for exploring spur trails on the return trip. As an addition to the Monitor & Merrimac Trail, allow at least three more hours because the trail is slow to travel.

ROUTE: Main Street (U.S. 191) north, Blue Hills Road west, then the Hidden Canyon Trail south. Alternately, travel the Monitor & Merrimac Trail, then take the shortcut indicated in that trail log to the start of this trail.

NAVIGATION: Drive north on U.S. 191 for 17.6 miles, west on Blue Hills Road for 3.6 miles, then south on the Hidden Canyon Rim Trail. Or, to reach the trail start from the lower drainage of the Tusher Canyon loop on the Monitor & Merrimac Trail, turn left at the trail junction in the lower drainage, drive 2.4 miles to Blue Hills Road, then west for 1.3 miles on that road to the trail start.

GEOLOGY TRAIL LOG:

0.0 As this trail spurs south from Blue Hills Road, it crosses an expanse of alluvium and sand held in place by desert vegetation. The low ridge of gray sediments to the north of this junction is Mancos Shale. Within that ridge, the thin layer of sandstone near the top is the Ferron Sandstone Member of Mancos Shale. The first ridge of gray deposits to the south of the junction is the Cedar Mountain Formation, which is similar in appearance to the Brushy Basin Member of the Morrison Formation. The higher and more distant rubbly slopes are the Salt Wash Member of the Morrison Formation. These two deposits are misaligned by a fault between them. The smooth, white sandstone occasionally glimpsed farther south is the Moab Member of Entrada Sandstone.

0.3 At a four-way trail junction, this trail angles left and crosses a shallow wash. Note how decades of overgrazing in this area have accelerated alluvium-cutting in the wash.

0.7 The trail crosses a shallow drainage marked by a line of trees and large shrubs. These are supported by subsurface water that originates from springs below Lunar Canyon, which is noted at mile 2.2. During wetter times of the year, this narrow gorge is a delight to hike because of its tiny, flowing stream that shortly disappears underground in the more open valley.

1.0 Here the trail is entering a drainage that is fed by a water source called Brink Spring, which was named after an early settler. The spring was developed for use by cattle. The trail is still traveling through Cedar Mountain Formation, but the high rocky hill above Brink Spring is Salt Wash-Morrison. As the trail continues beyond the water tank just ahead, it climbs into higher terrain past the collapsing remnants of a small dugout cabin to the left of the trail.

1.3 As the trail climbs steeply to the left, the dugout cabin is in the adjacent fenced area. Study this interesting historic structure, but do not hasten its deterioration.

1.4 As the trail approaches an area of slickrock, the rocky hill to the right is Morrison Formation deposits, with sandstone ledges of Salt Wash-Morrison above red slopes of the Tidwell Member of the Morrison Formation. The low ridge to the left of the trail is Cedar Mountain Formation. A major spur of the Moab Fault travels along the base of that ridge, marking its interface with the white slickrock the trail crosses. This fault line has offset the local geologic strata so that the Cedar Mountain deposits are aligned with the much older Moab-Entrada. From this vicinity, the painted-desert slopes to the right-rear are also Cedar Mountain.

1.5 At a trail junction here, this trail angles right to continue across the white Moab-Entrada sandstone, traveling roughly parallel to the red slopes of Tidwell-Morrison to the right.

1.7 Here, as the trail is traveling along the base of the slope to the right, it reaches the rimlands of Brink Spring Canyon. Park in this vicinity and walk over to the canyon rim. Note that the faultline mentioned earlier slices across the canyon where the white slickrock ends, aligning the younger Cedar Mountain deposits with the Moab-Entrada and the red Slickrock Member of Entrada Sandstone below that. Beyond the canyon to the southwest, the conical hill is largely Tidwell-Morrison capped with remnants of Salt Wash-Morrison. The higher plateaus to the south are the same combination, but with more Salt Wash-Morrison remaining.

2.0 Park off the trail somewhere in this vicinity and walk over to the edge of Brink Spring Canyon. The rimlands and sheer cliffs of this short but lovely canyon are Moab-Entrada. The angled inner gorge of red sandstone is Slickrock-Entrada. These two members are separated by a relatively thin layer of sediments that is largely obscured here by fallen rock and vegetation, but that is obvious farther along this trail. This anomalous deposit between these two members of the Entrada occurs only west of U.S. 191 where the Moab-Entrada is pinching out. It probably represents interlayering with the overlying Tidwell-Morrison.

Note in the upper end of the canyon the large crescent of seeping springs that have undercut the Moab-Entrada, creating a lovely hanging garden of fern, columbine and other water-loving vegetation. Note also where the drainage pours into the upper end of this canyon through a water-carved grotto. Hiking this drainage above the canyon for the next mile or so, as it parallels the trail, provides interesting examples of water erosion in solid sandstone. Eventually this trail crosses that drainage.

As the trail continues across the slickrock, traveling closely parallel to the Morrison Formation slopes to the right, watch for large boulders of Morrison perched atop the softer Moab-Entrada. Here, erosion of the softer rock has left some boulders sitting on small pyramids of remaining stone. These delicately balanced rocks are called "tripod rocks" because many such rocks are balanced on three pyramidal legs. A few such strange erosional phenomena have been found with "legs" a foot or more long.

In stretches of the trail ahead, navigation is difficult as it crosses open expanses of slickrock. In general, the trail parallels the base of the red Tidwell-Morrison slopes to the right wherever feasible, until it reaches a trail fork.

2.2 A short spur trail to the right here crosses a low ridge and enters Lunar Canyon, an open expanse of white Moab-Entrada slickrock that has eroded into a surface reminiscent of a brightly lit lunar landscape. This spur can be explored on the return trip without affecting the mileages in this trail log. Note the shallow drainage line, to the left of the trail along this next stretch, that was mentioned earlier as a hiking route.

3.4 At an inconspicuous trail junction here, this trail angles sharply left to travel the crest of a sandy ridge. As the trail travels that ridge, to descend into the drainage noted earlier that pours into Brink Spring Canyon, study the distinct lower limit of the white Moab-Entrada. The layered, dark brown rock below that is the anomalous deposit mentioned earlier in this log, as well as in the chapter of this book that describes the Moab area's geologic strata.

3.7 Here, the trail crosses the drainage noted earlier. It is interesting for part of a group to hike the drainage here on the return trip, to be met by the vehicle farther down.

Beyond the wash, the trail climbs into rolling terrain that is partly dune sand and partly thin ledges of the dark layered rock of the anomalous deposit noted earlier. This deposit also appears along the Spring Canyon Point Trail, described elsewhere in this book.

4.0 At a four-way junction here, continue straight ahead toward the La Sal Mountains and the rim of Hidden Canyon. The other spur trails can be explored later, if time permits, but are not a part of this trail log.

Brink Spring Canyon

4.2 Here, lovely Hidden Canyon comes into view, just as the trail angles down toward its rim. The base of the first knob of white Moab-Entrada near the trail has eroded to form a natural rock opening, or arch. Similar erosion in this deposit has also produced four arches in the northern rim of nearby Bartlett Wash. Continue down to the Slickrock-Entrada rim of Hidden Canyon and park. Between here and that rim, the trail is descending through the anomalous deposit mentioned earlier.

4.3 From the rim, where this trail log ends, study the maze-like contours and colorful Slickrock-Entrada slopes and ridges of Hidden Canyon. The canyon itself is entirely color-banded Slickrock-Entrada, with dune sand, wash sand and alluvium held in place by native vegetation on the canyon floor. There are two off-road vehicle trails that provide access to the floor of Hidden Canyon, neither of which is documented in the referenced book and map because they are frequently very difficult or impossible to travel due to drifting sand, steep sand-hills and eroding wash sand.

The ornate, colorful slopes and terraces of this canyon and adjacent Bartlett Wash offer endless hiking opportunities, some of which are described in the book ***Canyon Country* SLICKROCK HIKING & BIKING.** Just above the red and white layered Slickrock-Entrada of Hidden Canyon, and Bartlett Wash beyond to the south, the vegetated slopes are the anomalous deposit noted earlier. Above that, the white slickrock cliffline is Moab-Entrada topped with red slopes of Tidwell-Morrison and rocky ledges of Salt Wash-Morrison. The La Sal Mountains dominate the skyline to the southeast.

From this rim, it is possible to hike or bike down onto intermediate levels of the slickrock. The wrap-around cover photograph on the book, ***Canyon Country* MOUNTAIN BIKING,** was taken there. A few intrepid hikers have found ways to continue down to the canyon floor from farther along the rim to the left, but this is not advisable for the average hiker. The sandy main drainage in Hidden Canyon heads in the general direction of the mountains, then winds and twists through a deep, narrow gorge before emerging into Bartlett Wash near an off-road vehicle trail described in the referenced book and map. The mouth of the gorge is not easy to spot from that trail, hence the name Hidden Canyon.

Return to U.S. 191 via the same route, or spend additional time in the area hiking and exploring some of the spur trails noted in this log. If this trail was reached via the Monitor & Merrimac Trail, it is much quicker to return to U.S. 191 via Blue Hills Road than via the Monitor & Merrimac Trail.

Hidden Canyon

TRAIL NAME: **HURRAH PASS - CHICKEN CORNERS TRAIL**

SUMMARY: This trail log combines two named ORV trails that are usually traveled in sequence with Cane Creek Road, described elsewhere in this book. While the route goes through much the same combination of geologic formations all the way, it also crosses part of the Cane Creek Anticline and a little-known syncline, and offers outstanding views of the Colorado River gorge and other nearby erosional features. The various strata of ancient, colorful Cutler Sandstone through which the trails travel are outstandingly beautiful. The route starts from the end of Cane Creek Road, climbs to the summit of historic Hurrah Pass, then descends to the scenic rimlands of the river gorge. The trail closely parallels the river to famous Chicken Corners, then continues to breathtaking Gooseneck Overlook, high above the river yet far below nearby Dead Horse Point. The route lies entirely within the Lowlands District of Canyon Rims Recreation Area.

TRAIL SURFACE: dune and wash sand, broken and ledgy rock, alluvium and rough marine limestone.

VEHICLES: carefully-driven low-geared, high-clearance highway vehicles can normally go as far as the summit of Hurrah Pass, but only various kinds of off-road vehicles should continue beyond that scenic viewpoint.

MAPS: ***Canyon Country* OFF-ROAD VEHICLE TRAIL MAP - Canyon Rims & Needles Areas,** or ***Canyon Country* OFF-ROAD VEHICLE TRAIL MAP - Canyon Rims Recreation Area.**

GUIDEBOOKS: ***Canyon Country* OFF-ROAD VEHICLE TRAILS - Canyon Rims & Needles Areas,** or ***Canyon Country* OFF-ROAD VEHICLE TRAILS - Canyon Rims Recreation Area,** or ***Canyon Country's* CANYON RIMS RECREATION AREA.**

MILEAGES:

TO START: 12.0 miles, from Main and Center, Moab.
TRAIL: 15.5 miles, one-way to end.
RETURN: 27.5 miles, from end of described trail.
TOTAL: 55.0 miles, from Main and Center, round trip.

TIME: allow a full day for this route, a long day if any of its spurs are explored on the return trip.

GEOLOGIC STRATA VISIBLE FROM THIS TRAIL

Formation	Member	Page
Navajo Sandstone		Page 13
Kayenta Sandstone		Page 13
Wingate Sandstone		Page 12
Chinle Formation		Page 12
Moenkopi Formation		Page 12
Cutler Group	White Rim	Page 11
	Organ Rock	
	Cedar Mesa	
	Lower Cutler	
Hermosa Group	Upper Hermosa	Page 9
	Paradox Salts	

ROUTE: south on Main Street (U.S. 191), then west on Cane Creek Road to where it ends and this route begins.

NAVIGATION: Drive south on U.S. 191 for 0.7 miles, west on Cane Creek Road for 11.2 miles to where it ends, as defined, at a ford of Cane Creek. Continue beyond the ford on the Hurrah Pass Trail to its junction with the Chicken Corners Trail and on to where that trail ends at Gooseneck Overlook. Return via the same route, exploring any of several spur trails if time permits.

GEOLOGY TRAIL LOG:

0.0 This trail continues beyond the creek ford where Cane Creek Road ends. For a complete description of the geologic panorama visible from the vicinity of the ford, refer to the final description in the log for Cane Creek Road.

As the Hurrah Pass Trail starts, beyond the creek ford, it travels through canyon-floor alluvium and dune sand, with several nearby ridges of dark red Cutler Sandstone. As the trail continues, it heads toward sloping Cutler strata that provide easy access to lofty Hurrah Pass. The Cane Creek Anticline arched the strata in this vicinity. Accelerated erosion on its spine created the "pass" toward which the trail goes, which is the lowest point on the western skyline.

Cutler Sandstone strata are deposits made in the vicinity of a low-lying marine coastline, as the region's sea level and climate changed many times. These changes produced intermittent layers of shallow-marine limestone, tidal-flat muds and dune sand. In time, these became hard gray layering, thin water-affected layers, and thicker masses of cross-bedded red sandstone, respectively. See the sidebar, ROCK COLORS.

0.3 The Cane Creek Canyon ORV trail goes left here. The lofty cliffline of Hatch Point dominates the western skyline, with its layered Kayenta Sandstone, sheer Wingate Sandstone cliffs, slopes of gray Chinle Formation, and reddish layered Moenkopi Formation deposits below that almost to the valley floor.

1.8 On this stretch of the trail, note the safety fencing around Anticline Overlook, on the high promontory to the left. That lofty viewpoint is at the end of Anticline Overlook Road, described elsewhere in this book. The tall promontory to the right-front is topped with remnants of gray Chinle Formation. The lower part of the vertical eroded cliff is Moenkopi Formation, with the entire mass standing on a terraced base of Cutler Sandstone. As the trail continues beyond here toward Hurrah Pass, it travels at approximately the same stratigraphic level within the Cutler Sandstone, even though the strata have been arched upward by the Cane Creek Anticline. Along the trail, the fantastic shapes into which the Cutler Sandstone has eroded are a highlight of the trip. Some of these shapes have even been given unofficial names that will be noted as they become obvious ahead. Still other shapes are more prominent on the return trip.

2.7 Stop at a pullout on the right and walk out onto the promontory that juts outward from the ornately-eroded cliffline. Be extremely cautious of the crumbling left edge of this promontory because of the sheer drop below it. The view from the end of the promontory includes an outstanding display of colorful erosional forms of Cutler Sandstone. From this viewpoint, it is apparent that the Cutler strata are sloping downward into lower Cane Creek Canyon. East of Hurrah Pass, the angled anticline continuing up the canyon helped promote its formation.

From this vicinity, rock shapes on the eroded sandstone ridge above the trail ahead are sometimes called "Penguin Pulling a Covered Wagon." To the west, a cluster of rock on the skyline is called "The Little Train." As the trail continues, all of the colorful, ornately-eroded canyons that drain to the right of the trail cut through ancient and colorful Cutler Sandstone.

3.0 As the trail is passing a huge fallen boulder on the right, notice the delicately balanced rock silhouetted against the sky to the left. Just ahead, the trail passes beneath the Penguin, with its fanciful "Covered Wagon" in tow. There are other delicately balanced rocks in the terraced cliff to the left on ahead.

View of Cane Creek Canyon from Hurrah Pass summit.

3.5 Park here, just before the summit of Hurrah Pass, and walk toward the rim of Cane Creek Canyon for a breathtaking view of that spectacular and colorful gorge. From the pullout, look back along the trail and note the "Little Chinaman" figure on the eroded cliffline, and "The Footstool," a large, flat rock with a very thin base, below and to the left.

Exercise extreme caution at the crumbling rim. From this viewpoint, the fantasyland of eroding Cutler Sandstone below Hurrah Pass takes on another perspective, with its towers and ornate cliffs and terraces viewed from above. All of the sandstone outcrops in the valley floor are also Cutler, in some cases covered with alluvial fans of water-borne sediments. The steeply sloping cliffline that walls the canyon to the left is Wingate Sandstone above Chinle and Moenkopi deposits, with various Cutler strata on down to the canyon floor. Note the old mine trail that ascends from the canyon through the gray Chinle deposits toward a series of uranium mines. There are other mines out of sight at the summit of that grade. The narrow, downcanyon portal through which Cane Creek Road travels provides a glimpse of rounded Navajo Sandstone in the Behind-the-Rocks wilderness area.

The tips of the La Sal Mountains can be seen above the opposite canyon wall, where another mine trail angles steeply upward. To the left of that trail, just above the canyon floor, the balanced rock called "The Happy Turk" is conspicuous with its white turban. The convoluted, receding upcanyon cliffs of Cane Creek Canyon are the same combination of geologic strata, that is, sheer Wingate Sandstone cliffs topped with layered Kayenta Sandstone, then Chinle and Moenkopi deposits on down to the canyon floor, with a few outcrops of Cutler Sandstone in mid-canyon. All of the varicolored deposits at and below Hurrah Pass are also Cutler. From here, drive the last few yards to the summit of Hurrah Pass, where the view to the west is described next. At the summit, pull right off of the trail into a parking area on the ridgeline to avoid blocking the main trail.

3.6 From the summit of Hurrah Pass, the entire panorama to the west is spectacular, with the redrock fantasyland spoiled by only two human intrusions. See the sidebar, POTASH, for an explanation of the sprawling solar evaporation ponds beyond the river. To view the mill associated with these ponds, walk down onto the lower projecting point and look upriver, to the right. The small development beside the river on the far side is a private development. The towering column of Wingate Sandstone above the ponds is Airport Tower. The jutting peninsula beyond it is Dead Horse Point.

From this vantage point, the arching strata of Cane Creek Anticline are visible beyond the river, beginning below the ponds and continuing out of sight to the right. On both sides of the river, this anticline has raised and exposed much older Cutler strata than are visible in Cane Creek Canyon. The gray layer that tops the Cutler on both sides of the river is a shallow-marine limestone that is rich in fossilized mollusks, crinoids and other marine creatures with hard body parts.

To the southwest, the skyline is dominated by the sheer Wingate Sandstone cliffs of Island-in-the-Sky, with some of The Island topped with light-hued Navajo Sandstone. In the lower terrain, beyond the lefthand group of evaporation ponds, note how the gray-hued limestone is angled upward by the Shafer Dome anticline. For a more complete description of the geology on the far side of the river, see the log for the Potash - Shafer Trail, elsewhere in this book. Pyramid Butte tops the elongated ridge of Cutler deposits to the left of the southernmost ponds. Anticline Overlook, with its protective fenceline and shade pavilion, looms high on the skyline to the south.

Hurrah Pass was given its name during historic times when it was part of a twice-yearly cattle-drive route between distant Island-in-the-Sky and Cane Creek Canyon. In the fall, cattle were driven from their summer pasture on The Island, down the Shafer Trail which was a narrow stock trail at that time, across a ford of the Colorado River, then up to Hurrah Pass. Here, the cattle were counted to determine how many had been lost along the way. With this done, so the story goes, the cowboys let out a loud "hurrah" of relief that the worst of the arduous trek was over.

Two-wheel drive vehicles should not continue beyond the summit of Hurrah Pass as the trail descends the steep terracing of the Cane Creek Anticline toward the lower benchlands above the river. Drivers continuing beyond the summit should exercise caution along stretches of trail with adjacent sheer drops.

4.0 Note the tall tower of rock to the left of the trail, capped with an angled balanced rock. From this stretch of the trail, the other side of "The Little Train" is visible on the skyline above the tower.

4.2 Along this stretch of the trail, the industrial mill beside the river comes into view.

5.0 From here on down the grade, the trail is traveling rocky slopes of alluvium from the higher cliffline to the left, leaving the terraced benchlands of Cutler Sandstone it has been traveling. Note the varied coloration in the Cutler Sandstone ledges still below the trail level.

6.1 Here, the descending trail has reached the stratigraphic level of the gray marine limestone. Until recently, this was considered to be the top of the "Elephant Canyon Formation," but that designation was found to be based on faulty research. See the sidebar,

YES - GEOLOGISTS ARE HUMAN. These deposits are now considered to be lower Cutler strata.

As the trail crosses a shallow wash that cuts more deeply to the right, the Jackson Hole ORV trail goes down that drainage through a fenceline. In this vicinity, a few small marine fossils can be found in the limestone deposits. The Hurrah Pass Trail, as defined, ends at this junction, with the Chicken Corners Trail continuing ahead.

6.3 Here, the Chicken Corners Trail splits, with a shortcut spur going up the shallow, sandy drainage. The described route climbs steeply to the right, up the gray limestone slope, to reach and travel closely parallel to the rim of the scenic inner gorge.

6.5 Park here at a pullout near the river gorge rim and walk a few yards from the trail down onto the colorful Cutler Sandstone ledges for a view of the river and the private development across the river. Notice upriver to the right, how the Cutler strata are angled upward toward the crest of the Cane Creek Anticline. The broad expanse of white sandstone exposed through the cliffline gap to the north is in the Gold Bar Rim area. That light-hued rock is remnant layers of eroded Kayenta Sandstone just above the upper surface of Wingate Sandstone.

All of the sheer cliffs visible in this 360-degree panorama are Wingate Sandstone topped with layered Kayenta Sandstone. Below the Wingate, the terraced ledgy slopes are, in descending order, Chinle, Moenkopi and Cutler, with the interfaces between the Chinle and Moenkopi, and the Moenkopi and Cutler, generally obscured by rubble. Pyramid Butte looms above the river to the left. Dead Horse Point and associated highlands lie on the skyline to the west, and the convoluted cliffline of Hatch Point soars above the river gorge to the east.

As the trail continues, it travels roughly parallel to the rivergorge rim, traveling on or just above the gray limestone layer of the Cutler. Just ahead, that layer becomes less tilted as the trail leaves the vicinity of Cane Creek Anticline. The open meadows to the left of the trail for the next mile or so are largely dune sand barely held in place by native vegetation.

7.1 Ahead, as the trail is paralleling an expanse of exposed sandstone, note the abandoned river gravel on that red slickrock. When this river gravel is exposed for any length of time under the right conditions, it becomes darkened with a mineral coating called desert varnish. See the sidebar, DESERT VARNISH. Note that only river gravel in contact with the slickrock, or with sand eroded from the slickrock, has turned black. This is because these rocks can pick up minerals from the sandstone or sand. All of the low, rounded hills between the trail and the river are piles of gravel abandoned as the river cut more deeply.

7.6 Here, the described trail is joined by the shortcut spur noted earlier. Make note of this junction for the return trip.

8.0 At a junction here, the Old Mines Trail goes left toward the immense canyon system in that direction that cuts deeply into Hatch Point. The formation of that canyon has been promoted by an unnamed syncline that comes from that direction, crosses the river, then goes between the Cane Creek and Shafer Dome anticlines. The low, redrock mass to the left-front is called The Caves because of the series of caverns eroded into its eastern side. The Old Mines Trail provides access to these caves, to Dripping Springs and several old mines. This spur can be explored on the return trip without affecting trail-log mileages.

8.2 An alternate spur trail to The Caves, the Old Mines Trail and Dripping Springs goes left here.

8.8 Here, a short spur to the right goes down a narrow canyon toward the river. There are a few crude prehistoric petroglyphs on the red sandstone bluffs there.

10.0 Stop here and walk a few feet toward the rim of a small canyon. One undercut stretch of the rim there has a large gap behind it, creating a natural bridge.

Since this gray limestone deposit, and the strata below it to some depth, was once called the "Rico Formation," this span of rock was named Rico Bridge. For a photogenic view of the bridge and a better look at its size, walk around the rim to the left and descend into the lower terrain beneath the span.

10.6 Note the small reservoir damming the drainage to the left that was built to provide water for domestic livestock.

10.7 The Lockhart Basin ORV trail goes left here up a shallow drainage toward "Abe Lincoln," an angled spire at the right end of the Wingate Sandstone cliffline in that direction. Note the interesting erosional effects where this drainage pours into its lower canyon, cutting through Cutler limestone, then on down into lower Cutler deposits.

If time permits, on the return trip adventurous drivers might like to accept the challenge of the first mile or so of this trail, or even continue for the several additional miles to lofty Cutler Pass, with its spectacular panoramic view.

Rico Bridge from below.

11.0 Note how the prevailing winds have dumped dune sand over the crest of the cliffline to the left.

13.9 Here, the Chicken Corners Trail is approaching the narrow stretch after which it was named. Just ahead, as the trail starts around the base of the redrock cliff, it travels a sloping ledge one vehicle wide, right on the rim of the river gorge. At one time some large boulders made this narrow stretch even more hazardous. These were removed, but careful driving is still required. Just past the narrowest stretch, the cliff provides excellent shade for a lunch or rest stop.

Chicken Corners

14.1 The inner gorge of the Colorado River from Chicken Corners is breathtaking, with the multiple laminations of varicolored Cutler Sandstone starting not far above the river and extending to high above the trail level, then younger strata continuing on up to the top of Dead Horse Point. The strata at and just above the river are upper Hermosa marine shoreline deposits that closely resemble the lower Cutler strata in this area. The broad benchland at a lower level opposite Chicken Corners is reached by an off-road vehicle trail, part of which is visible. That trail is a spur from the Potash - Shafer Trail, described elsewhere in this book. An old foot trail descends from that benchland to the brushy riverbottom below that was once the site of an oil-drilling operation. There are exposures of petrified wood at several levels on the benchland. The logs that were deposited in the mudstone there washed in from where they grew, and are some of the earliest species of trees to evolve. The soft, gray layers of sediments between two levels of that benchland are rich in shallow-marine fossils.

The shade pavilion at the end of Dead Horse Point is visible from Chicken Corners, and from Gooseneck Overlook where this trail ends. The trail beyond Chicken Corners continues to travel on, or just above, the Cutler limestone layer and through dune sand in some areas.

14.8 Along this stretch of the trail, note that the limestone strata to the right beyond a shallow drainage are slanting upward toward Dead Horse Point and the missing summit of the Shafer Dome anticline, through which the Colorado River has cut its gorge.

15.5 The trail ends at spectacular Gooseneck Overlook, one of several named viewpoints in the Lowlands District of Canyon Rims Recreation Area. Here, the Colorado River is starting a broad sweep into a "gooseneck" entrenched meander. This immense loop is visible from Dead Horse Point, the high promontory to the north of here, and from one stretch of the Potash - Shafer Trail. Remnant layers of Kayenta Sandstone top Dead Horse Point and most of the higher rimlands visible from here, above sheer cliffs of Wingate Sandstone, with red and gray-layered Chinle Formation slopes below that. The thinner layering of Moenkopi Formation continues down to the more rounded, terraced bluffs of Cutler Sandstone.

This ancient marine tideland deposit continues downward almost to the river level, with each distinct layer representing a somewhat different environment in the region at that time. The ledged strata at and just above river level are upper Hermosa deposits. The more rounded, light red layers of Cutler and Hermosa represent coastal sand dunes, while the thinner water-affected layers represent periods of time when tides were frequently invading the coastal plains, or when fresh material was being carried into the region by runoff from the Uncompahgre Uplift to the east. The gray layers at the overlook level, and at intervals on down, represent periods of time when rising sea levels created shallow-marine conditions that were favorable to the kinds of sea life that deposited much of the gray limestone.

To the south and west, the high terrain is all part of Island-in-the-Sky in Canyonlands National Park. The southernmost parts of The Island are topped by cliffs of light-hued Navajo Sandstone.

In relatively recent times, since the exotic shrub tamarisk was introduced into the western United States, this hardy plant has invaded canyon country wherever there is enough water to support it. During historic explorations of the Colorado River by Major Powell and others, photographs taken showed no tamarisk, but plentiful river-willow and other native shrubs and trees. Recent photographs taken from the same vantage points as those on the Powell expeditions, show that tamarisk has helped stabilize river banks against excessive erosion during flood seasons.

View from the end of the Chicken Corners Trail.

TRAIL NAME: **KLONDIKE BLUFFS TRAIL**

SUMMARY: This fairly short trail travels largely along the western flanks of the Salt Valley anticline, providing a close look at the way in which such tectonic features tilt geologic strata and misalign them by faulting. It also offers an opportunity to study the unique erosional landforms produced by such tilting of ancient sandstone deposits. Other highlights are a number of petrified dinosaur tracks, the remnants of a small copper mining operation, and good rockhounding near the trail.

TRAIL SURFACE: graded dirt, wash sand, broken rock, slickrock and packed alluvium. Part of the trail is graded dirt, but the balance is very rough.

VEHICLE: off-road vehicles only, beyond the first junction in Little Valley.

MAP: ***Canyon Country* OFF-ROAD VEHICLE TRAIL MAP - Arches & La Sals Areas.**

GUIDEBOOK: ***Canyon Country* OFF-ROAD VEHICLE TRAILS - Arches & La Sals Areas.**

MILEAGES:

TO START: 17.0 miles, from Main and Center, Moab.
TRAIL: 7.1 miles, one-way to end.
RETURN: 24.1 miles, from end of described trail.
TOTAL: 48.2 miles, from Main and Center, round trip.

TIME: allow a full day, which will include time for viewstops, short hikes, exploring a few spur trails and some rockhounding in Little Valley on the return trip.

ROUTE: Main Street (U.S. 191) north, then east on the Klondike Bluffs Trail.

NAVIGATION: Drive north on U.S. 191 for the noted distance, then turn right on the graded dirt road there, within sight of the Grand County air field. Navigation along one slickrock stretch of the trail is confusing.

GEOLOGY TRAIL LOG:

0.0 The deposits immediately adjacent to the highway are alluvium and sand held in place by native vegetation. Over the next two miles, the trail travels through tilted and eroded deposits of Cedar Mountain Formation.

2.2 Here, the trail begins to drop through older sediments in its descent into Little Valley. From here to the valley floor, the trail is going through the Brushy Basin Member of the Morrison Formation. The Klondike Bluffs ahead are Entrada Sandstone, with the red part being the Slickrock Member and the white caprock being the Moab Member.

GEOLOGIC STRATA VISIBLE FROM THIS TRAIL

Formation	Member	Page
Burro Canyon/Cedar Mountain Formations		Page 19
Morrison Formation	Brushy Basin	Page 18
	Salt Wash	
	Tidwell	
Entrada Sandstone	Moab	Page 16
	Slickrock	
	Dewey Bridge	

2.6 At a trail junction here, the Klondike Bluffs trail goes left, dropping into the bottom of Little Valley. The rocky ridge and softer sediments to the left of the trail here are Brushy Basin-Morrison, as are the low hills to the right. Note that these strata, as well as the slickrock strata that dominate the eastern horizon, are all tilted by the Salt Valley Anticline that is centered in Arches National Park. See the sidebar, SALT ANTICLINES. This tilting has fractured the Entrada Sandstone on the anticline slopes, causing the parallel cracking clearly seen in the top of the white Moab-Entrada, and creating conditions which produce slender fins and spires of sandstone with extensive erosion. Such jointing and fin-forming, together with erosion, tends to create large natural rock openings, such as those in Arches National Park and elsewhere in this region. For more information about the complex processes of arch-forming, refer to the book, ***Canyon Country* ARCHES & BRIDGES.**

Most of the deposits in the lower part of Salt Valley are dune sand and alluvium held in place by vegetation. There is enough subsurface water, however, to support a few cottonwood trees in the wash, itself. Little Valley is popular for rockhounding because of the minerals in the Brushy Basin-Morrison deposits along its western slopes. See the side bar, ROCKHOUNDING. The vehicle trail to the right here provides access to the southern stretches of this valley. In the past, sites in Little Valley have been used in the making of western movies.

3.5 Note the painted-desert Brushy Basin-Morrison sediments in the bluff to the left of the trail. These sediments were deposited in a vast, shallow inland lake over a long period of time, and contain a variety of minerals as well as fossilized plants and animals. This member of the Morrison Formation has been the origin of a large number of discoveries of petrified dinosaur bones, both in the Moab area and elsewhere. The low, rocky bluffs on the east side of the drainage here are the upper layers of the Salt Wash Member of the Morrison Formation.

3.6 The trail crosses the main drainage in Little Valley near some old cottonwood trees. Beyond the wash, the trail enters the upper deposits of the older Salt Wash-Morrison.

3.7 At a trail junction here, the trail left continues up Little Valley, while the Klondike Bluffs Trail goes right.

3.9 At this junction, the trail left goes to a mining area. The Klondike Bluffs Trail continues ahead, dropping into a tributary drainage that cuts through rocky, tilted strata of Salt Wash-Morrison. At certain times of the year, springs in this small canyon keep its lower end wet.

4.3 The Klondike Bluffs Trail continues ahead across the canyon drainage here, while a short spur trail ascends to the left toward a small uranium mine. The Morrison Formation is one of several in this region that are uranium-bearing. It is unsafe to enter this and other uranium mines. The danger from falling rock is minor compared to the potential hazard of accumulated radioactive radon gas. See the sidebar, URANIUM. All of the rock in this vicinity is the lower part the Salt Wash-Morrison.

Beyond the drainage here, the trail ascends the steeply sloped Salt Wash-Morrison strata before dropping down through the red sediments of the Tidwell Member of the Morrison Formation, then onto the top of the white Moab-Entrada.

4.4 The trail is now traveling through the red sediments of the Tidwell-Morrison, which are visible on both sides of the drainage here just below the white sandstones and colored sediments of the Salt Wash-Morrison.

In a few more yards, the trail reaches the white sandstone of the Moab-Entrada. In this vicinity there are a few eroded petrified dinosaur tracks, with many more beside or near the trail ahead. Note that the red shale of the Tidwell-Morrison is only a few feet thick in this vicinity.

Beyond the point where the trail first descends to the Moab-Entrada, it climbs that sloping stratum at its interface with the Tidwell-Morrison, and continues near that interface to just beyond mile 5.2.

Beyond the uranium mine.

4.7 Here the trail begins to level out after emerging from the V-shaped upper end of the drainage line. It continues left, close to the interface between the red Tidwell-Morrison and white Moab-Entrada. Watch for a series of large petrified dinosaur tracks to the left of the trail along this first stretch. Apparently, dinosaurs traveled along the beach sands which constitute the Moab-Entrada. There are additional track sites farther along this trail, made by other species of dinosaurs.

5.0 Additional dinosaur tracks of a smaller species appear to the left of the trail here. Take care to avoid driving or walking on these and other tracks in this area, to help preserve them for others to study and enjoy. The tracks in this vicinity slowly erode away after prolonged exposure. That is why most of them still visible along this trail are close to the embankment of covering soils, indicating that they were exposed fairly recently.

5.1 Note the precariously balanced rock to the left of the trail.

5.2 Here, logs and rocks mark off an area of small dinosaur tracks in the sandstone adjacent to the bank of red sediments. The trail detours to the right around this valuable site. A few yards ahead, the vehicle trail splits. The Klondike Bluffs Trail, as described, continues straight ahead out onto the Moab-Entrada slickrock. An alternate but longer route that reaches the same destination goes left, continuing close to the geologic interface. On that alternate trail, there are a few additional tracks close to this junction and still more in the bottom of the drainage line a few yards west of this junction.

From this junction, the main trail travels on the upper surface of the Moab-Entrada, becoming obscure in several places. Watch for tire scuff-marks on the sandstone and places where tire tracks cross short patches of sandy sediments. If in doubt about the route, scout around on foot so as not to further confuse the trail with tire tracks indicating the wrong direction. If there are small rock cairns marking the route, please take care to leave these standing as guides for the next traveler.

To the left of the trail along this stretch, the red sediments just above the white Moab-Entrada continue to be the Tidwell-Morrison, capped with the lower sandstone layers of the Salt Wash-Morrison. The Moab-Entrada continues to ascend toward the northeast, tilted by the Salt Valley Anticline.

5.8 Here, the most indistinct part of the trail ends, as it drops into a mixture of sediments and slickrock.

6.2 The main trail continues straight ahead here, while another trail angles right. Between here and where the Klondike Bluff Trail ends, the several branching alternate trails are best left for nature to restore. The trail described in this log is the same one described in the referenced guidebook and map.

At another junction a few yards ahead, angle to the right. That junction is where the alternate route noted earlier rejoins the main trail.

6.4 The main trail goes right at the junction here, while the eroded trail climbing to the left goes to an abandoned uranium-mining complex. This stretch of the trail is traveling on the remnants of the red Tidwell-Morrison, while the rocky bluffs to the north and east of the valley that the trail is entering are Salt Wash-Morrison.

6.7 Stay left at the trail junction here.

7.0 The main trail continues ahead at this junction for a short distance. The trail to the right goes to a long-abandoned copper-mining site. This operation was constructed to take advantage of a pocket of copper mineral in the Moab-Entrada bluff above it. This mineral probably seeped downward from the higher Morrison deposits during past ages. The large V-shaped wooden troughs were used at one time in a first-step refining process in which the copper in solution was precipitated as a fine black powder onto scrap iron, in this case, apparently, crumbled tin cans because there are a number of these at the site.

The water for this operation was taken from a very deep well at the site. This shaft penetrated hundreds of feet of solid rock before entering a large underground cavern partly filled with water. Until recently, a small rock dropped into this drill hole could be heard to rattle for 15 or 20 seconds before going silent as it entered the cavern, then splashing into the water there. Now, someone has plugged this wondrous hole with a large rock. It is highly unlikely that this mining operation was ever economically feasible, but that is the story of all too many mining ventures in canyon country.

7.1 The trail, as described, ends here at the base of a steep grade. At one time, this trail could be traveled for about another half mile, but is now too eroded for four-wheeled vehicles. Mountain bikes, however, can still continue up this grade but should not go beyond the boundary of Arches National Park, as noted by a sign.

It is well worth hiking beyond the end of the feasible trail up onto the elevated plateauland that the trail once entered. From the southernmost rim of that plateau of Moab-Entrada, there are spectacular views to the south, with the white-capped red sandstone fins of Klondike Bluffs in the foreground below, and the panorama of more distant mountains and plateaus on south. With a little scrambling, it is possible to hike out onto a narrow projection of this Moab-Entrada mesa. This affords even better views of the Klondike Bluffs, and a panoramic view across Salt Valley and Arches National Park. This panorama includes Eagle Park, as well as the Devils Garden and Fiery Furnace areas of the park. Some of these features can be viewed more closely from the Arches Road, described elsewhere in this book.

Note from the vicinity of the described trail's end that the cliffs to the south are Moab-Entrada, while to the north and east the Tidwell-Morrison is at and above trail level. This is because a fault line along the base of the cliff to the south has downthrown the younger Morrison deposits to the north of the fault, leaving the Moab-Entrada aligned with Salt Wash-Morrison deposits in this immediate vicinity.

View from the summit of Klondike Bluffs from beyond where the trail ends.

ROCK COLORS

Fundamentally, the color of rock depends on its base material, its additional mineral content, and the forms these take within the rock, although there are many less obvious variables.

In the canyon country of southeastern Utah, the base material in most deposits is sand -- silicon dioxide -- which is white or pale amber. Non-sandstones have other base materials, such as silts, clays or shales, each with its own characteristic colors.

Thus, the colors of canyon country rock -- applying the general term "rock" to all geologic deposits whether hard or soft -- depend on several variables. These are:

1. the basic material, which may or may not have a characteristic color of its own;

2. any additional mineral content, and how it was transformed over time;

3. biological material deposited with the rock's mineral content, even though long-since decomposed or fossilized;

4. the biological environment during the time of deposit, which affected the deposit's subsequent chemical processes;

5. the climate and weather during the time of deposit, which strongly affected the biological environment during deposition;

6. the oxygen content of the atmosphere during and subsequent to the time of deposit, and the amount and duration of its contact with the deposit, which also affected subsequent rock chemistry;

7. the soluble minerals in the waters that saturated the deposit during and subsequent to its deposition, which often became part of the rock;

8. the chemical reactions and physical changes that took place following deposition, over a very long period of time;

9. the subsurface temperatures and pressures to which the deposit was subjected while it was lithifying;

10. the duration of exposure of the deposit on the surface, and the weather to which it has been exposed.

Each of these major variables interacted with some or all of the others, making it very difficult to trace the exact color-history of any particular exposure of rock. Still, certain general conclusions can be reached that apply to the colorful rocks of canyon country, with the understanding that there are always exceptions.

For one, rock colors depend on their present chemical and mineral content, after millions of years of slow changes below ground. They may not be the same colors now as they were when first deposited.

For another, the colors of some rocks change once they are exposed to weathering. Minerals are leached out entirely by rain and snow-melt, are changed by exposure to the hot desert sun, or are concentrated on rock surfaces, by slow and subtle chemistries -- some organic, some not -- in the form of brown or black "desert varnish." Further, there are vast exposures of sandstone in canyon country that have very little of their actual rock surfaces exposed, because these surfaces are almost totally covered with lichens. These primitive plants thrive on sunlight and the minerals they extract from the rock surface, plus moisture and carbon dioxide from the atmosphere. Where they grow thickly, they obscure the natural color of the rock.

Tracing the exact color-history of most canyon country rock formations would require immense effort and extensive research, in order to determine their exact physical and chemical histories. The easiest to trace would seem to be formations with the simplest mineral content, such as those that are predominantly red from iron content, but these present problems too.

Deposits with iron content that accumulated where life was prolific do not take on the characteristic rusty look of those laid down where life was scarce. The reducing chemistry of organic decay apparently affected the mineral content so as to preclude the formation of the iron minerals that are red-hued. Thus, geologic strata that are predominantly red-hued were almost certainly deposited during periods with climates too arid to support much plant and animal life. Red areas within formations that are mostly some other color probably indicate areas within those deposits that were inhospitable to life.

Several of the red-bed sandstones in this region were once vast tidelands, which would not seem to qualify as "arid," yet such areas were only intermittently wetted by sea water, and the rest of the time these highly saline, iron-rich tidelands were exposed to atmospheric oxygenation and intense equatorial heat and aridity, conditions not favorable to the growth of land vegetation and the animal life it supported.

The great red-bed deposits that dominate canyon country are all rich in iron, which forms red-hued minerals under the right circumstances, and they all came from the Uncompahgre Uplift during one or another of the several periods during which it rose above the surrounding terrain and contributed its erosional products. Other regional deposits, none of them red, came from various other highland sources, most of which are known.

There is, however, a major puzzle concerning the dominantly red color of these deposits that

continued ----

range from the oldest exposed -- the Hermosa Group -- through the Cutler, Moenkopi, Chinle, Wingate, Kayenta, Navajo, and lower two members of the Entrada. The existence of this puzzle is not obvious from casual observation, yet it is an important part of this region's geologic history, and the mineralogy of its dominant deposits. The puzzle is:

Were these iron-red deposits that color when they first accumulated and, if not, when and how did they acquire their present reddish color?

The largest school of thought on this question believes that these rocks and sediments have been red since they were first deposited. This is a natural assumption, largely based on very limited study of the matter.

A second group of geologists contends that few if any of these red-bed rocks were red when first deposited, that they acquired their characteristic red hue slowly, over a long period of time, as part of some very subtle chemistry.

Professor Emeritus William Lee Stokes, a former University of Utah geologist, made a case for this concept in his 1986 book, GEOLOGY OF UTAH, and in personal letters to this author. In his book, Stokes noted that two major factors were involved in creating the great red-beds that dominated the canyon country region for so long. These were *"a source of iron-bearing sediments and an arid or semi-arid climate under which oxygen or oxygenated waters could penetrate deeply and remain in contact with the previously deposited sediments as they lithified."* This idea is further supported by recent findings indicating that Earth's atmosphere has in the prehistoric past had a far higher oxygen content than it does now.

Stokes went on to point out that the Uncompahgre Uplift was a good source of iron-bearing minerals that were not red when deposited, minerals such as hornblend, biotite, pyroxine, chlorite and magnetite, which are dark green to black. In a letter to this author, he went on to explain that *"at first, one would see only light-colored sand with scattered dark grains. Bury this in a dry environment or loose condition so that air and water could penetrate deeply. Then the iron minerals gradually and slowly begin to oxidize in the pore spaces so that iron is released into the cementing material,"* in a form that is red in color -- hematite and lemonite. *"The color of red sandstone is in the cement materials between the grains,"* he concluded.

In his letters, Stokes went on to explain that while some of the younger rock in the Rockies might be reworked red sediments, such as those commonly found as loose sand in canyon country drywashes that drain its red-bed deposits, *"all the major red rock of southeastern Utah has undergone the conversion from light-colored strata,"* that is, none of canyon country's present red rock was red when deposited.

While there is little doubt that this concept is generally correct, this author has made observations in the field that do not fit this all-inclusive position. The chemistry of change, from dark-colored iron minerals in granules to red-colored iron minerals coating the rock's sand granules, requires two factors. One is the presence of oxygen, the other the presence of water. The chemical processes involved could not take place without both of these -- plus a lot of time.

Yet a careful observer of canyon country redrock will soon spot many places where red-hued sandstone is in intimate contact with light-hued sandstone in the same geologic deposit, with distinct boundaries between the two. This author has seen Navajo Sandstone with sharply defined alternating red and white layers barely two millimeters wide (about 3/32"), and both Navajo and Entrada sandstones with very cleanly defined lines between alternating red and white areas.

Chemical processes that take place in the presence of water and that depend on osmosis for the distribution of the chemical products -- the red iron minerals produced -- do not confine themselves to such sharp, narrow boundaries. In fact, the products of such chemical changes, in the presence of any moisture at all, would seep almost uniformly throughout the sandstone, leaving no distinct boundaries between red and white areas, and probably no white areas at all.

Thus, while the concept of slow change from non-red deposits to red rock may be applicable to some canyon country strata, it cannot be universally applied. There is more than enough evidence to support the alternate idea, that at least some of the "great red beds" of the region were red or partially red when deposited, and that they were not subsequently subjected to significant chemical change. It could be said, then, that both sides of the argument are correct -- some deposits were red when deposited, others were not -- and that in some cases it would take a lot of detailed study to determine which is which.

TRAIL NAME: **LABYRINTH OVERLOOK TRAIL**

SUMMARY: This short ORV trail ends at one of the many breathtaking overlooks in Canyon Rims Recreation Area. After crossing open high-desert meadows through a pass between immense ridges of red-hued sandstone, the trail drops into rugged sandstone tributary drainages of the narrows stretch of deep and spectacular Harts Draw, then climbs to the summit of a lofty, wooded mesa on the rim of that canyon. The undeveloped overlook at the south end of this mesa is almost 1,000 feet above the floor of the deep, twisting narrows of Harts Draw, and also offers distant views of the lower canyon and other colorful erosional features to the south and west. While mileages are given for driving from Moab, taking this trail, then returning directly to Moab, the trail is generally explored as a side trip from Needles Overlook Road, which is described elsewhere in this book.

TRAIL SURFACE: The first 2.3 miles of this trail are graded dirt that can be safely traveled by any vehicle, but the rest of the trail requires some type of off-road vehicle. This more difficult part of the trail is packed sediments and sand, rough slickrock, wash sand and one short but steep grade with loose rubble and rock ledges.

VEHICLE: Any vehicle for the first 2.3 miles, off-road vehicles beyond that. Those with standard highway vehicles who wish to visit this outstanding overlook can drive to where the trail is no longer suitable for their vehicles, then park there and hike the remaining 1.7 miles to the overlook.

MAP: ***Canyon Country*** **OFF-ROAD VEHICLE TRAIL MAP - Canyon Rims Recreation Area.**

GUIDEBOOKS: ***Canyon Country*** **OFF-ROAD VEHICLE TRAILS - Canyon Rims Recreation Area.** For a general overview of the geology of Canyon Rims Recreation Area, refer to the book, ***Canyon Country's*** **CANYON RIMS RECREATION AREA.**

MILEAGES:

TO START: 38.7 miles, from Main and Center, Moab.
TRAIL: 4.0 miles, one-way to end.
RETURN: 42.7 miles, from end of described trail.
TOTAL: 85.4 miles, from Main and Center, round trip.

TIME: Allow at least a half-day to explore just this trail and return directly to Moab, but visiting Labyrinth Overlook as a side trip from Needles Overlook Road takes only an hour or so longer. Those who have only highway vehicles, and who intend to drive the first stretch of graded dirt then hike on to the overlook, should allow an additional four hours.

GEOLOGIC STRATA VISIBLE FROM THIS TRAIL

Morrison Formation	Brushy Basin	Page 18
	Salt Wash	
	Tidwell	
Entrada Sandstone	Moab	Page 16
	Slickrock	
	Dewey Bridge	
Navajo Sandstone		Page 13
Kayenta Sandstone		Page 13
Wingate Sandstone		Page 12
Chinle Formation		Page 12

ROUTE: South from Moab on Main Street (U.S. 191), northwest on Needles Overlook Road, west on Jail Rock Road, south on Dry Valley Road, west on Dry Valley Loop Road, then southwest on the Labyrinth Overlook Trail to its end.

NAVIGATION: Drive south from Moab on U.S. 191 for 33.6 miles, northwest on Needles Overlook Road for 1.0 miles, west on Jail Rock Road for 0.9 miles to its junction with Dry Valley Road, south on that road for 1.0 miles, west then south on Dry Valley Road for 2.1 miles, to where the Labyrinth Overlook Trail spurs southwest. Use the referenced book and map to aid navigation, whether exploring this trail alone, or as a side trip from Needles Overlook Road. The road system in this part of Dry Valley can be confusing.

GEOLOGY TRAIL LOG:

0.0 From where this trail starts, by spurring southwest off of Dry Valley Loop Road, it first crosses open sage flats heading toward a gap in the nearby sandstone ridge. The sloping, rounded cliffs of that ridge are the Slickrock Member of Entrada Sandstone topped by red-hued deposits of the Tidwell Member of the Morrison Formation. Some of the larger domes also have white rubbly remnants of the Salt Wash Member of the Morrison Formation just above the red Tidwell. The open terrain is almost entirely decomposed rock, dune sand and sandy soils held in place by remnants of native vegetation.

Dry Valley, of which this is a part, was once "belly high to a horse" with lush grass, according to historic accounts, but these native grasses have long since been grazed to extinction by the tens of thousands of cattle and sheep that swarmed over every bit of Dry Valley for more than a century. The sparse remaining grasses seen in a few places are exotic species introduced in a futile attempt to improve grazing, but these non-native grasses do not thrive in this unique environment.

0.7 Here, the trail reaches the summit of the pass through the sandstone ridge, offering a first panoramic glimpse toward the west. The fence and water development just passed are used in support of livestock grazing, which continues despite the extreme sparsity of the exotic grasses here now. Destruction of the native ecosystem continues. See the sidebar, HUMANITY AS A GEOLOGIC FORCE. Dry Valley provides a vivid illustration of how specific human activities are severely affecting erosional rates.

To the right of the trail here, note that the large patches of desert varnish on the cliff face are brown in color. This indicates that the varnish in this location contains iron but little or no manganese. When both are present, the desert varnish tends to be dark brown or black. See the sidebar, DESERT VARNISH.

The panorama ahead, from left to right, begins with more low bluffs of Slickrock-Entrada sandstone topped with remnants of Morrison Formation deposits. The Abajo Mountains dominate the skyline to the south. The long, flat-topped ridge to the left of the Abajos is Peters Point Ridge. The isolated peak of the Abajos is Shay Mountain, one of the several central stocks of this intrusive igneous range. See the sidebar, LACCOLITHIC MOUNTAINS. The slopes of bare, white rock between the mountains and this viewpoint are Navajo Sandstone topped in a few places with thin remnants of the reddish Dewey Bridge Member of Entrada Sandstone. This iron-rich deposit has contributed to the red coloration of the soils found in many places in Dry Valley.

The distant highlands to the right of the Abajos are part of the spine of the Monument Uplift. See the sidebar, MONUMENT UPLIFT. The white sandstone slopes to the west are also Navajo Sandstone topped with remnants of Dewey Bridge-Entrada. The cliff to the right is Slickrock-Entrada topped with layered red Tidwell-Morrison and white rubbly remnants of Salt Wash-Morrison.

1.1 The trail spur to the right here, at a low point on the main trail, provides access to hiking around the base of Rone Bailey Mesa, to the right rear, and to the unnamed tributary of Bobbys Hole Canyon that contains Harts Draw Bridge. Details about this side trip and a picture of this strange natural bridge appear in the book, ***Canyon Country's* CANYON RIMS RECREATION AREA.**

2.3 The Narrows Overlook Trail spurs left here toward the inner gorge of Harts Draw, which is the main drainage to the left. Those who have time may wish to visit Narrows Overlook on the return trip. The two-wheel drive part of the Labyrinth Overlook Trail ends here, as it swings to the right beyond this junction. At a confusing junction just around this curve, continue straight ahead then angle left toward, then along, the ridgeline topped by pinyon and juniper trees.

2.6 Here, as the trail is traveling northward on the crest of the ridge, the Harts Draw drainage dominates the lower terrain to the left. The sloping white sandstone that walls that drainage here is Navajo Sandstone topped in some places with remnants of reddish Dewey Bridge-Entrada. The inner gorge of the canyon, which is barely visible, is rimmed with layered red-hued Kayenta Sandstone.

2.8 As the trail descends from the wooded ridgeline, the view to the north is colorful and spectacular. The peaks of the La Sal Mountains loom above red-walled Rone Bailey Mesa. As noted earlier, this mesa is largely Entrada Sandstone. The vertical and rounded cliffs are Slickrock-Entrada. The horizontal layering at the base of the cliffs is Dewey Bridge-Entrada, with the foundation of the complex plateau being white Navajo Sandstone. It is topped with Morrison Formation deposits. The elongated detached ridge of red sandstone to the left of Rone Bailey Mesa is Wind Whistle Rock. The smaller mass of red slickrock to the left of that is the top of Soup Rock, which is described in the log for Needles Overlook Road, elsewhere in this book. The deep drainage to the right of the trail is a tributary canyon that eventually joins Bobbys Hole Canyon, which in turn joins Harts Draw in its deepest narrows. Beyond here, the trail drops into lower terrain, traveling on sparse soils and rough, white Navajo Sandstone.

3.7 Here, the trail begins to ascend a flat-topped mesa of Navajo Sandstone. Labyrinth Overlook is at the south end of this mesa. The deep gorge to the right of the trail is Harts Draw. Its geology is described from Labyrinth Overlook.

4.0 The trail ends here at the rim of the mesa and Labyrinth Overlook. Park here and walk clockwise around the rim while reading the following description. Originally, this mineral-search trail was bulldozed on around the rim of the mesa, but that stretch goes nowhere. The high-desert vegetation is now recovering, so the trail should not be driven beyond the point where it first reaches the southeastern rim of the mesa.

The rocky rim of the overlook is a water-affected layer near the top of the Navajo Sandstone, just below its interface with the Dewey Bridge-Entrada, which has eroded away from this mesa. The panorama, beginning to the left and going clockwise around the overlook, begins with the sloping, white Navajo Sandstone of this mesa and another one to the east. The Abajo Mountains and Peters Point Ridge dominate the southern skyline. The Harts Draw drainage begins on the northern slopes of the Abajos.

All of the barren white sandstone visible from here is Navajo Sandstone topped with thin remnants of Entrada Sandstone. Below the Navajo, on down into the deep labyrinth of upper Harts Draw, the horizontal red layering is Kayenta Sandstone. The eroded cliffs of the inner gorge are Wingate Sandstone, streaked in places with desert varnish and rusty stains from the overlying iron-rich Kayenta. Note the serpentine course of the inner gorge, which is characteristic of very old drainage lines as they slowly cut downward through ancient rock deposits. See the sidebar, ENTRENCHED MEANDERS.

Continuing around the panorama toward the west, the inner gorge of Harts Draw wraps around the base of the mesa, then departs to continue in a northwesterly direction. The bottom of the gorge here is nearly one thousand feet below the overlook. Note the tributary drainage from beyond the gorge that drains into Harts Draw, creating a series of plunge-pools in its pouroff. Note also a large seep-cave in the inner canyon wall, just beyond another tributary drainage.

The deep gorge continues its serpentine course for several miles beyond the overlook, then opens out into the main part of Harts Draw. The white sandstone slopes above this more open stretch of the canyon are also Navajo Sandstone topped with thin remnants of red Dewey Bridge-Entrada. Layered Kayenta Sandstone extends downward from the Navajo. The canyon's sheer walls are Wingate Sandstone. Sloping, rubble-covered Chinle Formation deposits continue down to the valley floor below the cliffs of Harts Draw in the distance.

In the extreme distance to the north, the vertical Navajo Sandstone fins of Behind-the-Rocks are silhouetted against the sky, while the red sandstone buttes of Eightmile Rock, Soup Rock, Wind Whistle Rock and Rone Bailey Mesa lie below and to the left of the soaring La Sal Mountains. Note that in the bottom of the nearby drainage, in the direction of Wind Whistle Rock, the dune-sand Navajo Sandstone just above the red-layered Kayenta is red-banded by horizontal water-affected strata. This interlayering indicates that water deposits and dune sands were still fighting for dominance here during the early part of the Navajo Sandstone deposition period. See the sidebar, THE GREAT DESERT PERIOD.

On the return trip from this overlook, those who have the time will find it rewarding to visit Narrows Overlook, which is a point of land to the southeast of this overlook at a much lower level. Narrows Overlook provides a closer view of one stretch of the inner gorge of Harts Draw, and offers the opportunity to hike downward to the rim of the inner gorge.

TRAIL NAME: **LAVENDER CANYON TRAIL**

SUMMARY: This trail provides a close view of a canyon that has eroded into the tilted geologic strata of an enormous anticline, in this case, the Monument Uplift. The strata in the lower part of the long and very scenic canyon are relatively level and uniform, but in its upper end the strata are sharply tilted, with Cedar Mesa Sandstone rising from below the surface, then dominating the canyon's massive and colorful walls. This ancient seashore sandstone erodes into a fantasyland of sheer cliffs, fins, domes and abutments, with some walls carved into strange stone draperies by runoff water. Cedar Mesa Sandstone has a tendency to form natural arches, and Lavender Canyon offers many examples of this, including several multiple-opening spans.

TRAIL SURFACE: packed sediments for the first mile or so. Beyond there, the trail travels largely in or adjacent to the main drainage of the canyon, with wet and dry wash-sand, plus some rock, gravel, mud and flowing water. There are occasional small patches of quicksand that are easily avoided by staying in the tracks left by other vehicles.

VEHICLE: off-road vehicles only, although low-geared, high-clearance highway vehicles such as certain pickups and vans can travel most of the trail if skillfully driven and tire pressures are lowered to provide traction in the soft sand that is common on this trail.

MAPS: ***Canyon Country* OFF-ROAD VEHICLE TRAIL MAP - Canyon Rims Recreation Area,** or ***Canyon Country* OFF-ROAD VEHICLE TRAIL MAP - Canyon Rims & Needles Areas.**

GUIDEBOOKS: ***Canyon Country* OFF-ROAD VEHICLE TRAILS - Canyon Rims Recreation Area,** or ***Canyon Country* OFF-ROAD VEHICLE TRAILS - Canyon Rims & Needles Areas,** or ***Canyon Country's* CANYON RIMS RECREATION AREA.**

MILEAGES:

TO START: 64.4 miles, from Main and Center, Moab.
TRAIL: 17.2 miles, one-way to end.
RETURN: 81.5 miles, from end of described trail.
TOTAL: 163.1 miles, from Main and Center, round trip.

TIME: allow for a long day, or plan on camping overnight somewhere in the canyon or general vicinity before the return trip. It takes two days and overnight camping to explore the main canyon and a sampling of its many lovely tributary canyons.

ROUTE: south on Main Street (U.S. 191), west on Utah 211, then south on the Lavender Canyon Trail.

NAVIGATION: Drive south from Moab on U.S. 191 for 40.7 miles, then west on Utah 211 for 23.7 miles, to the start of the Lavender Canyon Trail. Drivers should be aware that mileages on this trail will vary more than usual because it travels in or closely adjacent to the canyon's broad, sandy drainage for most of the way. This will affect mileages because the actual route in the wide drainage varies from season to season, and also depends on how much short-cutting of its wide, sandy curves each driver takes.

GEOLOGY TRAIL LOG:

0.0 As this trail leaves Utah 211, it skirts around an irregular area of private land before entering the broad, open mouth of Lavender Canyon. The upper sheer cliffs of the canyon on both sides are Wingate Sandstone topped with layered Kayenta Sandstone, with one remnant of Navajo Sandstone above that in the right-hand cliff. Below the Wingate, Chinle Formation deposits slope down to and include a wide terrace of harder sandstone. Below that ledge, sloping sediments of Moenkopi Formation are partly obscured under rubble. The base of the cliffline on the right, in the vicinity of the canyon mouth, is the dark red Organ Rock Shale member of the Cutler Group. The canyon floor is alluvium and dune sand held in place by the native vegetation that remains after more than a century of overgrazing.

GEOLOGIC STRATA VISIBLE FROM THIS TRAIL

Formation	Member	Page
Navajo Sandstone		Page 13
Kayenta Sandstone		Page 13
Wingate Sandstone		Page 12
Chinle Formation		Page 12
Moenkopi Formation		Page 12
Cutler Group	White Rim	Page 11
	Organ Rock	
	Cedar Mesa	
	Lower Cutler	

Note that the geologic strata in the canyon are sloping gently upward toward the south. This is due to the Monument Uplift, some of whose highlands are visible beyond the canyon on the skyline. Farther up the canyon, the strata are tilted more sharply, exposing still older deposits.

1.0 Turn left through a gate in a fenceline and onto the trail that first heads toward the Six-shooter Peaks, then angles back toward Lavender Canyon.

1.1 Here, the trail is passing some large cottonwood trees and approaching the Indian Creek drainage. It is nearing the base of the righthand wall of Lavender Canyon. The interface between the dark red Organ Rock Shale, and the rubble-covered Moenkopi Formation sediments just above it, marks the time of a major extinction of life on Earth. The intersection between the base of the higher Wingate Sandstone cliff, and the Chinle Formation slope just below it, marks another such extinction. See the sidebar, GREAT EXTINCTIONS.

1.3 Here, the trail reaches Indian Creek, just below its confluence with the broader drainage of Lavender Canyon. There is generally some water here and intermittently on up the canyon drainage most of the year, but during the more arid months most of the drainage is dry. The trail travels in or close to the Lavender Canyon drainage for the rest of its described length. A few yards ahead, Indian Creek comes in from the left, while the trail swings right up the Lavender Canyon drainage.

1.9 Along this stretch of trail, as it is traveling in the wash, the adjacent brown-and-white-layered deposits are Moenkopi Formation. As the trail continues up the wash, note that even during times of the year when stretches of the drainage are damp, these wet areas end after a short distance. At those points, the water that always flows beneath the sandy surface has been forced upward by underground rock, to sink beneath the surface again when the sand is deep enough.

3.3 As the drainage curves to the right, notice in the low bluff to the left how alluvium once accumulated on top of the layered Moenkopi deposits. More than a century of overgrazing in Lavender Canyon has promoted accelerated erosion and this kind of cutting along its drainage, with the resulting massive loss of native soils. This erosion has been further hastened by the destruction of the canyon's higher watershed, which has been severely damaged by chaining operations in support of grazing. Chaining is uprooting native trees and shrubs by dragging large anchor chains or steel rails between two bulldozers, an environmentally destructive public land "range-management" practice common to this region.

5.5 Along this stretch of the wash, note the remnant domes of white Navajo Sandstone high above the left cliffline.

8.0 Here is another stretch of the drainage where seeping underground water has been forced upward by rock strata. Note the thriving water-loving vegetation wherever the water surfaces.

8.4 The cottonwood trees growing in the canyon bottom in this vicinity indicate the presence of permanent subsurface water. To the right, the light-hued rock that is appearing just above the wash level is Cedar Mesa Sandstone, a member of the Cutler Group. From here on up the canyon, the tilt of the strata caused by the Monument Uplift becomes more pronounced, raising older rock into view. Above the Cedar Mesa, which has not yet appeared on the left side, the layered dark red rock is Organ Rock Shale, then smooth slopes of Moenkopi terminated by a ledge of white Chinle Formation sandstone, with softer Chinle deposits above that. The sheer Wingate Sandstone cliffs are topped by layered Kayenta Sandstone, with a few remnants of Navajo Sandstone and younger deposits above that.

8.6 Here, the Cedar Mesa Sandstone is appearing at the base of the left canyon wall as a highly eroded ridge of light-hued rock. Note ahead, how the strata begin rising more abruptly, with the drainage becoming dominated by walls of Cedar Mesa Sandstone.

8.9 As the wash passes close to a low cliff on the left, note how water erosion has sculpted the Cedar Mesa Sandstone into a work of art.

9.4 Here, at a cottonwood tree beside the wash, the drainage forks. The main trail continues ahead. The smaller wash to the left comes from the Dry Fork of Lavender Canyon, which is well worth exploring. This trail log first goes left, up the Dry Fork, then returns and continues up the main canyon.

Going left up the Dry Fork drainage, the trail travels in the wash for a short time below an undercut rain-streaked cliff, then climbs out. It crosses the wash several times farther on.

9.6 Note the graceful multiple arches in the low cliff to the right of the trail. The walls of the Dry Fork are all Cedar Mesa Sandstone topped with remnants of younger deposits.

10.2 Note the two openings of Dry Fork Double Arch in the lowest part of the cliffline to the right, beyond the wash.

10.3 The trail climbs out of the wash to the left here. Do not continue driving up the wash.

Dry Fork Double Arch

10.4 The description of the Dry Fork of Lavender Canyon ends here in a giant cave with part of its roof collapsed, leaving an opening called Cave Window. Park here and walk on up the drainage for about one-third mile, watching for lovely twin Culebra Arches above the wash to the right. The serpentine ridge that defines these side-by-side natural spans gave them this name. "Culebra" is the Spanish word for "snake."

From Cave Window, return to the main drainage of Lavender Canyon.

11.3 Back at the main wash of Lavender Canyon, go left along another colorful rain-streaked wall.

12.2 Turn right on a short spur trail that goes toward a tributary canyon.

12.4 Stop at the base of the cliff beneath a large natural cave and study the prehistoric ruins in that cave. Return to the main wash by the same route.

12.5 Turn right in the main drainage. As the trail continues through another area where subsurface water has been forced upward, study the many undercut clifflines with colorful rain streaking. This streaking is partly desert varnish and partly lichens. See the sidebars, DESERT VARNISH and LICHENS.

13.1 Notice the large arch in the righthand cliffline. There is another arch higher in the same cliffline in the rear of the alcove just beyond this one.

13.2 Through a fenceline here, the trail enters the part of Lavender Canyon that is in Canyonlands National Park.

13.3 Note a large pothole arch high in the cliffline to the right.

13.5 Stop and look to the right-rear at Hand Holt Arch, a name given the span by early stockmen. Just

Culebra Arches, Dry Fork of Lavender Canyon

Hand Holt Arch

ahead, watch for another arch in the same canyon rim, and still another in the upper rim of the lefthand canyon wall.

13.6 Stop just before reaching the base of a high promontory on the right and look at immense Caterpillar Arch in the cliff rim to the right-rear.

13.7 A large tributary of Lavender Canyon goes right here, but this trail continues in the main canyon, which goes up the left drainage toward spectacular Cleft Arch.

14.0 Note just ahead, as the trail climbs out of the drainage, the banks of alluvium that are eroding away from the lower ridgeline of Cedar Mesa Sandstone.

14.4 As the trail is bending sharp left around a cottonwood tree, note the prehistoric ruins in the cliffline to the left.

Caterpillar Arch. For scale, note human figures.

14.8 As the trail angles left away from one cliff straight toward another, notice the arch in the skyline straight ahead.

15.1 Notice the seep line low in the cliff to the right.

16.1 Here, just beyond where another major tributary angles left, and the trail and main wash are angling right past an immense sandstone monolith, note the large natural arch in the skyline to the left-front. Although it does not show from the canyon floor, this is a triple arch, that is, it has three large legs rather than the one that is clearly visible from here. From this vicinity, part of the opening of enormous Cleft Arch is visible on up the canyon, but the span is not obvious when viewed edgewise.

16.5 To hike up under Cleft Arch, walk up the narrow drainage to the right here into the immense alcove in the righthand cliff, then ascend the steep slopes behind the span.

16.9 Here, the trail is directly below Cleft Arch, but continue ahead for a better view.

17.2 Stop in this vicinity for a photogenic view of Cleft Arch. Although it is possible to drive on up two additional forks of the canyon for short distances, this trail log ends here, very near the deepest part of the canyon as it is carved into lovely Cedar Mesa Sandstone.

On the return trip down the canyon, watch for prehistoric cliff-dwellings in the righthand (eastern) wall of the canyon. There are at least three within the first mile or so that were not apparent on the trip upcanyon. There are still more such structures in other tributaries of Lavender Canyon, most of them accessible by hiking. Do not disturb prehistoric remnants in any way. They are all protected by federal law.

The triple arch as seen from the trail.

Cleft Arch

GREAT EXTINCTIONS

Ever since life began on Earth more than 3.5 billion years ago, it has been subject to changes in its environment. Some changes were slow, and various species had time to evolve to fit the new conditions. Other changes were fast, too fast for some forms of life to adapt to, especially those species that were too specialized to fit the new environmental conditions. These species went extinct, leaving the future to less specialized species that could then slowly branch out to fill the newly-emptied ecological niches -- only to be faced later with further changes, both slow and rapid.

These changes in the planet's environment were what drove life to evolve, to develop, to branch into the multitude of orders, families, genuses and species that now inhabit its land, air and water. As Earth has evolved geologically, so has it evolved biologically, and the two processes have always been intimately linked.

Paleontologists searching the fossil record have found evidence of several major life extinctions and quite a few lesser extinctions that have taken place over the last 600 million years, since life has grown more complex than a broth of microorganisms living in warm, shallow seas. Three of the major extinctions occurred during the period of time represented by the geologic deposits exposed in the general Moab area. Since some geologic definitions of time are based on biology, it is not surprising that these three major extinctions took place at the interfaces between certain geologic eras and periods. The extinctions helped create the definitions.

It is also not just coincidental that these extinctions happened at the interfaces between drastically different geologic deposits, indicating that whatever caused the extinctions also caused major geologic changes, and that both effects were related and probably resulted from the same causes.

The first major life extinction that took place while presently exposed canyon country geologic strata were being deposited was at the Permian-Triassic interface, about 245 million years ago, between the Cutler and Moenkopi deposits. This extinction killed approximately 95% of the species of life on Earth. At present, scientists are not certain what caused this drastic extinction and its attendant change in deposition. Some claim the extinction was brought on by slow climate changes. Others hypothesize that some kind of cataclysmic event changed the world environment so rapidly and drastically that most life was wiped out, and that it also changed the form of deposition, from Cutler to Moenkopi, with a major period of erosion intervening.

Unfortunately, while that gap in the stratigraphic record covers more than 20 million years, there is no way to determine whether erosion took place over this entire time, or whether the Cutler was deposited well into the Triassic, then eroded away rapidly due to some catastrophic event. This lack of solid information about what happened during the long periods of time represented by stratigraphic unconformities creates a major gap in geologic and paleontologic knowledge.

The second major life extinction within exposed canyon country strata occurred at the Triassic-Jurassic interface, about 208 million years ago, between the Chinle and Wingate deposits. Again, the interface was obscured by a major unconformity, a gap in the stratigraphic record of more than 10 million years. But in this case, there is convincing evidence that the extinction and change in form of deposition had a catastrophic cause, and that erosion during the period of time represented by the unconformity was very rapid. Geologists have found the remains of an immense meteor crater in Canada that is more than 40 miles in diameter, and radiometric dating has placed its impact almost exactly at the Triassic-Jurassic interface. This event destroyed some 60% of all animal species, but had the greatest effect on marine invertebrate species, 75% of which did not survive. Land species were more fortunate. Both mammalian and dinosaurian predecessors survived to continue their development.

The third major life extinction occurred at the Cretaceous-Tertiary interface, about 66 million years ago, between the Mesaverde and Wasatch deposits. Again, the stratigraphic and biologic record is obscured by a major unconformity, a stratigraphic gap of at least 15 million years. This famous extinction, which killed some 75% of the species on Earth, including all the land-dwelling dinosaurs and almost all of the marine reptiles, plus many other less conspicuous species, is also controversial. Recent studies and discoveries, however, have pointed firmly to another even larger meteor impact, whose undersea crater is in the Gulf of Mexico, centered near the Yucatan Peninsula.

The mechanisms through which such natural catastrophes cause extinctions of life are very complex, and are still the subject of much debate among various groups of scientists, but the end result is beyond doubt -- a great number of plant and animal species ceased to exist after these interfaces between geologic eras and periods, and the types of geologic deposits changed drastically, indicating correspondingly drastic changes in climate, weather and perhaps even topography. It is unlikely that such changes are coincidental. It is highly probable that they had a common cause, whether it be meteor impact or some other natural calamity of major magnitude. Unconformities, whether they represent long or short periods of erosion, leave great gaps in the evidence.

Two of the three major extinctions that occurred during the deposit of exposed canyon country strata represent a curious paleontological factor, one very important to humanity.

Prior to the Permian-Triassic extinction, life

continued ----

forms on this planet were relatively simple and lagely aquatic. During the Triassic, after the first major extinction described, the reptile family developed two major branches. One eventually evolved into dinosaurs and other well-known reptiles, including those that still exist. The other evolved into what scientists call "mammal-like reptiles," which ultimately evolved into the mammal family. These long-range developments, however, were not smooth. The two major groups competed for the same niches in the same ecosystems, with first one dominating, then the other, over some 245 million years.

At first, during the Triassic, the mammal-like reptiles dominated, evolving large herbivores and predators, thus forcing the reptiles to remain small and less specialized. The Triassic-Jurassic extinction changed this balance. Whatever caused the extinction eliminated the larger, more specialized mammal-like reptiles, leaving only the smaller species and small reptiles. During the Jurassic, the reptiles won the competition for niches in the changed ecosystems, leaving the predecessors of the mammals to fit in wherever they could, which meant staying small and inconspicuous, as the reptiles had been forced to do earlier. Being conspicuous meant death, when the world was teeming with reptilian predators of all sizes. The roles had reversed. The hunted became the hunters, and large reptilian carnivores and herbivores evolved.

This balance reversed again with the Cretaceous-Tertiary extinction. The big, specialized reptiles, the dinosaurs and almost all other reptiles, went extinct, allowing what had by then become mammals to evolve to fill the ecological niches left vacant, and to become once again the dominant vertebrates. Once again, the hunted became the hunters, and once again big carnivores and herbivores evolved, this time from mammalian stocks. It was this second change in balance between reptiles and mammal-predecessors that led to the evolution of primates -- and the human species.

Thus, until about 2 million years ago, this planet's life faced only natural environmental changes, both slow and catastrophic. Then mankind came along, and many scientists contend that humanity is becoming the geologic force that is bringing the Cenozoic -- the biological-geological era we are presently in -- to an end by creating a major life extinction ourselves. See the sidebar, HUMANITY AS A GEOLOGIC FORCE.

Between 60,000 and 10,000 years ago, more than 50% of Earth's genera of large mammals were extinguished, largely by humans, and we are either working on the rest, or have domesticated them. On the land and in the sea, the remaining larger species are still going extinct rapidly, again due to human activities. We have decimated all the larger land predators and many of the large herbivores -- witness the American Bison. In the seas, we have driven most cetacean species to near-extinction, and are working on the next largest marine species, sharks.

Once humanity has finished its self-generated mass extinction by wiping out the Earth's large specialized mammals, including itself, will the cycle of dominance repeat? Will the reptiles once again evolve to fill the vacant ecological niches? Will the age of dinosaurs return, or will some other form of animal life take over the wrecked ecosystems that humanity leaves behind?

Thus, as canyon country visitors stand at a spectacular scenic overlook, such as Grand View Point in Canyonlands National Park or Needles Overlook in Canyon Rims Recreation Area, and study the broad spectrum of geologic strata visible, they should take special note of two geologic interfaces:

1. between the Cutler and Moenkopi deposits, where 95% of this planet's species went extinct, and our mammal-like reptile predecessors got their start and dominated the reptiles for some 235 million years, and

2. between the Chinle and Wingate deposits, where 60% of the planet's species went extinct, and our dominant mammal-like reptile forebears took a back seat to dinosaurs and other reptiles for some 140 million years, probably because of a major meteor impact.

Then, from wherever the Book Cliffs are visible, visitors should recall that:

3. beginning at the interface between the beige sandstone cliffs of the Mesaverde deposits and the overlying Wasatch deposits, the dinosaurs and many other species were forced into oblivion by an even bigger meteor impact, allowing mammals, and ultimately humanity, to rise to dominance in the world's animal hierarchy.

And while studying this region's magnificent erosional land forms, visitors should take special note that we -- humanity -- are now creating our own custom-built extinction, and that like each of the previous major extinction periods revealed in canyon country geologic strata, the present one is accompanied by a major period of erosion. The entire Colorado Plateau region is in an unconformity-in-the-making, one that has already lasted some 10 million years, with no end in sight.

There is a lot more than colorful rock and geologic history in view from this region's spectacular overlooks. There is also biologic history on a grand scale.

TRAIL NAME: **LONG CANYON TRAIL**

SUMMARY: This short but steep and rough ORV trail provides a close look at a deep and narrow canyon that slashes through five major rock formations that have been tilted by anticlines and synclines, with one of the most spectacular panoramas in the Moab area at its summit. The trail travels between the Colorado River gorge and the lofty meadowlands of Island-in-the-Sky in just 4.5 miles, gaining 1,860 feet in elevation along the way, with most of this in the last two miles. The entire canyon is an outstanding display of beautiful erosional forms. For those with off-road vehicles, this trail is usually explored as a loop together with Utah 279, plus all or parts of Utah 313 and the Island Road beyond the end of the trail. All of these roads are described elsewhere in this book.

TRAIL SURFACE: graded dirt, eroded rocky sediments and fractured rock, with about half of its length very steep and rough.

VEHICLE: off-road vehicles only, beyond the first 2 miles.

MAP: ***Canyon Country* OFF-ROAD VEHICLE TRAIL MAP - Island Area.**

GUIDEBOOK: ***Canyon Country* OFF-ROAD VEHICLE TRAILS - Island Area.**

MILEAGES:

TO START: 18.0 miles, from Main and Center, Moab.
TRAIL: 7.7 miles, one-way to end.
RETURN: 28.3 miles, from end of described trail.
TOTAL: 54.0 miles, from Main and Center, round trip.

TIME: allow a half-day for a direct return to Moab, plus time taken for stops along the way. Allow for a long day if this trail is taken as part of a loop trip incorporating Utah 279, the Long Canyon Trail, a side trip to Dead Horse Point via Utah 313, and all or part of the Island Road.

ROUTE: U.S. 191 north from Moab, Utah 279 down the Colorado River gorge, up Long Canyon, then return to Moab via Utah 313 and U.S. 191, with possible side trips to Dead Horse Point State Park and on the Island Road if a full day is available.

NAVIGATION: For a half-day trip with no time for exploring side-roads, drive north on U.S. 191, downriver on Utah 279, then return to the lower end of the Long Canyon Trail. At its upper end, turn right on Utah 313 to its junction with U.S. 191, then right from there to Moab.

GEOLOGIC STRATA VISIBLE FROM THIS TRAIL

Formation	Member	Page
Kayenta Sandstone		Page 13
Wingate Sandstone		Page 12
Chinle Formation		Page 12
Moenkopi Formation		Page 12
Cutler Group	White Rim	Page 11
	Organ Rock	
	Cedar Mesa	
	Lower Cutler	

GEOLOGY TRAIL LOG:

0.0 This off-road vehicle trail spurs from Utah 279 just downriver from immense Jughandle Arch. From the turn-around loop just beyond the railroad tracks, note a slender jughandle arch high in the cliffline to the left. Another natural rock opening is visible in that same promontory from just ahead.

As the trail starts up Long Canyon, the strata that wall the canyon ascend as well. Thus, while the trail climbs, it stays at about the same stratigraphic level until it begins its final steep ascent. The geologic strata all the way up the canyon are Wingate cliffs sometimes topped with remnants of layered Kayenta Sandstone, with the sheer cliffs standing on slopes of red and gray Chinle Formation and Moenkopi Formation deposits. For the first mile or so, the trail travels in Moenkopi deposits, wash alluvium and rubble from higher up. Then the trail and drainage cut into dark red Cutler Group sandstones.

0.3 The second arch noted earlier is visible high above the trail here in the cliff to the left. As the trail continues up the canyon, note how the Wingate Sandstone cliffs are fractured along vertical joints so that they erode into what appear to be rows of rock columns. This unusual fracturing is related to the strata-warping caused by the Cane Creek Anticline and other local tectonic activity.

0.6 Note how a seeping spring in the slope to the right of the trail supports a community of water-loving grasses. The seeping water has left a thin patina of white minerals on the surrounding surfaces as it evaporates. The Chinle Formation is rich in minerals, including those deposited from volcanic ash fallout. Watch for other such springs. Just ahead to the left and slightly above the trail, note the strange coloration and erosional surface of a layer of deposits beneath a red sandstone ledge.

0.7 The rounded slopes ahead are Moenkopi deposits. As the trail continues, this deposit is exposed as ledges and low bluffs on both sides of the trail, but is often covered by debris from the Chinle and Wingate above.

1.15 Note the finely laminated strata of Moenkopi deposits to the left of the trail, as it dips almost to the drainage level. This distinctive kind of layering is characteristic of Moenkopi in this area.

1.4 Note the gaps in the Wingate Sandstone in the lofty cliffline to the left, where erosion from both sides of a wall of Wingate has collapsed sections of the wall. Note also here the immense anvil-shaped rock perched on a spire in the righthand cliffline. Just ahead, the trail enters a narrow canyon walled with rounded knobs of dark red Cutler Group sandstone. The trail next travels through this deposit for some distance before ascending steeply through younger strata.

1.9 A dead-end spur trail branches left here. This short trail, originally built for mineral prospecting, is now used only for access to several primitive campsites. Beyond here, the trail begins to climb steeply.

2.4 Here, as the trail curves around steep switchbacks and crosses a drainage, it reaches the Moenkopi-Chinle interface. From this stretch of trail, the upper peaks of the La Sal Mountains first become visible downcanyon. Note ahead how the Wingate Sandstone cliffline has eroded into narrow, closely spaced columns. As erosion continues, these will be undercut and collapse, slowly destroying that wall.

2.5 At a narrow pullout on the left in a relatively level switchback, park and study the canyon below. An interesting balanced-rock column stands within a few yards of this curve. Adjacent to the drainage line far below is a pipe that marks the drill site for which the spur trail noted earlier was built. From here, there is a first good view back down the canyon. The downward plunging cliffline of Long Canyon recedes toward the rounded fins of Navajo Sandstone in the Behind-the-Rocks wilderness area, with the lofty peaks and slopes of the La Sal Mountains on the skyline beyond. From this viewpoint, the rock sequence described earlier is spread out below and beyond like a gigantic three-dimensional stratigraphic map.

Note in the high Wingate cliffline to the left a large arc-topped alcove. Unless that arc of rock collapses before then, it will eventually become an immense arch, as the thin opening behind it slowly enlarges. Beyond here, the trail continues to ascend steeply through the Chinle Formation with its alternating layers of soft gray deposits and harder ledges of red-tinged sandstone.

2.9 Note at this point that the trail is climbing steeply in or close to a drainage line. Frequent flash-flooding thus damages the trail and sometimes makes it impassable until repaired with heavy machinery. In the past, this trail has been completely closed for long periods of time because of rock collapses and flood damage here and along two other stretches ahead.

3.5 From a pullout on the right just beyond a steep switchback curve, park for another breathtaking view back down Long Canyon. From here, the entire canyon down to the Colorado River is visible, with the sheer Wingate cliffs of the river gorge beyond the green band of vegetation that marks the river. Some of the ridge of rounded Navajo Sandstone that dominates Poison Spider Mesa appears beyond the Long Canyon cliffs to the left, while domes and fins of the same rock show beyond the canyon's right wall. The entire range of the La Sal Mountains extends across the southeastern horizon. Note again how the relatively thin southern wall of Long Canyon is eroding away into spires and columns, then gaps in the cliffline.

3.8 At this point, in the late 1960s, an enormous rockfall from the cliff to the left completely buried this trail, closing it for a year or more. Note the newly exposed rock in the cliff, with no desert varnish on its surface, just above the broken rubble to the left of the trail. More of the same rockfall was eventually pushed over the side of the trail to the right into the deep canyon. Just beyond this stretch of trail, note the gap in the high cliffline to the left, where undercutting has collapsed the Wingate wall that splits the upper canyon.

4.1 Just beyond here, the trail reaches the Wingate-Chinle interface, then climbs still more steeply through the Wingate. Note to the left of the trail here, how surface-water run-off has carved the cliffs just above trail level into strange shapes and forms, including tiny arches.

4.2 From a narrow pullout on the left here, notice how the extensive drilling and blasting above and left of the trail done during its construction has accelerated further collapse, creating a continuous maintenance problem. See the sidebar, ROCK EROSION. Just ahead, the trail enters a very narrow gorge in the drainage line. Occasional flooding also makes this stretch of trail difficult to maintain. The Long Canyon Trail is an outstanding example of how challenging it is to build roads and trails in the rough, vertical terrain of canyon country, and how expensive these are to maintain when they do not conform to the natural erosional contours of the land.

This particular off-road vehicle trail was constructed during one of the past uranium booms to make a convenient connection between the river and the highlands of Island-in-the-Sky. It has since been maintained for scenic and recreational purposes, although it also provides shortcut access to petroleum exploration and pumping operations in the higher terrain.

4.5 Here, where a short spur trail goes left, the Long Canyon Trail tops out approximately at the interface between the Wingate and overlying Kayenta Sandstone. For additional spectacular views of Long Canyon and the panorama to the south and east, walk or drive to the end of the spur trail, then walk beyond where it ends. This short trail travels the top of the gigantic wedge of rock that divides the two upper branches of Long Canyon. Its walls are sheer and hundreds of feet high. Some crevices in the Wingate Sandstone point plunge to breathtaking depths. Hikers who go beyond the end of the spur trail should exercise extreme caution. The panoramic view from this promontory is one of the best in the general Moab area. It has been the site for at least one sunset wedding.

From this junction, the distant Book Cliffs are visible to the northeast, high above the red cliffs of Klondike Bluffs in Arches National Park, with nearer red sandstone bluffs below these. The immense sloping mesa of lightly vegetated red-brown sandstone a little farther east is Arths Rim. It is separated from the white sandstone slopes of Gold Bar Rim farther east by the deep slash of upper Little Canyon. The distant slopes of the Uncompahgre Uplift are visible on the skyline to the left of the La Sal Mountains.

Beyond this junction, the trail closely parallels the rim of the other upper tributary of Long Canyon, offering spectacular views down into and across that deeply cut gorge.

4.6 Note the flat-topped rock column just beyond the canyon rim to the left. From here on to the trail's end, it travels through the lower levels of Kayenta Sandstone, although this deposit is mostly obscured beneath sand, decomposed rock, soils and native vegetation. In the past, some wag placed a sign beside the trail here, visible to those approaching the trail from the other direction. The sign read PUCKER PASS, which supposedly meant that certain parts of the human anatomy would tend to pucker during the descent of the steep and hazardous trail down Long Canyon.

5.0 Along the next stretch of the trail, it closely parallels the northern boundary of Dead Horse Point State Park.

6.2 Note The Knoll, the rounded prominence on the horizon to the right-front. For more information about this novel remnant of Navajo Sandstone, refer to the road log for Utah 313. In this vicinity, the trail crosses open meadows of decomposed rock, soils and sand held in place by remnants of native vegetation. The higher elevation here affords more moisture for the growth of vegetation communities, which in turn support more wildlife. In the past, the native grasses in such meadowlands were thick and lush. These grasses have since been destroyed by more than a century of overgrazing by domestic livestock. The sparse grasses here now are exotic species that have been sown in a futile attempt to replace the native grasses.

6.7 The trail to the right here in the open meadow provides vehicle and hiking access to the spectacular upper rimlands of Bull and Day canyons, although the trail and its various spurs were actually constructed for mineral exploration and for access to a small operating oil well.

7.7 Here, the Long Canyon Trail joins paved Utah 313. Turn left to go to Dead Horse Point, turn right to return to Moab or to reach the beginning of the Island Road. Both of these roads are described elsewhere in this book. From this junction it is 1.7 miles to the junction where the paved Island Road goes left into the Island-in-the-Sky District of Canyonlands National Park, and Utah 313 goes right for the return trip to Moab.

MONUMENT UPLIFT

Some geologists call this immense geologic deformation an "upwarp," rather than an "uplift," but the meaning is the same. In the distant geologic past, stupendous natural forces caused the crust of the Earth to bulge upward in an immense, elongated blister. The remnants of this physical deformation of the region's sedimentary strata are still highly visible today as major landforms in the region, some of them in Canyonlands National Park and adjacent Canyon Rims Recreation Area and described in several geology logs in this book.

Monument Uplift is about 100 miles long on the surface but, as with most such anticlines, it is probably even longer if the distorted strata still underground are considered. The uplift formed during a long period of severe geologic activity called the Laramide Orogeny, which took place during the late Cretaceous and early Tertiary, about 45 to 75 million years ago, although geologists differ somewhat about the time.

This long interval of crustal deformation was caused by continental drift activity. It produced the present Rocky Mountains, the Kaibab Uplift in the Grand Canyon vicinity, the Defiance Uplift in Arizona, the Zuni Uplift in New Mexico, and the San Rafael Swell and Circle Cliffs uplifts in central Utah, as well as other major tectonic features such as subsiding basins.

Some geologists think that during this period of severe tectonic activity, the Colorado Plateau, the immense oval of continental crust that is the heartland of the Four Corners States, was rotating somewhat in relation to the surrounding North American continental plate, and that this contributed to the deformation processes. Although this concept has yet to be widely accepted, it is entirely possible because during that period the continental plate changed its direction and rate of travel at least three times. These directional changes may also have triggered the Laramide Orogeny and provided the stupendous amounts of energy involved in the resulting tectonic activity.

Regardless of the causes of its formation, and despite millions of years of subsequent erosion, there are still many prominent remains of the Monument Uplift. During its erosional phase -- which is still going on -- the more fractured rock in the arching dome of the uplift eroded away more easily, although parts of it still remain as highlands or cliffs and spires.

The northern nose of the uplift, or immense anticline, dives downward in the Needles District of Canyonlands National Park and adjacent areas of the Beef Basin District of Canyon Rims Recreation Area. The spectacular sandstone spires and fins of The Needles were formed by the erosion of sandstone strata cracked in an intricate checkerboard pattern by the upward bulging of the uplift. The numerous grabens there and in the adjacent Beef Basin District are also a part of that faulting and jointing.

The Colorado River formed Cataract Canyon along the northwestern flank of the uplift. The Elk Ridge highlands in the Dark Canyon District of Canyon Rims Recreation Area and on south are part of the uplift's remaining spine. In this vicinity, ancient geologic strata have been raised more than 4,000 feet by the uplift. The region's deepest gorge, Dark Canyon, formed in the uplifted strata.

Farther south, Comb Ridge, Lime Ridge and Raplee Anticline are part of the eastern flank of the Monument Uplift, and the spectacular spires and cliffs of Monument Valley are part of its central heartland.

South of the Abajo Mountains, the San Juan River cuts through the uplift, just as the Gunnison River once cut through the Uncompahgre Uplift, and the Colorado River still goes through the Kaibab Uplift in the Grand Canyon and across the Moab Valley salt anticline. Similarly, the Green River slices through the Uinta Mountains and the Dolores River cuts through the Gypsum and Paradox salt anticlines. How these rivers were able to cut through these huge ranges and anticlines is fascinating, but too complex for discussion here.

The Abajo Mountains thrust upward through the Monument Uplift millions of years later, adding to the deformation of the rock strata in that vicinity. See the sidebar, LACCOLITHIC MOUNTAINS, for more details about the several island mountain ranges in the general Four Corners region.

TRAIL NAME: **MONITOR & MERRIMAC TRAIL**

SUMMARY: This popular ORV trail travels through a variety of geologic strata and erosional landforms and provides a close look at several places where major faults have offset these strata. Along the way, the trail goes near massive buttes, including the Monitor & Merrimac Buttes, plus towers, mesas and great slopes of colorful sandstone, a large natural bridge and exposures of "Dewey Bridge agate." There are also lofty canyon-rim overlooks that provide spectacular panoramic views of the major geologic features in Arches National Park, plus both upper and lower Sevenmile Canyon. The Monitor & Merrimac Buttes were named after the two Civil War ironclad ships that bore these names.

TRAIL SURFACE: a mixture of packed sediments, dune sand, wash sand, a small flowing stream, rough slickrock, and broken, ledgy rock.

VEHICLE: off-road vehicles only.

MAP: ***Canyon Country* OFF-ROAD VEHICLE TRAIL MAP - Island Area.**

GUIDEBOOKS: ***Canyon Country* OFF-ROAD VEHICLE TRAILS - Island Area** and ***Canyon Country* SLICKROCK HIKING & BIKING.**

MILEAGES:

TO START: 15.8 miles, from Main and Center, Moab.
TRAIL: 17.2 miles, one-way to end.
RETURN: 15.8 miles, from end of described trail.
TOTAL: 48.8 miles, from Main and Center, round trip.

TIME: allow a full day to travel the described segments of this interconnected series of loop trails, plus a couple of short trail spurs to historic and prehistoric sites near the described route.

GEOLOGIC STRATA VISIBLE FROM THIS TRAIL

Formation	Member	Page
Dakota Sandstone		Page 20
Burro Canyon/Cedar Mountain Formations		Page 19
Morrison Formation	Brushy Basin	Page 18
	Salt Wash	
	Tidwell	
Entrada Sandstone	Moab	Page 16
	Slickrock	
	Dewey Bridge	
Navajo Sandstone		Page 13
Kayenta Sandstone		Page 13
Wingate Sandstone		Page 12

ROUTE: Main Street (U.S. 191) north to the beginning of the described trail.

NAVIGATION: Drive north from Moab on U.S. 191 for the noted distance, then west on the described trail just before the highway crosses the bridge over Courthouse Wash. This ORV trail is a series of interconnected loops. This geology log travels the outer perimeter of these loops, following the route that provides the best geologic overview. Navigation of the described loops of

this complex ORV trail route is complicated by several factors, making it essential to use the referenced guidebook and map. Additional navigational guidance is given in this trail log.

GEOLOGY TRAIL LOG:

0.0 As the trail heads west from U.S. 191, it first crosses railroad tracks, then enters a broad exposure of Cedar Mountain Formation which closely resembles the Brushy Basin Member of the Morrison Formation. In some places the Cedar Mountain is topped with ledges or ridges of Dakota Sandstone. The main drainage area of Courthouse Wash, just north of this junction, is the general Courthouse Pastures area north of the Monitor & Merrimac. East of the highway, Courthouse Wash travels across several miles of open desert before plunging into a canyon system that travels the length of Arches National Park before joining the Colorado River near the U.S. 191 bridge across the river.

0.2 The trail crosses the Courthouse Wash drainage. Note the trees and water-loving vegetation growing near this drainage, indicating considerable subsurface moisture even though the wash surface is usually dry.

Note in the distance to the left, just beyond the drainage, the red Entrada Sandstone monoliths on the skyline. Beyond the wash, the trail continues to travel in decomposing Cedar Mountain Formation deposits.

0.6 The trail loop described in this log closes at this junction, with the description starting by going left here and returning via the trail to the right. Cedar Mountain and Dakota deposits continue to dominate the foreground. The geology of the red-walled canyons to the south will be described further on.

1.1 The spur of this ORV trail that goes through Mill Canyon angles right here, but the route covered in this log continues straight ahead. Those who wish to walk the BLM's Mill Canyon Dinosaur Trail, and the partly restored Halfway Stage Station just ahead, can do so after completing the described geology loop without affecting the mileages in this log.

One branch of the Moab Fault angles across the landscape ahead, cutting along the northern walls of the two mesas there. The Courthouse drainage funnels between these two mesas. This narrow canyon makes a good, short hike beyond the historic stage station. Other hikes in the vicinity, including one that goes up onto Mill-Courthouse Mesa, to the right-front, are described in the referenced hiking book.

The thick white layer of rock on top of the sloping mesas in this area is the Moab Member of Entrada Sandstone. The color-banded cliffs below that are the Slickrock Member of Entrada Sandstone. In the lefthand mesa, the more distant lower slopes are the reddish Dewey Bridge Member of Entrada. The white sandstone outcrop at the base of the nearest end of that same ridge is on this side of the fault noted earlier, and is the Salt Wash Member of the Morrison Formation. The eroding gray slopes to the east and southeast are Cedar Mountain deposits.

1.2 The spur trail to the Halfway Stage Station goes right here. The described trail continues straight ahead across the Courthouse drainage. Note the window-like opening behind the partially detached spire of sandstone on the left side of the butte ahead. There is another large arch in the top of that butte that is only visible from the eastern rim of Mill-Courthouse Mesa. A picture of that span appears in the referenced hiking book.

1.9 The fault noted earlier is now running along the base of the cliff to the west, offsetting the strata so that Entrada deposits are aligned with the much younger Cedar Mountain Formation. On ahead, this trail climbs to the right and crosses the fault.

2.1 At a trail junction, the described route angles right and climbs steeply into higher terrain, cutting through the upper level of the Cedar Mountain deposits and a thin ledge of Dakota Sandstone. The trail to the left descends through a mining area, then returns to U.S. 191. Note how the strata in the gray hills south of this junction were tilted and warped by the faulting. As the trail climbs the grade ahead, note to the left across the drainage how the faulting has aligned the rubbly Cedar Mountain deposits with white Navajo Sandstone, thus clearly indicating the faultline in this vicinity.

2.3 From the summit of the grade, and for the next mile or so, the trail travels at the stratigraphic level of Navajo Sandstone, although this rock is sometimes covered with sand and soil held in place by sparse vegetation. Note the immense sand dune lying against the cliff ahead, deposited there over a long period of time by prevailing winds. Note also the vehicle tracks ascending the sand dune in several places. The area is popular for hill-climbing activities.

2.6 Organized sand-hill climbs, which in the past started at this point, may be resumed again in the future. Beyond here, the trail travels on solid Navajo Sandstone for some distance. Note how the soil that was once placed here to create a parking area at the base of the sand-hill climb is eroding away from the surface of the slickrock, thus reverting the area to its natural erosional surface. Note also how all the strata in this vicinity are tilted sharply upward toward the south.

Beyond here, note the cliff to the right. The lower horizontal terraces of white rock are Navajo Sandstone. The ornately eroded redrock level above that is the Dewey Bridge-Entrada, which is somewhat interlayered with the next higher Slickrock-Entrada along this stretch of cliff. The fault noted earlier can be seen to the left of the trail here as a clear junction between white Navajo Sandstone and the ridge of dark rubble beyond.

Note on the northeast skyline the red and white Entrada Sandstone cliffs and spires of the Klondike Bluffs area in Arches National Park. Other park features can be seen to the east, including Skyline Arch.

3.2 Park here as the trail crosses a stretch of slickrock and walk onto the higher slopes to the right for an excellent panoramic view of the Courthouse Pastures area, described next.

Due south, a low ridge of Navajo Sandstone dominates the skyline. The Monitor & Merrimac Buttes farther to the right are Entrada Sandstone. Big Mesa dominates the skyline to the southwest and west, with an intervening ridge of Entrada Sandstone obscuring part of the mesa. Note the sand dune piled against that ridge by prevailing winds. Beginning at its base, the strata in Big Mesa are red-banded Dewey Bridge-Entrada, sheer cliffs of Slickrock-Entrada, a white cliff of Moab-Entrada, rubble-covered slopes of the red Tidwell Member of the Morrison Formation, then the ledgy Salt Wash-Morrison on up. All of the white rock in the lower terrain in this panorama is Navajo Sandstone. The same combination of strata applies to the other clifflines in this vicinity. The large, isolated butte that the trail has just passed is called Courthouse Rock. The high mesa just west of Courthouse Rock is Mill-Courthouse Mesa. The referenced hiking book describes how to reach the summit of that mesa and hike its breathtaking perimeter.

To the east on the skyline are the Klondike Bluffs and other Entrada Sandstone features in Arches National Park. The painted-desert hills in the near-distance to the northeast beyond the main highway, are the Brushy Basin Member of the Morrison Formation. The Book Cliffs are visible on the skyline to the north, while one stretch of the Uncompahgre Uplift is visible on the horizon to the east.

Beyond here, the trail skirts along the left of the white sandstone exposure for a few hundred feet, then drops down to cross a drywash before heading generally toward the left end of the Navajo Sandstone ridge on the south skyline.

3.4 Go left at the trail junction here.

NAVIGATIONAL NOTE: For the next hundred yards or so, the trail ascends a slope of white slickrock and is obscure. There are two trail junctions in sequence just a few yards beyond the upper end of this slope. The left spur at the first junction goes first east, then northeast. The described route angles left (southeast) from the main trail at the second junction, and travels toward low terrain between the junction and a high ridge to the east. Stay to the right while ascending the expanse of slickrock, thus bypassing the first junction and arriving just right of the second one, that is very near the low ridge of white sandstone.

3.7 At the above-described junction beyond the slickrock grade, angle left, following the trail across brushy sand dunes, then down into the lower terrain where powerline poles cross the sloping meadows. The trail straight ahead here goes directly to the Monitor & Merrimac Buttes but is not described in this log, which follows the outer perimeter of the several connecting loops of the Monitor & Merrimac Trail. Another spur here angles right, around the south side of the nearby sandstone ridge.

NAVIGATIONAL NOTE: The past mineral-search activity encountered ahead has made it difficult to find and follow the described trail, but this trail heads in the general direction of the highest rimland to the southeast. Several trails descend into the lower terrain between this trail junction and the rimland beyond that this route travels. Thus, any route that descends and crosses the two drainages ahead, then ascends toward the distant rim parallel to the deep canyon to the left, will be suitable and not vary significantly from this log's mileages.

3.9 The deep canyon to the left here that is fed by two separate drainage lines is Corral Canyon. Its vertical cliffs are Wingate Sandstone topped with layered Kayenta Sandstone. All of the slickrock on the slopes toward the La Sal Mountains is Kayenta Sandstone.

As the described trail descends into lower terrain it enters an area that has been severely disturbed by past mineral-search activities. Numerous holes were drilled in the slope beyond the drainages in an attempt to trace the vein of uranium ore in the Chinle Formation deposits far below. The antiquated 1872 federal law that still governs "hardrock" mineral claims and mining on public land permits almost any kind of environmental devastation of the land during mineral-search and mining, and does not require restoration or cleanup afterwards, thus leaving great areas of otherwise beautiful public land looking like an industrial junkyard. The now-defunct mines that this havoc supported are visible in the elevated Chinle Formation slopes to the west of U.S. 191 as it passes this stretch of cliffline north of the Utah 313 junction. These mines and associated damage are an example of this travesty of land administration.

4.2 At a trail junction, continue ahead to cross the wash. Beyond that, the trail climbs back toward the rim of Corral Canyon, then travels slickrock and sandy slopes roughly parallel to the rim, toward the distant ridgeline. Stay on what appears to be the most heavily traveled trail.

4.5 At another multiple trail junction, continue down toward the second drainage, then climb another slope, still following the most conspicuous trail.

4.8 At an obscure trail junction here, go left up a white slickrock slope.

5.0 The trail reaches the eastern rim of Courthouse Pasture here, but continue south along the rim toward a better viewpoint.

5.4 Stop here at an expanse of slickrock rim for an excellent viewpoint. The Entrada Sandstone cliffs, spires, ridges and buttes of Arches National Park stretch across the eastern panorama. For details about the geology of this park, see the log for the Arches Road elsewhere in this book. In general, the white sandstone expanses east of the highway are the Moab-Entrada, which tops the reddish Slickrock-Entrada.

Sinuous lower Sevenmile Canyon cuts through these deposits to the east of the highway for several miles, then joins the canyon of Courthouse Wash coming from the north. The combined drainages continue south through the Courthouse Towers section of the park. The painted-desert Brushy Basin-Morrison hills noted earlier lie beyond open meadows to the northeast. The Book Cliffs dominate the entire northern skyline.

Monitor & Merrimac Buttes and Big Mesa loom above the dune-sand pastures to the west. The detached spire of rock near the cliff-rim just north of this viewpoint is Wingate Sandstone topped with remnants of Kayenta Sandstone.

Walk down the slope here toward the west on an eroding spur trail toward beautiful Uranium Arch, which bridges a slickrock drainage to the left of the spur about 100 yards from the main trail. The span formed in layered Kayenta Sandstone above a large seep-cave.

Beyond this viewpoint, continue southeast along the lofty rim. The trail is obscure in places, but watch for signs of tire tracks and crumbled rock across and beyond expanses of slickrock.

5.6 Another spur trail to the right here goes toward Uranium Arch. Note in the cliff to the left how the reddish, upper surface of the Wingate Sandstone has been exposed by erosion.

5.8 The trail climbs left onto a mass of bare Kayenta Sandstone, then continues past some old drill holes toward the rim. It then descends from the sandstone exposure into lower, vegetated terrain just back from the rim.

6.1 At a trail junction here, turn right. The rough and steep trail straight ahead climbs onto a higher butte that the trail to the right skirts around.

6.2 Stop somewhere along this stretch of slickrock trail for another perspective of Courthouse Pasture and Monitor & Merrimac Buttes, which are left and right, respectively. The large circular shape on the near side of Merrimac Butte is reminiscent of the sidewheeler housing on that Civil War ironclad ship.

6.25 At what appears to be a trail junction here, the trail angles steeply to the left up a sandstone slope. After a short climb, watch for cairns marking a right turn toward a level stretch of gravel and slickrock.

Uranium Arch

6.4 Here, the trail starts descending to the right, traveling white Kayenta Sandstone slopes along the rim of Sevenmile Canyon. Watch just ahead for where this trail drops off of the slickrock and angles sharply left. As soon as the trail crosses a patch of red sediments, it drops down a ledge, then swings sharply to the right again, to continue downward on white Kayenta Sandstone. Beyond this short double switchback, the trail travels closer to the canyon rim.

6.6 Park here and walk to the nearby rim for a spectacular view of Sevenmile Canyon below, Arths Rim and Gold Bar Rim to the southeast, with the La Sal Mountains looming above Moab Valley. The Abajo Mountains are on the skyline to the south. All of the sloping, eroded terrain to the southwest beyond Sevenmile Canyon is Kayenta Sandstone, with major drainages cutting down into Wingate Sandstone. Below, Utah 313, described elsewhere in this book, goes right up Sevenmile Canyon, then climbs into the higher terrain to the southwest through remnants of rounded white Navajo Sandstone. Beyond this canyon-rim viewpoint, follow the red-dirt tire tracks on the white slickrock and occasional small rock cairns for guidance.

6.7 Stop here and study the short "hanging canyon" to the left of the trail, where a drainage cuts down into red Wingate Sandstone. Watch just ahead for where this colorful canyon angles left and ends abruptly. This hanging canyon is visible when traveling Utah 313 as an angled notch in the cliffline high above. Beyond here, the trail continues to drop down sloping terraces of Kayenta Sandstone set with sparse pinyon-juniper forest.

7.1 Stop here and hike a few yards left for a view of two more hanging canyons that cut deeply into the red Wingate Sandstone that walls Sevenmile Canyon. One of these lovely canyons is accessible to hiking, but the other is not.

7.4 Here, the trail is climbing from the Kayenta Sandstone rimlands of Sevenmile Canyon toward the brushy meadows adjacent to the Monitor & Merrimac Buttes. The white rocky ridge just ahead is a remnant of Navajo Sandstone.

7.6. As the trail enters the brushy meadows, pause and enjoy the colorful view to the southeast, with eroded crags of Navajo Sandstone on the far side of Sevenmile Canyon silhouetted against the La Sal Mountains. Straight ahead, note the remnant of red Dewey Bridge-Entrada on top of the ridge of white Navajo Sandstone. At a trail junction just ahead, either route will eventually reach the Monitor & Merrimac, but the described route goes to the right.

7.8 At a junction here, continue straight ahead. The network of old seismograph trails in Courthouse Pasture confuses navigation. Just ahead, angle to the left at another junction.

8.1 Angle left at another junction. Note that the broad brushy meadows of Courthouse Pasture are almost entirely dune-sand held in place by vegetation and exotic grasses planted in an attempt to improve grazing after the native grasses were grazed to extinction.

8.3 At another junction, continue straight ahead toward Monitor Butte. Continue straight ahead at another junction in a few yards.

8.5 At another junction, go left toward Big Mesa, with the Monitor & Merrimac to the left. In a few yards, angle left again at a juniper tree.

8.9 As the trail approaches a low sandstone ledge at the base of the Monitor, angle right toward the Merrimac for a few yards, turn left up onto the ledge of rock that lies between the two immense monoliths, then continue south between them toward the slickrock terrace rim and along the rim to the right.

9.1 Park on the rim, below the eastern end of Merrimac Butte and study the scene to the south and west. The deep drainage below is the North Fork of Sevenmile Canyon. The red-and-white banded slickrock that surrounds that drainage on all sides is Navajo Sandstone. This rock also forms the base under the Monitor & Merrimac Buttes, with the interface being a few yards down the slope. Note the dune-sand barely held in place by vegetation at the base of the Merrimac, and the accumulation of "Dewey Bridge agate" that has eroded out of the Dewey Bridge-Entrada in which it forms.

To the west, Big Mesa dominates the skyline, with its sequence of rock, from bottom to top, being white Navajo Sandstone, layered Dewey Bridge-Entrada, sheer cliffs of Slickrock-Entrada, broken bluffs of white Moab-Entrada, then the red sediments of Tidwell-Morrison with remnants of rocky Salt Wash-Morrison on its slopes. Both the Monitor and Merrimac have the same sequence except that no Morrison is left on top of the Merrimac, and only a little Moab-Entrada. There is neither Morrison nor Moab-Entrada on top of the Monitor.

At the base of Merrimac Butte.

Remnants of a petrified forest are in desert oasis sediments on top of the white Navajo Sandstone ridge due south of here beyond Sevenmile Canyon. A few fragments of that same deposit of petrified wood have been found on this side of the canyon. See the sidebar, PETRIFIED OASES.

The challenge spur trail described in the referenced book and map that goes clockwise around Merrimac Butte, is not described in this book. To continue this geology log, turn around and skirt the base of Merrimac Butte in a counterclockwise direction to find and follow the trail there.

9.3 As the trail crosses a patch of sediments close to the base of the Merrimac, note the beautiful desert varnish streaks on the "paddlewheel" housing of the Merrimac. See the sidebar, DESERT VARNISH.

9.4 Just ahead, as the trail descends a gentle slickrock slope near the base of the Merrimac, note how the red Dewey Bridge-Entrada on which the trail is traveling phases into white Navajo Sandstone. The trail continues to skirt counterclockwise around the Merrimac's colorful base, then goes left past a stock pond blasted out of solid sandstone.

9.6 From the stock pond, continue across the slickrock away from Merrimac Butte toward the center of the nearest arm of Big Mesa, to where the trail drops off of the slickrock and is visible in the sandy meadows beyond. Along this stretch of trail, note to the left the natural window through one low arm of rock that projects from Merrimac Butte.

9.8 Here, at a five-way junction, the challenge spur around the Merrimac comes in from the left, while two seismograph trails go to the right. The described route continues straight ahead, going between Big Mesa and a detached ridge of red Entrada Sandstone farther to the right. At this junction, the upper tips of the Determination Towers are visible above the dune-sand slopes straight ahead. The described route goes near the base of these towers. Note the lovely pattern of desert varnish on the vertical cliffs of Big Mesa.

10.8 At a junction here, the described route goes left toward the gap in the cliffline, with the far slopes of Tusher Canyon showing through the gap. From here, note the "slit-window" arch through the northernmost of the Determination Towers, and another oval window below that. At one time, enormous and beautiful Courthouse Arch spanned the exposure of white sandstone below the central saddle in the detached ridge of sandstone in mid-Courthouse Pasture to the right-front. The arch collapsed during the winter several years ago, probably from the freezing and expanding of snowmelt behind one of its abutments.

The trail straight ahead at this junction is another of the branches of the Monitor & Merrimac Trail that is not described in this log. It descends through Mill Canyon and provides easy access to the Mill Canyon Dinosaur Trail, then continues on to join the described route at mile 1.1. Angle left here toward the portal into Tusher Canyon.

11.1 The spur left here climbs onto the white Navajo Sandstone base upon which the Determination Towers stand. That spur can be used to circumnavigate the Towers, then return to this trail, but is not included in this description. Note that the lower halves of the larger towers are Dewey Bridge-Entrada while the upper halves are Slickrock-Entrada. The shorter towers are entirely Dewey Bridge-Entrada. Note also how thick the Moab-Entrada is on top of the arm of Big Mesa to the right of the towers.

11.2 Another short spur trail to the left provides access to the benchlands below the towers.

11.7 Here, the trail passes through the portal into Tusher Canyon. In that canyon, the geologic strata continue to be the three members of Entrada Sandstone, topped with remnants of Morrison Formation deposits. Note that as the canyon floor and the trail descend, the canyon walls descend even more abruptly because of the general tilt in the strata in this vicinity.

continued on page 208

PETRIFIED OASES

For tens of millions of years, the general Moab vicinity was an immense low-lying coastal desert. See the sidebar, GREAT DESERT PERIOD. During the peak of this long arid time, the entire region was one vast sea of light-hued sand dunes, dunes that were ultimately to become Navajo Sandstone. In places, this sand accumulated to depths greater than 2,500 feet before a long period of erosion set in, or at least sand deposition ceased. This period was followed by a brief wetter phase, then by the deposition of a different type of sandy desert. These last two deposits are now the red-hued lower members of Entrada Sandstone.

Like most present-day deserts, the Navajo Sandstone desert had its drylakes, its "ephemeral lakes" or "playas." See the sidebar, PETRIFIED PLAYAS. And like even the most arid of today's deserts, such as the Sahara in northern Africa, the Navajo Sandstone desert had occasional springs -- oases -- tiny islands of verdure surrounded by vast expanses of drifting, shifting dunes.

The widespread existence of the petrified remnants of such desert oases in Navajo Sandstone was discovered only recently, in the late 1980s, by southeastern Utah amateur geologists, including the author of this book. That the hard water-layered deposits were left by oases, rather than playas, was indicated by two factors.

First, the oasis deposits contained the fossilized remnants of trees -- petrified wood -- in some cases with the tree stumps still erect and in position, as the trees grew. Some of the trees were quite large, with trunks a foot or more in diameter and perhaps 40 feet tall. The only such tree identified to date was a species of conifer.

Second, the oasis deposits were always darker in color than playa limestones, which are white or light gray, indicating that the oasis deposits had been associated with the decay of organic matter, the decay commonly associated with dead vegetation.

These two factors alone indicated that the Navajo Sandstone desert oases were small islands of vegetation, each with its permanent source of water, its trees, shrubs and, beyond doubt, animal life. Some of the petrified oases found also had tangled masses of petrified algae, of the kind that grows in stringy clumps in freshwater springs and shallow ponds.

In two locations, dark oasis deposits with petrified wood in them were covered by light gray playa deposits, indicating that the oases had been flooded and completely buried by accumulations of the fine sediments and minerals of a newly-formed ephemeral lake, or playa, probably brought in by runoff from an unusually heavy series of storms. The highly mineralized playa deposits killed the trees and other vegetation at the oases and permanently sealed off the springs, although at one such buried oasis there is evidence that the spring broke through the overlying deposits for a time, leaving a stump-like column of petrified algae.

Altogether, over a period of several months, the amateur geologists who first discovered these curious petrified oases in Navajo Sandstone found more than a dozen such locations, and reported them to several professional geologists and paleontologists. To date, however, none of these professionals has made a systematic study of the reported sites, even though such studies would certainly reveal a wealth of paleontological information, help reconstruct the unique Navajo Sandstone paleoenvironment, and firmly resolve the major mystery of where Navajo Sandstone falls on the geologic time scale.

As of this writing, various geologic authorities differ over the age of Navajo Sandstone as much as 20 million years, a problem whose answer most certainly lies in the petrified oases of canyon country.

As with the playas in Navajo Sandstone, its oases also had animal residents and visitors, as indicated by the presence of petrified tracks in oasis deposits. Many small creatures, including insects, must have inhabited these small islands of water-loving vegetation amid vast expanses of barren dune-sand, while other animals doubtless came there for water, or to prey upon those who lived there. It is doubtful, however, that any large herbivores wandered such a vast desert. Certainly no oasis could long support such animals.

The discovery of both petrified playas and petrified oases in the great Navajo Sandstone desert creates a picture of an arid region that was not, after all, entirely barren of plant and animal life, as formerly believed by both geologists and paleontologists. Trees and shrubs grew around these scattered desert springs, algae and probably other freshwater life thrived in their pools, and larger animals, even large predators, visited these isolated islands of life now and then.

Perhaps some day, when professional scientists have taken the time to study the many petrified oases in canyon country, a more detailed picture of life in the middle of the Great Desert Period can be reconstructed, and Navajo Sandstone can finally be dated accurately. And maybe some of the other mysteries associated with Navajo Sandstone can be resolved, such as the existence of petrified logs buried in dune-sand, with no playa or oasis deposits below them, and whether the great Navajo desert actually eroded during the period shown on geology stratigraphic charts as a major unconformity, or simply did not accumulate additional sand during that period of time.

Strangely, the vast deserts that dominated the region before and after the Navajo did not have oases, or at least none have been found in the Moab area so far, in either Wingate or Entrada sandstones. Navajo Sandstone is unique in this way.

BLM archaeologist, left, examines a large slab of oasis tracks.

Small tracks in oasis sediments, with coin for scale.

Theropod tracks in oasis sediments on Hatch Point. The one below is a natural casting.

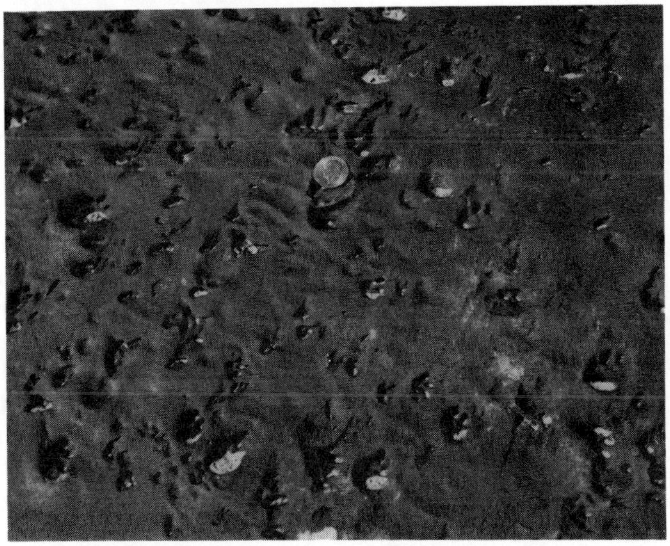

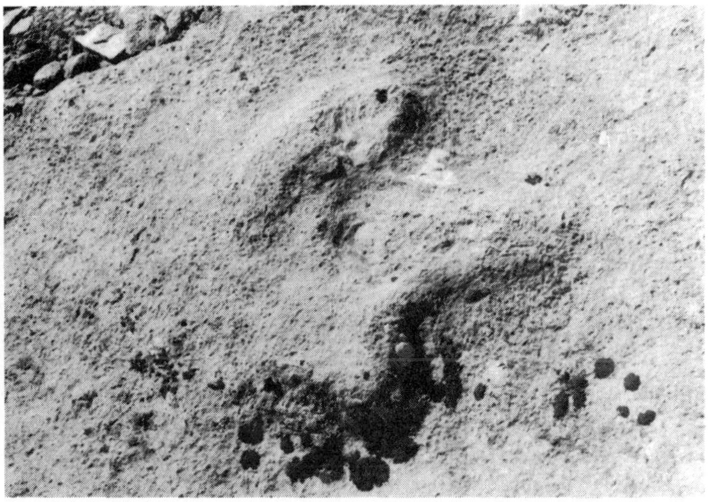

11.8 Here, the trail descends steeply into Tusher Canyon drainages. There is a confusion of trails between here and the main drainage, but any route that reaches that drainage is suitable and will not affect the log's mileages significantly. Note that the portal vicinity and the bottom of the canyon are deep dune-sand deposits left there over a long period of time by the prevailing winds, which are guided by the surrounding sandstone walls. To the right, the beautifully terraced Slickrock-Entrada cliffs and slopes high in the cliffline are accessible to hikers from farther down Tusher Canyon, as described in the referenced book on slickrock hiking.

12.1 At a trail junction here, continue down the drainage. The trail to the left goes into the upper end of Tusher Canyon, but is not covered in this description. Just ahead, the trail travels for some distance in a narrow wash-bottom.

12.5 Here, this trail drops into the main Tusher Canyon wash and continues down the drainage, while another spur trail goes back upcanyon toward an immense mass of dune-sand that is only partly vegetated. The low ridge of Entrada Sandstone just west of this junction hides a beautiful narrow valley that can only be reached by hiking over the ridge from on down the wash.

13.2 One of the routes described in the referenced hiking book begins here by ascending the color-banded Slickrock-Entrada terraces to the right, then continuing all the way up to the southern tip of Mill-Tusher Point with its fantasyland of mini-canyons, terraces and slopes eroded into bare Slickrock-Entrada sandstone. Note that where this trail came through the Tusher Canyon portal, the Slickrock-Entrada cliffs were undercut and had sheared away along vertical planes. Along this lower part of the canyon, where the Slickrock-Entrada has not been undercut, it has eroded into terraces and gentle slopes. Exposures of Slickrock-Entrada such as this make excellent slickrock hiking.

13.4 The cottonwood trees appearing adjacent to the narrowing drainage indicate the presence of permanent subsurface water which is forced to the surface just ahead by underground rock strata. Along this next stretch of trail, note that the descending Slickrock-Entrada is disappearing underground, leaving the canyon walled with white Moab-Entrada for a short distance.

13.7 A fault crosses the wash here at about right angles. Beyond here, for a short distance, the white sandstone on the right is Salt Wash-Morrison. A steep spur trail to the left here is a very rough and challenging connection between this main drainage of Tusher Canyon and a tributary canyon to the west. This trail log continues down the wash.

13.9 The rubbly hills to the left of the wash here are Cedar Mountain deposits. The white sandstone to the right of the wash is still Salt Wash-Morrison. Just ahead, note that the white sandstone deposits on the right end abruptly at another fault, with the Cedar Mountain deposits dominating on both sides of the drainage from there on.

14.4 At a junction here, the trail to the left goes on to connect with the lower end of the Bartlett Wash ORV trail in another 0.9 miles, then with Blue Hills Road in another 1.5 miles. From there, it is another 1.3 miles west to the start of the Hidden Canyon Rim Trail, described elsewhere in this book. That alternate route travels through deposits that are largely Cedar Mountain Formation. Those wishing to take the Hidden Canyon Rim Trail can go left here, but this description goes right, to close the loop begun earlier.

As the described route goes eastward from here, it crosses more exposures of Cedar Mountain Formation, typified by the pale blue-green slopes just beyond this junction.

14.9 Note along this stretch of the trail the large amount of dune-sand in the vicinity that is barely held in place by sparse vegetation. The low gray hills in the near distance to the east, north and west are all Cedar Mountain deposits, which closely resemble Brushy Basin-Morrison painted-desert deposits. As the trail continues, note that where it is light-hued or gray-green or white, it is cutting through Cedar Mountain deposits, while in places where the trail and surrounding soils are brown-hued, it is cutting through accumulated dune sand that has been enriched by centuries of decayed vegetation.

16.6 The described loop route closes here, at the junction first reached at mile 0.6, with Courthouse Rock and Mill-Courthouse Mesa to the right on the skyline. To return to U.S. 191, continue east at this junction. To visit the Mill Canyon Dinosaur Trail, turn right, then right again just ahead, toward Mill Canyon. To visit the Halfway Stage Station, go right also and continue to its spur trail at mile 1.2 of this geology log.

17.0 The return route crosses Courthouse Wash again, with slopes and ledges of Cedar Mountain Formation to the right.

17.2 The described trail reaches U.S. 191. Turn right to return to Moab. Turn left to go toward Interstate 70.

TRAIL NAME: **ONION CREEK - THOMPSON CANYON - POLAR MESA TRAIL**

SUMMARY: This geology trail log is actually a continuous route that combines all of two named off-road vehicle trails, part of another, and part of an improved road. Together, these make a spectacular and geologically fascinating loop route that includes traveling the length of a salt anticline valley that is still active, across and along some of the massive faulting caused by such activity, a broad overview of the stair-step geology of the Dolores River gorge and the northern flanks of the Uncompahgre Uplift, skirting the rimlands of breathtakingly deep and narrow Thompson Canyon, and many different lofty viewpoints of the colorful, geologically distorted scenery of Fisher Valley, the Fisher Towers, Richardson Amphitheater and the open desert terrain on to the north. The route begins and ends near river level, but climbs high into the cool, wooded foothills and associated plateaulands of the La Sal Mountains in between.

TRAIL SURFACE: graded dirt, numerous stream fords, steep and rocky grades, sand, packed and eroded soils and sediments, rough slickrock, graveled and paved road, broken rubble and, during or following rain, slippery mud.

VEHICLES: Except during or following rain, which causes fresh cuts in the route's many stream fords, the first 9 miles can be traveled by carefully-driven highway vehicles, as can the final 8 miles, but the rest of the loop route can only be traveled by skillfully-driven off-road vehicles.

MAP: ***Canyon Country* OFF-ROAD VEHICLE TRAIL MAP – Arches & La Sals Areas.**

GEOLOGIC STRATA VISIBLE FROM THIS TRAIL

Formation	Member	Page
Morrison Formation	Brushy Basin	Page 18
	Salt Wash	
	Tidwell	
Entrada Sandstone	Moab	Page 16
	Slickrock	
	Dewey Bridge	
Navajo Sandstone		Page 13
Kayenta Sandstone		Page 13
Wingate Sandstone		Page 12
Chinle Formation		Page 12
Moenkopi Formation		Page 12
Cutler Group	White Rim	Page 11
	Organ Rock	
	Cedar Mesa	
	Lower Cutler	
Hermosa Group	Upper Hermosa	Page 9
	Paradox Salts	

GUIDEBOOKS: ***Canyon Country* OFF-ROAD VEHICLE TRAILS -- Arches & La Sals Areas,** and **The Utah-Colorado MOUNTAIN BIKE TRAIL SYSTEM – Route 1 - Moab to Loma – *KOKOPELLI'S TRAIL.***

The trail into Onion Creek.

MILEAGES:

TO START: 23.3 miles, from Main and Center, Moab.
TRAIL: 32.6 miles, one-way to end.
RETURN: 31.1 miles, from end of described route.
TOTAL: 87.0 miles, from Main and Center, round trip.

TIME: This loop route requires a full day to travel, an even longer day if combined with the full described length of Utah 128.

ROUTE: Main Street (U.S. 191) north, Utah 128 upriver, all of the Onion Creek and Thompson Canyon ORV trails, part of the Polar Mesa ORV trail, then return via the Castleton-Gateway Road, Castle Valley Road, Utah 128 and U.S. 191.

NAVIGATION: Drive 2.6 miles north from Moab on U.S. 191 then 20.7 miles upriver on Utah 128 to where the Onion Creek Trail goes right toward Fisher Valley. From there, follow the navigational instructions in this log and the referenced ORV trail guidebook and map for the Onion Creek, Thompson Canyon, and Polar Mesa ORV trails, then return via the Castleton-Gateway and Castle Valley roads, Utah 128 and U.S. 191.

GEOLOGY TRAIL LOG:

0.0 As this trail leaves Utah 128 as a graded road passable to most vehicles, it crosses rolling terrain that is largely alluvium, decomposed Cutler Sandstone and both dune and wash sand. The occasional dark red rock outcrops on each side of the trail are also Cutler Sandstone. The trail travels southeastward, first toward the Fisher Towers, then toward the erosional fantasyland of lower Onion Creek, with several rock towers looming against the more distant higher terrain.

To the right of the Onion Creek drainage, the high plateau is the tip of Fisher Mesa. Note in the base of the mesa the detached butte and angled strata that are related to the massive faulting that connects the Salt Valley and Fisher Valley salt anticlines. The conspicuous spire in the Fisher Towers area is The Titan, whose height is 900 feet.

This first short stretch of the trail affords good views of Power Pole Rim to the extreme left, the rimlands of Dome Plateau to the rear, Castle Tower, the Priest and Nuns and the tip of Adobe Mesa to the right, with the more distant western wall of Castle Valley beyond those prominences. The La Sal Mountains loom against the skyline between Fisher and Adobe mesas.

The lowest dark red deposits in the Onion Creek - Fisher Towers area are Cutler Sandstone, with Moenkopi Formation deposits continuing upward. In the cliffs on each side, rubble-covered Chinle Formation sediments slope upward from the red-hued Moenkopi to the sheer cliffs of Wingate Sandstone that are topped with layered Kayenta Sandstone.

0.8 Here, the trail reaches its first view of Onion Creek, as that seasonal stream curves along an eroded bluff of Cutler Sandstone to the left of the trail. For the next several miles, the trail closely parallels this stream, fording it many times. As the trail continues up the drainage, more outcroppings of Cutler Sandstone appear on both sides of the trail, together with occasional remnants of alluvial gravels perched on rocky terraces.

1.0 Here, the trail reaches the first of many fords of Onion Creek. Standard highway vehicles should not continue beyond here if the stream has recently cut into the trail, making low banks at the fords.

From here on through the Onion Creek labyrinth, watch the ornate Cutler Sandstone bluffs and low cliffs for interesting spires, balanced rocks and small arches that are characteristic erosional phenomena of this ancient rock. As the trail continues, note the white crust on the sediments adjacent to the stream in many places. This is minerals that have been dissolved in the stream from where it flows through exposures of Paradox Formation salts. Evaporation from wetted streambank sediments then redeposits these salts.

1.8 Here, after several more fords of Onion Creek, the trail is entering the deeper stretch of the Onion Creek labyrinth. Note the interesting spire to the right of the trail and several ornate rock columns ahead, with The Titan on the more distant skyline. As the trail continues through the winding labyrinth, watch ahead and above the trail for strange erosional forms. Some of these have been given such names as The Camel, The Poodle, Donald Duck, Mr. Magoo, The Chess Men, Grandma and Little Red Riding Hood, Humpty Dumpty, and the Candlestick. Any children in the vehicle might like to see if they can spot the rock shapes these names represent. The Cutler Sandstone in the Onion Creek labyrinth readily forms small natural arches, or holes through solid rock, many of which are visible from the trail.

2.5 As the trail crosses the stream here, The Poodle is on the skyline straight ahead. In a few yards, as the trail curves right along the base of a cliff, The Camel is high in the skyline just beyond The Poodle, and Donald Duck is in between, visible to the left-rear from below The Camel.

3.0 Here, just beyond another stream ford, as the trail is traveling several yards higher than the stream, The Candlestick and Grandma and Little Red Riding Hood come into view on the skyline straight ahead.

Note that along this stretch of trail, the Cutler Sandstone strata are sloping steeply upward toward the east. This stratigraphic distortion is part of the Fisher Valley salt anticline.

3.3 Here, as the trail starts to climb above the narrowest stretch of the Onion Creek labyrinth, The Chess Men are silhouetted against the skyline below Little Red Riding Hood.

From this point it is possible for passengers in a vehicle to hike through the Onion Creek narrows and be met by the vehicle about 3/4 mile on ahead, where the trail once again descends to cross the creek. Hikers should be alert for soft patches of sand beside the creek.

3.7 Here, the trail crosses the creek narrows over a small wooden bridge. Along this stretch of the trail, the Cutler Sandstone cliffs soar to magnificent heights. Note the characteristic way this ancient marine-tideland rock weathers, with rounded, laminated surfaces that tend to erode into columns and spires, and everything patinaed with dark desert varnish. See the sidebar, DESERT VARNISH. Shortly beyond the bridge, the canyon begins to open out into the central part of Fisher Valley.

The Candlestick

4.1 Here, the trail reaches Onion Creek again at an appropriate place for picking up hikers who have walked through the narrows. Beyond here, the trail enters the more open stretches in the middle of Fisher Valley, leaving the narrow confines of the Onion Creek labyrinth, although crenelated towers and walls of Cutler Sandstone continue to dominate to the left. Above these red-hued cliffs and ridges, brush-covered slopes of layered Moenkopi Formation phase upward into Chinle Formation deposits and the still higher Wingate Sandstone cliffs of Fisher Mesa and the Top-of-the-World cliffline. Vehicle trails described in the referenced ORV guidebook and map reach these lofty rimlands.

4.4 The lighter, brick-red outcrops to the right of the trail beyond the creek are Chinle Formation deposits topped with yellowish Wingate Sandstone rubble. This stratigraphic displacement is due to massive faulting and slumping in this part of the salt anticline valley. Note here and on ahead how all of the strata on both sides of the trail have been severely tilted and warped by the same anticline activity.

5.1 Here, the colorful Paradox Formation that dominates this part of Fisher Valley first comes into view as sparsely vegetated color-banded hills and slopes of mineral salts. These ancient marine salts are the exposed spine of the ridge of such deposits that make up the Fisher Valley salt anticline. See the sidebar, SALT ANTICLINES. In Fisher Valley, erosion has exposed an enormous mass of such multicolored evaporite salts. Onion Creek, which originates from springs within and just above the salt exposure, has cut deeply through this upthrust. Tests made within recent decades suggest that Fisher Valley Paradox salts are still extruding upward, with measurable movement, indicating that this salt anticline is still active.

For the next mile or so, the trail travels through this immense surface exposure of Paradox salts, from which Onion Creek acquires a heavy burden of soluble minerals. Particularly malodorous springs that originate from these upthrust salts up ahead gave Onion Creek its name, although in times of heavier water flow, its odor is less discernible. The characteristic sulfurous odor of the creek becomes more pronounced as the trail nears the area of the springs.

As the trail continues up the canyon, note how the original horizontal layers of the Paradox deposits have been drastically distorted by the extrusion process that forced them upward. Each of these layers of salts has its own characteristic color, texture and mineral content, including common salt, gypsum, anhydrite, black shale, gray sandstone and limestone.

6.2 Seeping springs in this vicinity, beginning just upstream of a tributary drainage to Onion Creek coming in from the left, contribute their malodorous content to Onion Creek. If time permits, the drainage to the left here provides an excellent hiking route that ends in deep and narrow grottoes in dark red Cutler Sandstone, beyond an old mine that was active at the turn of the century. Documents found in the collapsed remnants of a cabin near the mine indicate that it was occupied by a man and his wife in 1904, when this obscure tributary canyon was far more remote and inaccessible than it is now.

7.2 Here, as the trail is emerging from the cut that Onion Creek has made through the Paradox salt extrusion, heavily faulted and distorted dark red Cutler Sandstone deposits appear again to the left of the trail, although Paradox salts continue to be visible to the right of the trail for some distance yet. Just ahead, the trail climbs steeply away from Onion Creek, along a narrow ridge between Onion Creek and tributary drainages to the left. The trail cuts through Cutler Sandstone and more recent alluvial deposits as it ascends toward the level of upper Fisher Valley, which will be discussed further ahead in this log.

7.7 Stop at a safe place along this stretch of trail and look back down the colorful and distorted Onion Creek drainage.

8.0 Along this stretch of the trail, as it climbs through eroded Cutler sediments, note to the right of the trail beyond Onion Creek the sharp interface between the red-hued Cutler deposits and the gray Paradox salts. The lofty rimland of Fisher Mesa dominates the southwestern skyline. The high plateau above Fisher Valley is one arm of immense North Beaver Mesa. The light-hued Wingate Sandstone rimland to the left along this stretch of the trail is Top-of-the-World.

First appearance of Paradox salts.

8.7 From here, just before the trail reaches the relatively flat meadowlands of upper Fisher Valley, stop and study the eroded terrain to the right of the trail. Earlier in Fisher Valley's geologic history, the upthrusting Paradox salts formed a barrier to drainage from the upper end of the valley. Light red sediments from still higher terrain accumulated behind this barrier to form the horizontal strata of relatively young deposits visible in the slopes and hills to the right, including those through which the trail is now ascending.

During the current phase of erosion, Onion Creek, which originates from numerous springs in the hilly maze to the right, and local rainfall and snowmelt, have slowly eroded these younger deposits into their present configuration, carrying them away to the Colorado River through the narrow gorge that Onion Creek has cut through the Paradox salts. This erosion is still eating into the rim of the upper basin of Fisher Valley.

In the high cliffline to the northeast, note the large arch near the upper rim. While the opening behind this arch is not very wide, it is quite long, leaving the immense arc of rock supported only by its massive abutments. This arch can be reached by a short hike from the end of the Top-of-the-World Trail, which is described in the referenced ORV guidebook and map.

9.0 Here, the trail tops out from the grade and reaches the level of upper Fisher Valley, with its broad meadows of alluvium. Much of this valley is a privately-owned ranch, but the mapped trails that cross it are open to the public. Note ahead, the drastically tilted strata that have been distorted by the salt anticline activity of Fisher Valley. The sloping mesas there are fractured Wingate sandstone with traces of layered Kayenta Sandstone above that. The larger bluff has slopes of rubble-covered Chinle Formation deposits continuing down to the valley floor.

9.7 At a trail junction here, the Cottonwood Canyon ORV trail (Kokopelli's Trail) goes left, while this trail swings right toward distant ranch buildings on another segment of Kokopelli's Trail. Kokopelli's Trail, described in the referenced book with this subtitle, is a mountain bike route that begins in Moab, travels on existing roads and ORV trails, then ends near Grand Junction, Colorado.

10.1 Here, the main trail enters private ranchland. Stay strictly on the described trail, and shut any closed gates after passing through them. Do not drive the graded road that goes toward the ranch buildings.

10.7 Here, the described trail angles sharply left toward a notch in the ridgeline ahead. That low ridge is entirely Moenkopi Formation sediments. To the right, the higher cliffs of North Beaver Mesa are Wingate Sandstone topped with a thin remnant of Kayenta Sandstone, with vegetated slopes of Chinle Formation angling down to the protruding bluffs of dark red Moenkopi. The entire valley floor is relatively recent deposits of alluvium.

Standard highway vehicles and low-geared, high-clearance highway vehicles such as trucks and vans that have come this far, should not continue beyond this point. Only skillfully-driven off-road vehicles of several types can travel the next 16 miles of steep, rough and eroded trail.

11.5 Here, the trail tops out of a steep grade and descends into lower terrain beyond a gate, which should be left as found, either open or closed. The slopes to the right along this next stretch of trail are Moenkopi deposits, while the rounded cliffs to the left are Wingate Sandstone. This mismatch was caused by the major faultline on which the trail is traveling as it descends. The distant high cliffs are still Kayenta, Wingate and Chinle, in descending order.

12.5 Here, a spur trail goes to a designated camping area that has shaded campsites, picnic tables and primitive toilet facilities. The main trail descends to the left to cross a drainage. A few yards up this drainage, a small dugout log structure juts from the wall of sediments and rock. The cabin was built here to take advantage of the perennial springs in the drainage. Please do nothing to accelerate the slow collapse of this historic cabin. The cottonwood trees here are taking advantage of the permanent source of water. As the trail continues, note the many cottonwoods thriving in this vicinity. These indicate that there is considerable subsurface water here, even in and near the seemingly arid drainages.

Historic cabin near the trail.

12.6 Stop here and study the scene ahead and to the left. The massive faulting through which this trail is traveling has severely displaced the geologic strata in this area. The outcrops of white rock to the left and ahead are the Slickrock Member of Entrada Sandstone that is many hundreds of feet below its normal stratigraphic level, which is high above the sheer Wingate Sandstone cliffs that surround this descending drainage. Farther on, the trail climbs steeply onto part of the downthrust ridge ahead and travels closely parallel to the faultline that displaced it. The faults along this trail are part of a continuous series of faults that connect the Fisher Valley and Sinbad Valley salt anticlines. Just ahead, the trail crosses another trickling stream.

12.9 Here, as the trail is traveling between the two light-hued Slickrock-Entrada outcrops, note the large cave eroded into the one on the left. Beyond the cliffline with the cave, the lower ridge of light-hued rock is Navajo Sandstone that has also been faulted downward from its normal position, high above the Wingate cliffs.

13.1 Here, the trail starts climbing steeply into higher terrain just beyond a small drainage. The upper part of the cliff ahead is Slickrock-Entrada, while the lower light-colored layers are Navajo Sandstone.

13.8 Stop somewhere along the trail here and study the scene through which it is traveling. The high cliffline to the northwest is an eastward extension of Top-of-the-World. The distant cliffline to the left of that is Fisher Mesa. The tilted outcrop of rock silhouetted against Fisher Mesa is Wingate Sandstone with a remnant of layered Kayenta Sandstone on top of that. In the lower terrain, toward the gap in the cliffline, the low ridges are drastically displaced Navajo Sandstone, while a few of the outcrops nearer the trail are displaced Slickrock-Entrada. On ahead, to the right of the trail, the smooth-walled cliffline is also Slickrock-Entrada that has been displaced downward to a lesser extent by faulting between it and the higher cliff of fractured Wingate Sandstone behind it. Beyond here, the trail continues to travel on top of the lower ridge of Slickrock-Entrada.

14.0 Note to the right of the trail here, the sharp line of demarcation between the light-hued Slickrock-Entrada cliff and the darker Wingate and Kayenta sandstones at the same level. That interface marks the faultline that displaced the Slickrock-Entrada downward. The uppermost layered light-hued rock in that rubbly cliffline is Navajo Sandstone.

14.5 Just before the trail begins to descend for a short stretch, the panorama to the east opens up. The highest ridgeline is the northwestern end of the immense and ancient Uncompahgre Uplift. The major drainage line this side of the Uncompahgre that cuts through numerous colorful rock strata is the lower Dolores River gorge. This entire colorful panorama is described in more detail from a point farther along the trail. Beyond this viewpoint, note how the strata to the left and right are descending toward the Dolores River. Both clifflines are predominantly Wingate and Kayenta sandstones topped with thin remnants of light-hued Navajo Sandstone.

15.1 Here, just after it has crossed a grassy meadow, the trail begins another steep ascent.

15.4 From a pullout at a hairpin curve to the right, study the breathtaking panorama to the east. First, note that all the geologic strata on both sides of the Dolores River far below slope toward the river. This general slope was created by the Uncompahgre Uplift on the far side, and by the La Sal Mountains and Fisher Valley salt anticline on this side, with the Dolores River drainage developing in the lower terrain. The color-banded sandstone cliffs on both sides of the river are Slickrock-Entrada with layered rubble-covered slopes of the Salt Wash Member of the Morrison Formation continuing upward. The upper slopes of the Uncompahgre are highly vegetated but are also Salt Wash-Morrison, which was deposited completely over the Uncompahgre before its more recent uplift periods. See the sidebar, UNCOMPAHGRE UPLIFT.

Below the Slickrock-Entrada cliffline on the far side of the river, the terraced darker deposits are Kayenta Sandstone, with a short stretch of vertical Wingate Sandstone cliffline visible upriver to the right. Beyond this sheer cliffline, the steeply eroded gray slopes are Chinle Formation. Note that while there is some Navajo Sandstone on this side of the river, as noted earlier and as visible a few places from this viewpoint, the Navajo does not exist beyond the Dolores in this vicinity except for a small patch in the distance toward the left. The eastern edge of Navajo Sandstone lies in the vicinity of the Colorado-Utah border. It apparently ended at the base of the Uncompahgre in this area. Farther southeast, it ended on the western slopes of the Paradox Valley salt anticline, as noted in the log for Utah 46 - Colorado 90, described elsewhere in this book.

The canyon immediately to the southeast of this viewpoint drains into deep and narrow Cottonwood Canyon farther north, between here and the Slickrock-Entrada bluffs in that direction. The light-hued rock that tops the drainage ridge to the southeast is a thin remnant of Navajo Sandstone. Beyond this spectacular viewpoint, the trail ascends still more, then travels at about the Navajo-Kayenta interface for about 3 miles.

16.1 Beginning here, the deep inner gorge of Thompson Canyon comes into view beyond the trees. The lower walls of the canyon are Wingate Sandstone with layered Kayenta Sandstone above that, and remnant

continued on page 216

UNCOMPAHGRE UPLIFT

"Uncompahgre" is a Ute Indian expression meaning "red water springs," which probably referred to colorful iron-rich springs now in the Uncompahgre National Forest & Primitive Area.

Most historical accounts attribute the first explorations of the Uncompahgre region in western Colorado to members of the 1776 Dominguez-Escalante Expedition, but recent research by this author and others has determined that explorer Juan Maria Antonio Rivera was searching the region for precious metals in the service of Spanish commercial sponsors as early as 1765.

Modern geologists have applied the name Uncompahgre Uplift to the ancient spine of an immense northwest-trending ridge that began to form more than 300 million years ago in what are now adjacent areas of Colorado and Utah. At present, the above-ground part of the uplift appears just south of Grand Junction, Colorado, and ends to the southeast by blending into the San Juan range of the Rocky Mountains.

The Uncompahgre Uplift first appeared in the distant past, about 10 million years after several other regional uplifts. These uprising ridges of ancient Precambrian granite were the first land to appear above the shallow seas that had covered the region for millions of years. As they rose, they isolated a huge bay from the surrounding ocean, turning it into a "dead sea" of sorts. See the sidebar, SALT ANTICLINES, for more details about the interaction of the Uncompahgre Uplift with the adjacent marine deposits that accumulated to great depths along its southwestern flank.

The first upthrusting of the Uncompahgre barely broke the surface of the entrapped embayment and was re-submerged not long afterward, geologically speaking. About 2 million years later, however, regional tectonic forces thrust the uplift upward again, this time to considerable heights. The Uncompahgre Uplift and others in the general region were active during the major mountain-building period that produced the Ancestral Rocky Mountains, the predecessors of the present Rockies that eroded away and were largely covered by Morrison Formation and later deposits before the present range developed.

Erosion from the Uncompahgre's lofty red-granite spine then began. Its sediments gradually filled adjacent lower terrain, including the basin of Paradox salts that had accumulated to the south and west. These were the upper Hermosa and Cutler Group deposits, the first of the famous canyon country "red-beds."

Geologists disagree about the timing, but do agree that the Uncompahgre continued to uplift at intervals from then on, and some have found evidence suggesting that it is still rising, very slowly. At least one geologist believes that at one point the uplift reached heights of over 12,000 feet above sea level.

There is evidence to support a second major rise about 280 million years ago, some 23 million years after the initial low uplift and during early Cutler deposition. Another rise occurred at about the Permian-Triassic interface, 245 MYA, near the end of Cutler deposition, and corresponded to a major extinction of life. See the sidebar, GREAT EXTINCTIONS. If that extinction was due to a meteor impact, as others were, the impact may also have triggered regional tectonic activity, including this particular rise in the Uncompahgre.

Another small uplift occurred at about the interface between the Moenkopi and Chinle deposits, some 225 MYA. This ended Uncompahgre uplifting for a long time, although one authority indicates that there was another small rise during the Moenkopi deposition. Either way, its higher terrain soon eroded down to where Chinle Formation deposits covered the uplift.

For many millions of years the Uncompahgre was quiescent, as canyon country strata continued to accumulate, erode and accumulate again. Morrison Formation deposits also covered the Uncompahgre, as did Burro Canyon, Dakota, Mancos Shale, and some Mesaverde sediments. By then, the Ancestral Rockies had also eroded down to a few low hills, and the canyon country region was largely low-elevation coastal plains.

Then, a prolonged period of tectonic activity began. See the sidebar, LARAMIDE OROGENY. Some of the older uplifts, like the Uncompahgre, rose again, and new uplifts such as the San Rafael Swell began. Later, entire new ranges of mountains were created, such as the Uintas, Henrys and the present Rockies. The newly created subranges of the Rockies tended to obscure the southeastern end of the ancient Uncompahgre.

In the midst of this period of millions of years of slow or sporadic tectonic activity, at least two other meteors struck, one in canyon country. See the sidebar, UPHEAVAL DOME. The other annihilated the dinosaurs and most of their associated ecosystem, thus giving mammals an opportunity to dominate, which in turn led to the evolution of humanity. Toward the end of this tectonic period, both the Uncompahgre Uplift and the San Rafael Swell rose still higher.

Although geologists disagree about the timing, they do agree that the Uncompahgre uplifted twice again during geologically recent times, once about 10 million years ago, when the entire region between the Sierra and Rocky mountains rose from 7 to 10 thousand feet, triggering the current period of erosion in canyon country, and again about 1 million years ago. A few geologists are convinced that this latest rise is not over yet.

This last uplifting of the Uncompahgre was associated with a strange sequence of events. Geologists who have studied the geologic history of this region's intricate drainage system have attempted to reconstruct the creation of Unaweep Canyon, the deep granite-walled gorge that cuts completely through the Uncompahgre between Gateway and Grand Junction in Colorado.

continued ----

According to one scenario of events, as the young Colorado River was slowly cutting down through the sediments that had accumulated on the uplift, it abraded the uplift's Precambrian core, creating Unaweep Canyon, with the Gunnison joining the Colorado east of the uplift. Later, for reasons as yet unknown, the Colorado was supposedly diverted into a drainage that was curving around the northern nose of the uplift, perhaps by conventional "stream piracy," perhaps in some other manner, leaving the Gunnison to continue flowing through Unaweep Canyon until it was diverted into its present channel.

More recent research, however, firmly establishes that Unaweep Canyon was cut into the spine of the Uncompahgre by the ancestral Gunnison River, which then joined the Dolores River near the present town of Gateway, Colorado, while the Colorado River flowed approximately along its present route. Later, the Gunnison was also diverted, this time by the most recent rise of the Uncompahgre, which stopped its flow through Unaweep Canyon. The Gunnison then joined the Colorado. This left Unaweep Canyon drained by two small streams -- East Creek and West Creek -- that flow both directions from its higher center.

Today, the Colorado River is making its own cut down through the Precambrian granite spine of the Uncompahgre Uplift. This is visible as the very dark, ornately water-carved rock that river-runners see along the walls of Westwater Canyon.

Thus, the Uncompahgre Uplift has played a major part in creating the present canyon country scene. It helped define the gigantic basin in which Paradox salts accumulated. With some help in the form of lateral squeezing from the uplift and eastward crustal pressures, these marine mineral deposits later formed the region's strange salt anticline valleys. The Uncompahgre also contributed stupendous amounts of iron-rich sediments to canyon country, its great "red-beds," as well as some of the run-off water that carved the region's complex canyon systems.

Its most recent rise caused one of the region's major rivers -- the Gunnison -- to change its course, forcing it to join the Colorado rather than the Dolores. Both the Dolores and Colorado continue to carry away great quantities of Uncompahgre and canyon country sediments, although now these accumulate behind Glen Canyon Dam rather than going on to the sea.

Today, the densely wooded spine of the uplift, itself, and most of its adjacent highlands, also offer endless scientific and recreational opportunities. The Morrison deposits on its eastern slopes have yielded the petrified bones of numerous dinosaurs, including famous Supersaurus and Ultrasaurus, and the high, wooded uplift with its numerous tiny lakes and network of roads and off-road vehicle trails is a scenic paradise for hikers and mountain bikers.

The primordial ridge of ancient granite that first thrust up from the sea more than 300 million years ago has outlived the vast mountain range that was born with it, and is still rising after the creation of a second companion mountain range. It shows every sign of surviving for another 300 megayears.

bluffs of light-hued Navajo Sandstone set back from the rimlands on both sides. Ahead to the left, the highest prominence with a base of Slickrock-Entrada is one end of Polar Mesa, which was probably named after the light-hued sandstone that walls the entire mesa, giving it the appearance of a gigantic iceberg floating in a green sea of vegetation.

Note that the Salt Wash-Morrison strata on the mesa have been cut by uranium mining trails. The Polar Mesa Trail, part of which this route travels on ahead, ends at a sizable mining community on top of Polar Mesa. That once-active settlement is now defunct, although some of the mineral claims there were "patented," and are hence now private land.

Beyond here, the trail travels on Navajo Sandstone remnants closely parallel to the Thompson Canyon drainage, which becomes more visible on ahead.

16.9 Pause here for a breathtaking view of Thompson Canyon where it is joined by a short spur canyon draining from the direction of Polar Mesa. Beyond here, Thompson Canyon becomes deeper and more spectacular as the trail continues.

In another quarter mile, the trail climbs away from Thompson Canyon to skirt along the upper rimlands of North Beaver Mesa, high above upper Fisher Valley. As it ascends, the trail travels through fractured and heavily vegetated Navajo Sandstone.

17.9 At a trail fork here, go right. The trail straight ahead is a little-used alternate route that is eventually rejoined by the one to the right, which offers a spectacular view of Fisher Valley and the terrain beyond.

18.2 Stop here, beside a large slab of undercut sandstone on the rim to the right, for a breathtaking view of upper Fisher Valley, with parts of the Onion Creek labyrinth showing beyond the meadows. The Grandma and Little Red Riding Hood monoliths are visible from here, as is the top of The Titan in the Fisher Towers area. The lofty rimland to the right of Fisher Valley is Top-of-the-World. The opposite rimland barely visible from here is Fisher Mesa, with a detached butte of that promontory looming above Little Red Riding Hood. The nearby sheer-walled peninsula is an extension of North Beaver Mesa, on which this trail is traveling. Its vertical walls are Wingate Sandstone and its layered upper strata are Kayenta Sandstone. Above the trail on the left, the white rock is Navajo Sandstone. Beyond this overlook, the trail climbs steeply upward to rejoin the bypass trail.

18.8 Here, the scenic loop rejoins the bypass trail. Turn right. The next stretch of trail, as it climbs gradually toward another arm of Polar Mesa, crosses a number of outcrops of Navajo Sandstone that have been exposed by erosion.

20.2 The trail reaches a fenceline marking the National Forest boundary. Up to this point, the trail has been traveling through public land administered by the Bureau of Land Management. Close the gate after going through. The Slickrock-Entrada cliffline ahead walls another part of lofty Polar Mesa. Note the tall conifers growing along the spring-watered base of the cliff.

21.5 Here, the trail has reached brushy mountain meadows, while still traveling along the base of the Slickrock-Entrada cliffs of Polar Mesa. Other geologic deposits are not visible here because they are densely overgrown or buried under soils.

21.7 Here, the trail tops out, then begins to descend away from Polar Mesa.

22.0 At a junction here, the Thompson Canyon Trail ends by joining the Polar Mesa Trail. This log goes right to join the Castleton-Gateway Road in another 4.4 miles. Those who have the time and inclination may want to go left here first, to explore Polar Mesa and its historic community of uranium mines. That 15-mile side trip is not covered in this log.

As the trail continues toward the La Sal Mountains, the geologic strata are still too obscured by alluvium and dense vegetation for observation, but the rock beneath the soil is Entrada Sandstone for the first mile or so.

Between here and the Castleton-Gateway Road, note the numerous tall Ponderosa pines growing in the drainage to the left of the trail, and on both sides of the trail farther ahead. There are many good primitive campsites in these groves.

24.4 Here, as the trail begins descending again after going through several beautiful groves of tall Ponderosa pines, a broad panorama of the La Sal Mountains opens up ahead. This range is laccolithic. See the sidebar, LACCOLITHIC MOUNTAINS for more details about the geology of such ranges. As the trail descends, it passes outcrops of beautifully crossbedded Navajo Sandstone.

24.7 Here, the trail is descending steeply through Navajo Sandstone that has been fractured by several parallel faults. The lower of these faults has aligned the Navajo with Salt Wash-Morrison deposits.

25.9 Park here, just beyond an old corral, for another breathtaking view of upper Fisher Valley and the Top-of-the-World rimlands beyond. The small bridge here that the trail crosses spans a ditch through which streams from higher terrain have been diverted to flow into Fisher Valley for use by the ranch there.

26.4 Here, the Polar Mesa trail ends by joining the improved Castleton-Gateway road, about midway between Moab and the town of Gateway, Colorado, in the Dolores River gorge. Turn right to continue this description. Beyond here, the route continues to travel through Salt Wash-Morrison that is largely obscured by dense vegetation.

27.2 This stretch of road affords still more breathtaking views of Fisher Valley, the Top-of-the-World rimlands, Fisher Mesa, The Titan and other Fisher Towers spires, plus the more distant rimlands of Dome Plateau and the Book Cliffs in the extreme distance.

27.8 From a wide pullout here, note that although the road is climbing steeply, the tilted strata here above the Fisher Valley anticline are climbing even more steeply. At this pullout, the deposits along the road are Morrison Formation, but to the right the highest cliffline below the road-cut ahead is Entrada Sandstone. Below that, the two lower clifflines are Navajo Sandstone and Wingate Sandstone separated by sloping vegetated terraces of Kayenta Sandstone.

As the trail continues climbing, it offers almost aerial viewpoints of upper Fisher Valley.

28.7 Here, the trail tops out just beyond a stock-watering pond, and is traveling on top of the Moab Member of Entrada Sandstone. The spur ORV trail that goes right here crosses Moab-Entrada that is largely covered by soils and vegetation, but exposed in a few areas. Some of these exposed areas have petrified dinosaur tracks in them. The first tracks appear in the trail itself, a few yards beyond the fenceline, in sandstone darkened by desert varnish. Others are in the brushy terrain to the left of the trail. There are similar tracks at this same stratigraphic level, but at much lower elevation, on the eastern and western monoclines of the Salt Valley anticline, in the borderlands of Arches National Park. The Klondike Bluffs Trail, described elsewhere in this book, passes some of these tracks. See the sidebar, PETRIFIED TRACKS.

29.0 The Fisher Mesa ORV trail spurs right here. In about 100 yards, it reaches a primitive campground with tables, fire grills and a pit toilet, then continues to the end of immense Fisher Mesa. That long, difficult and spectacular trail is described in the referenced ORV map and guidebook, but not in this book.

30.6 Along this stretch of what is now a paved road, the scant exposures of slickrock to the right across a drainage are Entrada Sandstone, with this deposit angling downward as the road is traveling through Morrison Formation that is obscured by dense vegetation.

31.1 Here, upper Castle Valley comes into view. For a description of its geology, refer to the log for Castle Valley Road elsewhere in this book. Note the bare slopes on the mountains ahead that are too steep and slide-prone for vegetation to become established. Note also, to the right-front, a sharp ridge of light-hued sandstone that is a remnant of the tilted and faulted interface between the ancient salt anticline of Castle Valley and the much younger upthrusts of the La Sal Mountains.

32.6 Here, the road crosses Castle Creek. As defined in the referenced ORV guidebook and map, the Castleton-Gateway Road ends here, and the road that continues down the valley is Castle Valley Road. This geology log ends here, but the geology visible for the next 1.6 miles, between here and the La Sal Mountain Loop Road junction, is described from Castle Valley Road and, to some extent, from the first part of the La Sal Mountain Loop Road.

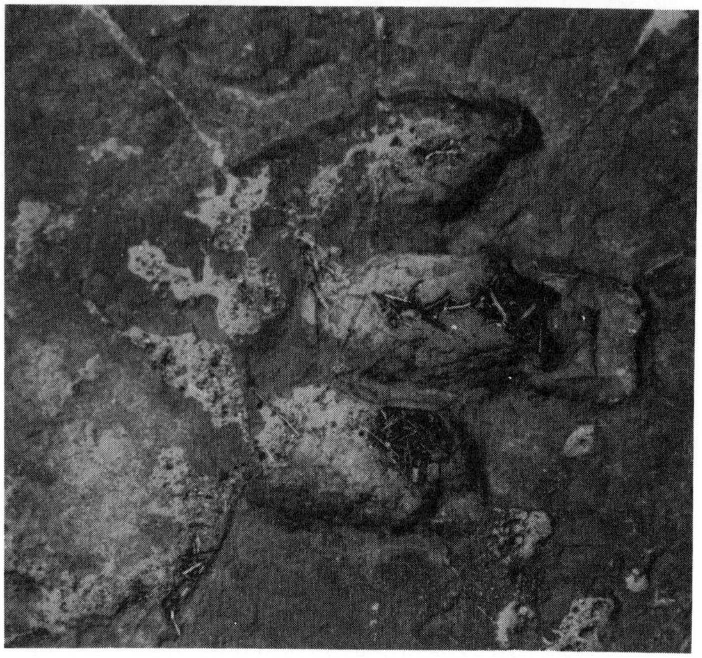

One of the dinosaur tracks at Mile 28.7.

Thompson Canyon

TRAIL NAME: **POTASH - SHAFER TRAIL**

SUMMARY: This popular ORV trail travels the colorful rimlands of the inner gorge of the Colorado River, through a fantasyland of ancient red Cutler Sandstone deposits, before ascending the breathtaking switchbacks of the historic Shafer Trail to the elevated meadows of Island-in-the-Sky, where the route ends by joining the paved Island Road. Although the stratigraphic sequence visible from the route is mostly uniform, along the way the trail cuts across two major anticlines and a small syncline, and five mega-years have carved a unique variety of erosional landforms from this combination of strata. As it crosses one anticline, the Potash Trail goes around the sprawling solar evaporation ponds that are so conspicuous from Dead Horse Point and Anticline Overlook, providing a close look at this environmental disaster area. Beyond the ponds, the trail travels through unspoiled red-rock terrain and skirts the rim of the river gorge, far below soaring Dead Horse Point but still high above the river. It then enters Canyonlands National Park and climbs to the level of the park's famous White Rim before ascending the Shafer Trail. It should be noted that individuals who suffer from acrophobia will not enjoy traveling the Shafer Trail switchbacks.

ROAD SURFACE: gravel, sand, packed sediments, rough marine limestone and slickrock, with three miles of very steep, rough and narrow switchbacks.

VEHICLES: off-road vehicles only for the entire trail, although when it is dry, carefully-driven, low-geared, high-clearance highway vehicles can travel the entire length. It should be noted that this route enters Canyonlands National Park, and that only street-licensed motor vehicles operated by licensed drivers are allowed on park roads and ORV trails. Thus, such non-street-legal off-road vehicles as ATVs can travel this route only to the park boundary and back.

GEOLOGIC STRATA VISIBLE FROM THIS TRAIL

Navajo Sandstone		Page 13
Kayenta Sandstone		Page 13
Wingate Sandstone		Page 12
Chinle Formation		Page 12
Moenkopi Formation		Page 12
Cutler Group	White Rim	Page 11
	Organ Rock	
	Cedar Mesa	
	Lower Cutler	

MAP: ***Canyon Country*** **OFF-ROAD VEHICLE TRAIL MAP - Island Area.**

GUIDEBOOK: ***Canyon Country*** **OFF-ROAD VEHICLE TRAILS - Island Area.**

MILEAGES:

TO START: 19.9 miles, from Main and Center, Moab.
TRAIL: 20.4 miles, one-way to end.
RETURN: 32.1 miles, from end of described trail.
TOTAL: 72.4 miles, from Main and Center, round trip.

View from near the trail beginning.

TIME: The trail as described requires a short day to travel, a full day if time is taken to explore one or two of the shorter spur trails noted along the way, a long day if either of the longer spurs is taken.

ROUTE: Main Street (U.S.191) north, then Utah 279 downriver to the trail start, where paved Utah 279 ends.

NAVIGATION: drive north from Moab on U.S. 191 for 4.2 miles, downriver on Utah 279 for 15.7 miles, continue on the Potash Trail to the base of the Shafer Trail grade, then up that trail to were it ends by joining the Island Road. Return to Moab via the Island Road, Utah 313 and U.S. 191, or via the Island Road, Utah 313, the Long Canyon ORV trail, Utah 279 and U.S. 191.

GEOLOGY TRAIL LOG:

0.0 From where paved Utah 279 ends at an industrial chemicals mill, this trail continues on downriver past other parts of this sprawling operation. For more information about the grim human and environmental history of this plant and the associated solar evaporation ponds that the trail passes ahead, see the sidebar, POTASH.

As the trail begins, it travels on riverbank sediments and gravels past gigantic Jackson Hole, on the other side of the river. This "hole" is an abandoned meander of the Colorado River. The high, isolated promontory to the left-front from here is the center of the abandoned loop in the river. See the sidebar, ENTRENCHED MEANDERS. Beginning here, the river cuts through the arching strata of the Cane Creek Anticline, and on ahead the trail travels its eroding slopes.

In this vicinity, the strata on the far side of the river are, from top to bottom, remnants of layered Kayenta Sandstone above vertical Wingate Sandstone cliffs, with red and gray slopes of Chinle Formation below that. Layered slopes of red Moenkopi Formation sediments continue down to more rounded clifflines of Cutler Sandstone, which continues down to the river level. Much of the lower terrain is covered with river gravel left there as the Jackson Hole meander was abandoned by the river.

1.2 Along this stretch of the trail, the terraced red cliffs on both sides of the river are lower Cutler Sandstone deposits, with most of the terraces and slopes covered with rounded river gravel. As the trail continues, note how the strata on both sides of the river are tilted by the Cane Creek Anticline. The high promontory to the left is part of Hatch Point. Some of the developments of Anticline Overlook are visible on the highest tip of Hatch Point. The Anticline Overlook Road, described elsewhere in this book, ends at that spectacular viewpoint.

1.3 The spur left here goes to a paved public launch ramp on the Colorado River that is used by river-runners. Ahead, the trail climbs parallel to the ascending strata to reach higher terrain, where it travels for many miles at roughly the same stratigraphic level, while continuing to ascend and descend the angled slopes of the Cane Creek Anticline. The highest sandstone abutments to the left across the river here are Wingate Sandstone.

1.6 The rock strata adjacent to the trail in this vicinity are shallow-marine deposits in the lower Cutler Sandstone. In some areas, this limestone deposit is rich in marine fossils, although in this local area the rock is so highly lithified that the fossils are difficult to extract for collecting. These fossils here are exposed on the surface of the rock by weathering, and are visible as fragments of shells, crinoid stalks, and other marine creatures that left hard body parts or shells. For the next 11 miles, almost anywhere that the trail goes through gray limestone deposits, these may be fossiliferous.

At one time, this marine limestone layer, and the sandstone strata below it to considerable depth, were called the "Rico Formation." This was later renamed the "Elephant Canyon Formation," but further recent study found this designation to be based on faulty research. That study suggested that "lower Cutler" beds was more appropriate, so that designation is used throughout this book. See the sidebar, YES -- GEOLOGISTS ARE HUMAN for more details.

As the trail continues, notice that the upward-arching strata of the Cane Creek Anticline on the other side of the river are leveling off on the summit of this salt anticline. See the sidebar, SALT ANTICLINES. In the next mile or so, watch for the drilling platforms and plugged holes of early oil explorations.

2.7 The dark red sandstone deposits through which the trail has been traveling, and which dominate the low bluff ahead, continue to be Cutler marine-tideland sediments. To the left, beyond the river, the high mesa is more of huge Hatch Point, while the lofty peninsula to the right is a part of Island-in-the-Sky. The deposits in these two elevated promontories are, from top to bottom, layered Kayenta Sandstone, sheer-walled Wingate Sandstone, red and gray slopes of Chinle Formation, layered Moenkopi Formation, then varied red-hued Cutler Sandstone deposits on down. This combination of rock strata continues for most of the rest of this trail, with minor variations along the way.

Note that, while the strata arched by the Cane Creek Anticline are now descending, the general tilt of the higher terrain on both sides of the river is upward toward the south. Because of this general upward slope, the south-trending Colorado River is cutting more and more deeply into ancient Cutler Sandstone strata. Some of these older strata are visible from stretches of the trail ahead, as it closely approaches the inner gorge of the Colorado River.

3.5 The spur left here is a dead-end trail that goes to a private leaseholding of Utah State land beside the river. Note in the low bluff ahead and to the left here, the color and texture variations in the Cutler Sandstone deposits. These are indicative of changes in the tides, sea level and climate, and of transient storms, during the long period of Cutler deposition. Some of these more laminated strata may be fossiliferous, while others that represent tideland dune-sands are not. A few such strata, still below this level but exposed in the inner gorge of the river, contain anomalous petrified wood, although the trees probably did not grow in the deposit in which they occur.

3.9 Here, the trail begins climbing steeply into higher levels of Cutler deposits. Note the beautiful color-banding and erosional forms in the rock visible from the trail as it ascends.

4.3 As the trail nears the grade summit, note the immense balanced rock to the right of the trail. Just ahead, as the terrain to the north and east opens up, note how the strata in that direction are arched upward by the Cane Creek Anticline. The gray-hued rock that dominates the lower terrain visible from here is the fossiliferous Cutler marine limestone through which this trail travels for many miles.

4.6 Here, the trail is skirting around a series of immense ponds, in which soluble salts pumped upward from the Paradox Formation are concentrated by solar evaporation before being transported to the mill at the beginning of this trail for separation. In earlier years, this trail followed a route directly through the terrain now occupied by these evaporation ponds. It was rerouted when this previous BLM-administered public land was transferred to the State of Utah and then to private ownership.

The crenelated dark red cliffs to the right of the trail are Cutler Sandstone deposits. The convoluted abutments of Hatch Point continue to dominate the southeastern skyline beyond the Colorado River gorge.

5.2 Note here to the right of the trail, how construction of the salt ponds has blocked a natural drainage and dammed rain runoff. Water-loving vegetation has taken advantage of this serendipitous pond.

5.4 From where the trail tops the gradual ascent while skirting around the salt ponds, many more ponds are visible on lower terraces above the river gorge. From this same higher vantage point, the interface between Cutler and higher Moenkopi deposits is just above the trail level to the right. To the left-front, the conical free-standing tower of rock is Pyramid Butte, which will be visible from the trail for several miles. The upper, finer layers of the butte are Moenkopi deposits, while its lower part and broad base are Cutler strata.

6.0 Just ahead, the drainage that the trail crosses was created by a small, unnamed syncline that lies between the two anticlines whose slopes the trail travels. Beyond the drainage, as the trail skirts closely around the base of a colorful outcrop of Cutler Sandstone, note the beautiful erosional forms this rock takes once exposed to extensive weathering, with balanced

rocks, towers, caves and many grotesque shapes. The relatively unlayered sandstone in this mass to the right-front represents a long period of time when the Cutler deposits were coastal dune-sands relatively unaffected by tides or new water-borne sediments. The higher strata to the right of the trail here were more frequently affected by brief tidal invasions, probably storm-surges.

6.5 As the trail tops out of a short grade, note to the right-front the horizontal water-affected layers of the darker sandstone in a low bluff near the trail, and the angled cross-stratification in the lighter red rock that is aeolian.

6.8 Here, the trail goes through a fenceline, reenters public land, then continues to travel through various levels of Cutler Sandstone deposits. The rounded gray slopes ahead in the near-distance are capped with the harder limestone layer in the Cutler that is rich in shallow-marine fossils. Two spur trails penetrate this sloping area and continue to the rimlands of the Colorado River. These spur trails are not described in this log but can be explored with adjustments to this log's mileages. The upward arching gray slope is the eastern end of Shafer Dome, an anticline that curves around the base of Dead Horse Point, angling the Cutler strata there. This stratigraphic tilting is visible from the trail for the next 6 miles.

7.0 As the trail leaves the solar evaporation ponds, note that the ground surface is largely covered by dune sand deposited here by prevailing winds guided by the higher terrain on each side of the trail. The tall promontory to the right against the skyline is locally known as Airport Tower.

7.3 The first of the two spur trails noted earlier goes left here up a shallow drainage, to cross the broad expanse of gray limestone then reach the rim of the Colorado River gorge. This spur travels fairly closely around the base of Pyramid Butte, ending on its southern side, at the rim of the gorge. It is possible to hike

View from end of spur trail at Mile 7.4.

beyond the end of this spur and on down to an old ranch development in the brushy riverbottom.

7.4 Just ahead, a few feet beyond a shallow drainage that the trail crosses, the second noted spur trail goes left to the rim of the river gorge, then travels along that breathtaking rim, where there is good hunting for fossilized mollusk shells. This spur does not connect with the one just noted.

7.8 Note the defunct reservoir to the right of the trail that once caught the water from springs in this vicinity for use by domestic livestock. The reservoir is now being overgrown by exotic weeds and shrubs, and the trickle of water has resumed flowing down its natural drainage. Beyond here, the trail parallels that moist drainage for a short distance, then crosses it. Note ahead, the sloping limestone strata that have been tilted by the Shafer Dome anticline.

8.2 Here, just beyond another fenceline, a spur trail goes left by a lone cottonwood tree. The other trees that once grew here in this shallow wash died when the reservoir just passed retained most of the natural water flow for years. This very rough and somewhat difficult spur trail continues down the drainage toward the Colorado River, ending on narrow benchlands of Cutler Sandstone overlooking the river. Short stretches of that spur are visible from the main trail ahead.

The anomalous petrified wood noted earlier in the lower Cutler are exposed on these benchlands by erosion. The large logs there, which represent some of the earliest true trees to evolve, evidently washed into that level during massive flooding from higher terrain, since they are in thick layers of what was mud deposited by fresh water. It is not legal to collect more than a few small pieces of petrified wood on public land.

Note to the left of this junction, the canyon cutting into the older red tideland deposits below the gray limestone that caps the anticline slopes.

8.6 Note in the drainage to the left of the trail, just after it has crossed a tributary wash, how the layered gray marine limestone is exposed. A short hike in that drainage will reveal other details about how different types of Cutler deposits erode. Here, the spur trail noted at mile 8.2 is traveling the slopes beyond the wash. Just ahead, that spur trail dives steeply into the developing canyon of the main drainage.

8.7 As the trail skirts around one head of the canyon to the left, the spur trail is now traveling down the canyon. Note that this drainage line angles downward, while the surrounding geologic strata angle upward. This indicates that the drainage began long before the rock strata now visible were exposed. Along this stretch of the trail, the lofty promontory to the right is Dead Horse Point. Developments on that state park overlook are visible from the trail ahead. Beyond here, for the next several miles, the main trail travels on or near the rough surface of the Cutler marine limestone.

9.7 Note to the right along the next stretch of trail, the sharp interface between the gray shallow-marine limestone and the dark red tideland sediments, both Cutler deposits. This interface represents a time when a shallow sea was buried under iron-rich, fresh-water sediments from the Uncompahgre Uplift. From then on, the area became marine tideland barely above sea-level that was mostly dune sand, only occasionally affected by storm surges and more fresh-water layering.

10.2 Stop here where the trail tops a grade and goes through another fenceline at the rim of the Colorado River inner gorge. The trail has just reached a high point on the Shafer Dome anticline. Exercise extreme caution while walking near the edge of this crumbling cliffline, here and at other rim viewpoints ahead.

More than 500 feet below, the deeply entrenched Colorado River meanders along through lower Cutler deposits, with some Upper Hermosa strata just above river level. This older rock has been thrust upward in this vicinity by the Shafer Dome anticline.

Pyramid Butte looms against the La Sal Mountains to the east, while Hatch Point dominates the complete southeastern skyline. The cliffs above the trail here are younger Cutler Sandstone strata. Note back along the trail, below and to the right of distant Airport Tower, the steeply sloping limestone strata through which the trail has passed. Beyond this viewpoint, the trail will continue to travel on Cutler strata angled by the Shafer Dome anticline.

The left end of the relatively small, isolated butte of red sandstone on the far side of the river at about the same level as this viewpoint, is Chicken Corners. The narrow ledge that the Chicken Corners ORV trail traverses there, between the butte and the river gorge, gave that trail its name. The Hurrah Pass - Chicken Corners Trail is described elsewhere in this book.

To the southwest, the lofty terrain of Island-in-the-Sky dominates the skyline. Note that in some places this elevated peninsula is topped by white Navajo Sandstone above layered, red-hued Kayenta Sandstone. To the south, through the notch cut by the river in the spectacular panorama, some of the Cedar Mesa Sandstone spires in the Needles District of Canyonlands National Park are visible on clear days in the extreme distance.

10.3 Here, the shade pavilion at the tip of Dead Horse Point comes into view high to the right-front. The trail continues skirting around this lofty promontory for the next mile or so, traveling the slopes of the Shafer

Dome anticline. Note that the distorted strata are now descending, with the trail continuing to travel at the stratigraphic level of the marine limestone.

11.2 Drive very carefully here and just ahead, as the trail travels a ledge between soaring cliffs and a sheer-walled tributary canyon of the Colorado River gorge. Note that the trail is still closely paralleling the interface between the gray marine limestone and red tideland Cutler deposits.

11.9 Stop here at a pullout for a spectacular panoramic view of a gooseneck bend in the Colorado River, with the river approaching from the left and leaving to the right. This is the same loop visible from Dead Horse Point at the end of Utah 313, described elsewhere in this book. The geologic strata both above and below the trail here are Cutler deposits. Note how strata arched by the Shafer Dome are descending in the downriver direction.

The higher strata continue to be Moenkopi, Chinle, Wingate and Kayenta, in ascending order, with a few exposures of the White Rim Sandstone Member of the Cutler Group appearing on the slopes through gaps in the cliffline ahead, and high on the promontory visible between the gaps. White Rim Sandstone is discussed ahead, where it is more conspicuous. Beyond here, the trail skirts along the rim of another spectacular tributary canyon, again traveling on the shallow-marine limestone layer.

12.3 Just beyond a shallow drainage that the trail crosses near its pouroff into the canyon, note the red-rock spires in the lower cliffline, silhouetted against the verdant riverbottom.

12.4 Here, just beyond the drainage, fallen slabs of White Rim Sandstone are conspicuous on the high slopes ahead. These slabs have fallen from the rim just above them. White Rim deposits are beach sand left during a brief encroachment of the sea from the west, followed by a long period of erosion. White Rim deposits thin to nothing in the vicinity of the Colorado River, but are quite thick in the rimlands of the Green River farther west. The only remaining trace of White Rim Sandstone east of the Colorado River is one exposure in lower Castle Valley that is visible from Castle Valley Road, described elsewhere in this book.

On ahead, the trail begins a gradual ascent through lower Cutler deposits to the White Rim level. Along that stretch of trail, study the series of columns that have eroded from the older Cutler deposits, protected by large cap-rocks of harder White Rim Sandstone. Such columns, spires and balanced rocks are characteristic of White Rim erosion, especially in the eastern part of the deposit, where it is thin enough to fracture into smaller blocks.

12.8 As the trail starts descending toward a shallow drainage, note the spectacular pouroff into the deeper canyon.

12.9 Here, as the trail approaches a narrow stretch of the drainage, it enters Canyonlands National Park. Park Service regulations apply from this point on, with special emphasis on keeping all vehicles on designated trails. Only street-licensed motor vehicles operated by licensed drivers are allowed on park roads and ORV trails.

13.2 Note the series of columns in the cliffline ahead, most of them capped with large slabs of White Rim Sandstone.

13.3 Just ahead, as the trail is traveling in a wash, a cliff to the left offers a shaded noontime rest stop. Beyond here, as the trail continues its ascent toward the White Rim, still more rock columns topped with White Rim Sandstone come into view on the left.

14.2 Here, the trail is traveling just below the White Rim, with photogenic views to the rear of several rock

columns capped with White Rim Sandstone. Note that in many places the harder White Rim Sandstone has been undercut by erosion and seeping springs. Watch for places where fallen, tilted boulders of White Rim have protected the sediments below them from rain-erosion, leaving columns with angled caprocks. There is one to the right here, beyond the drainage. Just ahead, the trail makes its final ascent to the upper surface of the White Rim, past walls of that sandstone beautifully streaked by water runoff.

14.7 Here, the trail reaches a junction. The described route goes right up the Shafer Trail, which connects to the Island Road. The White Rim Trail goes left. Those who have the time may wish to explore this spectacular trail as far as Musselman Arch, about 3-1/4 miles from here, then return and continue up the Shafer Trail with mileage adjustments. This 6-1/2-mile side trip requires as least 2 additional hours.

The historic Shafer Trail began early this century as a livestock trail, used for taking cattle from summer pasturage on Island-in-the-Sky, to a river ford, then across Hurrah Pass and into mid-Cane Creek Canyon. Some cattle were then taken on to market, while others wintered in the canyon before returning to the Island by the same route in the spring. With the first uranium prospecting boom, the stock trail was improved to allow wheeled traffic between the Island and the broad bench-lands the White Rim Trail travels, which in turn gave access to spur trails that climbed to the uranium-bearing Chinle Formation deposits around the base of Island-in-the-Sky. Since Canyonlands National Park was established, the Shafer Trail has been maintained for recreational purposes, and remains one access route to the White Rim Trail. The trail was named after the early cattlemen who first developed it, John and Frank Shafer.

The sheer cliffs all around this junction are part of Island-in-the-Sky, the enormous complex peninsula that angles up toward the south and juts into Canyonlands National Park. The sheer, mid-level cliffs are Wingate Sandstone, with slopes of red and gray Chinle deposits continuing down to where they phase into the thinner laminations of Moenkopi Formation. Above the Wingate cliffs are layered Kayenta Sandstone, with rounded light-hued cliffs of Navajo Sandstone against the skyline. The drainage that the trail ascends toward the west is a mixture of rubble, alluvium, dune sand and decomposing rock from higher strata.

The Shafer Trail continues up the narrowing drainage of one arm of Shafer Canyon, then climbs very steeply up a series of switchbacks through Moenkopi and Chinle deposits, then up a steep slope of rubble that lies against the Wingate Sandstone cliff. Above that, the trail zigzags through layered Kayenta strata, then levels off at the base of the Navajo Sandstone, where it travels for some distance on Kayenta ledges before climbing to the summit of Island-in-the-Sky. This switchback grade is dangerous and should be driven very carefully. If the trail is wet from recent rain, it should not be attempted at all.

Some of the trail's sharp switchback curves provide pullouts suitable for viewing and photography, but the main trail should not be blocked to other traffic. If a descending vehicle is met on a narrow stretch of the trail, safety and generally accepted rules-of-the-road require that the descending vehicle back up to the nearest stretch of trail wide enough for passing, but on this unique and unusually dangerous trail, local conditions may require exceptions to this. The best rule is -- do what is safest for both vehicles.

Drivers who have even a touch of acrophobia, or who are not experienced drivers of their off-road vehicles, may find it preferable to return to Moab via the Potash Trail, rather than ascend the Shafer Trail.

14.9 The red sediments on either side of the trail along this stretch are Moenkopi Formation.

15.5 At about this level, the trail reaches the Moenkopi-Chinle interface, where layered red-brown sediments are topped by a thick ledge of light-hued sandstone.

16.1 A pullout on the left here, just beyond the first switchback, affords a breathtaking view to the southeast. Two extensions of Island-in-the-Sky frame the panorama, which includes Dead Horse Point, with a large arch-topped alcove beneath its southern tip, and the La Sal Mountains looming above light-hued Navajo Sandstone in the Sand Flats area. Some of the White Rim colonnades noted earlier are visible to the left of the White Rim Trail far below. The trail is traveling through Chinle Formation deposits, with some of them being unlithified sediments and others being hard, ledgy sandstone.

16.4 Here, as the trail is traveling across a giant rubble slope, it meets the Chinle-Wingate interface, which is clearly discernible to the left where massive aeolian Wingate Sandstone meets the horizontal layering of water-deposited Chinle sediments.

16.8 Here, at another switchback pullout, the trail reaches the Wingate-Kayenta interface.

17.3 Pause at another switchback pullout and study the immense spring-seep alcove in the Navajo Sandstone that looms above the trail. Notice the long seep-line that is about thirty feet above the Kayenta-Navajo interface, and the beautiful water run-off streaks above the seep-cave.

17.4 Here, the trail reaches the Kayenta-Navajo interface, where it levels off and travels on top of the Kayenta for the next 2 miles.

19.1 Pause here, just beyond a narrow stretch of trail, and look back at where the trail has traveled across an undercut ledge of rock.

19.4 Pull into the parking area here and walk over to the rim for a view of the Shafer Trail. From here on to the end of the trail, it is traveling through Navajo Sandstone, but beyond the parking area this deposit is largely obscured by dune sand and decomposed rock, with only occasional Navajo outcrops visible.

19.7 Notice that the trail is traveling through a large area of dune sand held in place by native vegetation and exotic grasses that were introduced after the native grasses were grazed to extinction. The sand here was deposited by prevailing winds. Dune sand is common just back from many clifflines in this region, especially those that face south, southwest or west, the directions from which prevailing winds come. These winds pick up decomposing rock granules, carry them upward over the cliff, then deposit the sand on the relatively level ground just back from the cliffs. Throughout geologic history, wind has been an important geologic force that has shaped the deposits of many canyon country strata, and even brought the Navajo Sandstone materials from other regions.

20.4 Here, the Shafer Trail and this geology log end at the paved Island Road. Go left to continue into the Island-in-the-Sky District of Canyonlands National Park. Go right to return to Moab. For alternate routes back to Moab, refer to NAVIGATION at the beginning of this trail log.

Shafer Trail

CRYPTOGAMIC SOILS (MICROBIOTIC CRUST)

There is a form of life that thrives in the high desert terrain of the Colorado Plateau that is seldom noticed or even recognized by the casual visitor as being alive, yet it is not only one of the oldest forms of life on the planet Earth, but also one of the most important to its biosphere, its world-wide, integrated community of plants and animals.

Blue-green algae first appeared in the shallows of warm seas 3-1/2 billion years ago in the form of "stromatolites," or large clumps of microscopic cells that grew to several feet in diameter, like gigantic, odd-shaped underwater mushrooms. These communities of primitive algae, or "cyanobacteria," were first found by scientists in fossil form. For a long time they were thought to be extinct but, recently, living communities were found in remote locations, essentially unchanged in 3500 million years.

As scientists have slowly reconstructed the early history of life on Earth, they have concluded that cyanobacteria played a critical part in the spread of other forms of life, especially land life. These strange and very primitive bacteria have characteristics of both animals and plants. Under the right conditions they are mobile, yet their bodies bear chlorophyll, the substance that enables plantlife to synthesize nutrients.

According to current thinking, some cyanobacteria slowly adapted to other environments than the sea, first to the changing conditions in tide pools, which were subjected to periodic drying, occasional flooding with fresh water, then more inundation with sea water. In time, many millions of years, this led to cyanobacteria -- blue-green algae -- that were adapted to life on the land, under all conditions, from arid deserts to polar icecaps. This, in turn, set the stage for other forms of land life, both plants and animals, to develop.

Before algae had adapted to life on land, wind and water quickly carried any loose particles of decomposed rock either into the sea or into the nearest low basin, leaving little but barren rock which could not host life. The algae, however, managed to live on land by developing a protective cover, or sheath, that served to preserve its moisture and shield it from the harsh surrounding environment. It grew in silts, sands and fine gravels, binding the surrounding tiny rock particles to itself with its sticky sheath, for protection from the erosion that would otherwise carry these materials away.

In time, as land-dwelling cyanobacteria spread to occupy virtually every land environment, they created the first soils, and these soils eventually hosted other forms of plantlife, which in turn provided environments favorable for the development of animal life, in two different ways.

Blue-green algae were not only key to the establishment of all other forms of land life, they were also an important factor in changing the planet's atmosphere. For at least 3-1/2 billion years, chlorophyll-bearing algae in the seas, and later on the land, have been utilizing atmospheric carbon dioxide and emitting oxygen as a waste product.

This gradually converted the atmospheric composition from one with large percentages of carbon dioxide and very little oxygen, to one in which the balance was reversed. As land plants developed and spread across the world they, too, used chlorophyll and helped in the process of converting the atmospheric balance to one with a significant percentage of free oxygen.

This change in the atmosphere brought with it many other changes. It helped cool the entire planet, because atmospheric carbon dioxide traps solar heat, and the free oxygen, a rich source of energy, provided the basis for the high-energy metabolisms needed by mobile forms of life -- animals. The released oxygen also provided a protective canopy of ozone in the upper atmosphere. This form of oxygen blocks certain ultraviolet radiation that is so inimical to both plant and animal life that neither could have evolved without the ozone layer.

Thus, thanks to the adaptation of blue-green algae to life on land, all other land life, both plant and animal, was able to develop, living together in a harmony that depended on maintaining the changed balance of atmospheric gases. Plants and animals each depended on the waste products of the other. Animals intake oxygen and emit carbon dioxide, while plants utilize carbon dioxide and emit oxygen. Even now, the chlorophyll in marine algae and microplants produces most of the planet's oxygen and absorbs most of its atmospheric carbon dioxide.

This, however, was not the end of the story of blue-green algae, or cyanobacteria, but the beginning. This strange form of life is still everywhere on Earth, adapted to virtually every environmental niche on the planet -- including its high deserts such as those of the Colorado Plateau and its heartland, canyon country.

Here, a most ancient form of algae, the kind that evolved to thrive in the hot, arid sands and silts that dominated all land 3-1/2 billion years ago, still plays a vital part. It binds together the arid sands and sediments of the high deserts, protecting them from erosion by wind and water, holding the scant precipitation that does fall, helping to build stable soils, fixing vital nitrogen and forming a basis for all other kinds of life, both plant and animal.

In its desert-dwelling mode, blue-green algae grows in the sandy soils of the region in the form of masses of microscopic threads, with each tiny thread coated with a sticky sheath that binds the adjacent rock particles to it, just as its predecessor species did billions of years ago. Most of the time such communities of algae are dry and dormant, but each time the sands receive a bit of moisture, the algae revives and grows rapidly, leaving behind a mass of empty sheathing that continues to bind the sandy soil particles. This tangle of abandoned algal sheath also continues to absorb and hold the scant moisture, which can be used by the entire microplant community, and each time it is wetted the sheathing expands, aerating the soil.

This complex of microscopic processes initiated by blue-green algae creates a kind of dark, lumpy soil crust that endures, resisting both wind and water erosion, and providing an environment suitable for other tiny forms of plantlife -- among them fungi, mosses and lichens. Lichens are themselves a symbiotic combination of algae and

continued ----

fungi. See the sidebar, LICHENS.

This crusty surface on desert sands and silts was first called "cryptogamic soil," with "cryptogam" meaning "hidden, non-flowering plants." This definition depended on considering cyanobacteria to be plants because of their chlorophyll, which is not strictly true. Recently, however, the ambiguous cyanobacteria have been given a classification all their own, making the term, "cryptogam," obsolete. Now, the correct term for the crusty surface that binds perhaps 75% of Colorado Plateau soils, is "microbiotic crust," although it will doubtless be a long time before all literature sources incorporate this change.

Thus, a form of life that first developed billions of years ago still plays a vital part in the protection and preservation of the entire ecosystem of canyon country. Without its microbiotic crust, the sandy, arid soil of the region would quickly be carried away by wind and water. This would destroy the basis for other plantlife and, in turn, leave an environment in which animal life could not survive.

Left to itself, microbiotic crust can thrive, and easily repair the little natural damage it sustains from the scant wildlife that inhabits the region. But human activities do enormous damage to these fragile crusts. Domestic livestock grazing has seriously damaged the region's fragile microbiotic crusts, allowing erosion to carry away countless megatons of soils. Wheeled and tracked vehicles have caused even more ruination, and both of these types of human activities continue today, virtually unabated, creating ever-escalating damage to the high-desert ecosystem.

As long as this nation's very limited environmental laws do not recognize the value of cryptobiotic crusts, and fail to place severe limits on damage to them by human activities, there is little hope of stopping the continuing destruction. The federal agencies responsible for land administration in canyon country currently pay lip service to the protection of microbiotic crusts by urging recreationists to avoid damaging them, but continue to allow far greater damage by livestock grazing and mineral search and development activities.

Those who understand this environmental tragedy, however, and who do not wish to contribute to it, should not ride or walk across patches of this dark-colored crusty soil with its tiny hills and hummocks.

Vehicle drivers and riders can stay strictly on established roads and off-road vehicle trails. Hikers can usually avoid walking on the crusts by detouring around them on rock exposures, or on the soils beneath trees and shrubs that are generally free of these crusts. In terrain where microbiotic crusts thrive, conscientious hikers often add appreciably to the distances they hike, just by taking circuitous routes to avoid these crusts.

The question about how long it takes for damaged microbiotic crusts to recover and once again protect this region's sandy soils from wind and water erosion, is complex. Of course, if the damage is severe, the crust is beyond repair and erosion runs rampant. Evidence of this is everywhere that human activities have occurred. In Moab Valley, for example, when fierce winds blow, great clouds of dust rise from areas where the microbiotic crusts have been destroyed by human developments, while nearby areas that are still protected generate little or no dust. Rain that falls on disturbed areas runs off quickly, carrying with it heavy loads of sediments and cutting deeply into native soils.

These sediments then cloud the region's streams and rivers, turning them the colors of the various sediments, and the rivers bear countless megatons of soils into downriver reservoirs, eventually to fill them to the brim. Studies have shown that this is happening much faster than predicted by the dam-builders -- thanks to continuing human activities that promote erosion and ignore protection of fragile desert soil microbiotic crusts.

But what if these crusts receive only minor damage, say from the feet of deer, desert sheep or the occasional careless hiker? Conventional thinking indicates that replacement takes many years. Some authorities say five to seven years. This may be true, if recovery is defined as complete replacement of the entire original complex of microscopic plants. However, while some of the microplants that live communally with the blue-green algae do grow very slowly, the cyanobacteria themselves grow quite rapidly when moisture is available, such as during and following rainfall, or during the winter when snow brings long-lasting moisture.

Thus, limited damage can be partially repaired rather quickly by the cyanobacteria that form the basis of the microbiotic community, but complete restoration of the crust does take years, even decades -- providing the damage does not start erosion too severe to permit recovery.

Much of the damage that humanity has already done to the Colorado Plateau's microbiotic crusts could be repaired, given time, if the activities creating the damage were stopped. Today, uranium prospecting and mining, one of the major past causes of damage from vehicles, is effectively dead, but the seismographic and drilling activities associated with oil and gas explorations continue unabated, as does the extensive use of public land for domestic livestock grazing, which causes even more destruction to the soil communities. Careless recreationists cause some occasional damage, with their wheels and feet, but at least there is some hope that this can be minimized by education.

Certainly, preserving the vital microbiotic crusts that remain in this high-desert region is worth whatever it takes, because it is the basis for all life here, including that of humanity and its non-destructive activities.

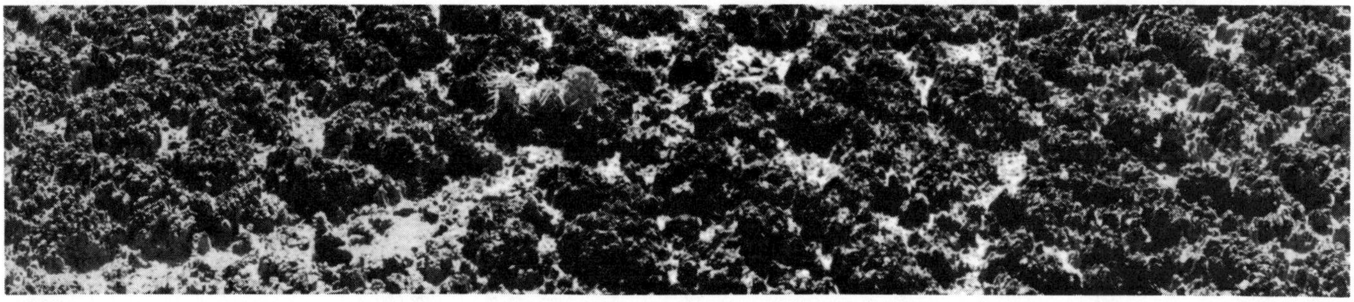

TRAIL NAME: **SAND FLATS TRAIL**

SUMMARY: This off-road vehicle trail is an extension of the road that provides access to the popular Sand Flats recreation area near Moab. The trail climbs steeply through spectacular terrain, eventually to join the La Sal Mountain Loop Road, described elsewhere in this book. Along the way, a fairly narrow spectrum of geologic strata are exposed, but the variety of erosional landforms is outstanding, and the trail crosses broad tablelands whose strata have been lifted high above their normal elevation by igneous intrusions associated with the formation of the La Sal Mountains. The trail also offers a number of outstanding panoramas of Moab Valley and the redrock canyon country that surrounds it.

TRAIL SURFACE: eroded graded dirt, rough slickrock, rocky grades, sand and packed sandy soils.

VEHICLE: off-road vehicles only, for the entire trip.

MAP: ***Canyon Country* OFF-ROAD VEHICLE TRAIL MAP - Arches & La Sals Areas.**

GUIDEBOOK: ***Canyon Country* OFF-ROAD VEHICLE TRAILS - Arches & La Sals Areas.**

MILEAGES:

TO START: 10.2 miles, from Main and Center, Moab.
TRAIL: 11.1 miles, one-way to end.
RETURN: from the end of the described trail:
21.3 miles, via the Sand Flats Trail and Road;
26.0 miles, via the La Sal Mountain Loop Road route.
TOTAL: 47.3 miles, via the La Sal Mountain Loop Road.

TIME: allow a full day for the round trip from Moab. This will include time for stops and short hikes along the way.

GEOLOGIC STRATA VISIBLE FROM THIS TRAIL

Formation	Member	Page
Morrison Formation	Brushy Basin	Page 18
	Salt Wash	
	Tidwell	
Entrada Sandstone	Moab	Page 16
	Slickrock	
	Dewey Bridge	
Navajo Sandstone		Page 13
Kayenta Sandstone		Page 13
Wingate Sandstone		Page 12

ROUTE: Moab city streets to the Sand Flats Road, then that road to where it ends at the water tanks on a ridge between Negro Bill and Rill canyons, where the graded dirt road becomes an ORV trail, then on the trail to where it connects with the paved La Sal Mountain Loop Road. Return either by the same route, or via the La Sal Mountain Loop Road in either direction.

NAVIGATION: drive east on Center Street, south of Fourth East, east on Mill Creek Drive then the Sand Flats Road up the hill past the cemetery. Continue beyond the water tanks that mark the end of the Sand Flats Road.

View from Mile 0.0, Sand Flats Trail.

GEOLOGY TRAIL LOG:

0.0 This ORV trail as described begins at the spring-fed water tanks and pullout where the Sand Flats Road, as described, ends. The trail begins traveling below a small remnant of Navajo Sandstone, above on the left, and within horizontal layering of Kayenta Sandstone. The cliffs and rock dome ahead across Rill Canyon are Entrada Sandstone. The Porcupine Rim ORV trail, not described in this book, spurs left at the water tanks.

0.2 From a pullout on a curve in the trail, the cliffs and dome ahead are Navajo Sandstone set above terraces of Kayenta. More exposures of the Slickrock Member of Entrada Sandstone are appearing on the righthand skyline. The knobs on top of these domes are remnants of the Moab Member of Entrada Sandstone. For the next mile or so, the trail skirts the rimlands of Rill Canyon. Watch for wet areas on the slopes beyond Rill Canyon, where springs reach the surface. In late fall and early winter, when this trail is still accessible, these springs, and others on up the canyon, become very obvious because of the immense ice cascades the seeping water forms. In some severe winters, these springs build up masses of ice so immense that there are still remnants of them in late spring.

0.7 The small pullout here provides a good view of mid-Rill Canyon, with Moab Valley and its rugged cliff-line to the west.

0.8 At the slickrock pullout here, the cliffs just above the trail-level on both sides of the canyon are Navajo Sandstone above ledgy Kayenta Sandstone. Note that the Navajo Sandstone is not as thick here as it is farther west. This deposit thins to nothing just east of the Utah-Colorado border. The higher cliffs to the right are the two members of Entrada Sandstone noted earlier. The bottom of the inner gorge of Rill Canyon below is Wingate Sandstone that has step-like pouroffs in its drainage and, in some seasons, pools of water from seeping springs in the terraced slopes above.

1.2 From this stretch of trail, seeping springs are visible as lines of dense vegetation in two steep drainages beyond Rill Canyon. Note here that, although the Sand Flats Road and Sand Flats Trail have been traveling within or near the base of the Navajo Sandstone, the road and trail have climbed considerably. This illustrates how the strata in this area have been severely tilted by the upthrusting of the La Sal Mountains. See the sidebar, LACCOLITHIC MOUNTAINS.

1.7 From a pullout here, note the slender spires and balanced rocks in the bare Slickrock-Entrada at the head of Rill Canyon. The cliff walls at trail-level here are still Navajo Sandstone. The canyon below is partly Navajo, partly layered Kayenta. The high dome across the canyon is Slickrock-Entrada topped by remnants of Moab-Entrada. Note below the slender spires, the two deep spring-seep caves at the head of Rill Canyon, and the dense vegetation that these springs support.

1.8 Note the seeping springs to the left of the trail, with their willow and cottonwood trees.

1.9 Here, the trail crosses another seeping spring that supports thriving cottonwood trees and exotic tamarisk shrubs. In a few more yards, there is another spring with reed grasses and other water-loving plants growing to the right of the trail.

2.1 Here, the trail tops out of the steep Rill Canyon grade and continues to travel through wooded, decomposing Navajo Sandstone. For the next mile or so, in the warmer seasons, watch for collared lizards sunning themselves on the rocks beside the trail. A few yards ahead, an alternate trail to Porcupine Rim spurs left.

2.9 Note the spires of Slickrock-Entrada across the canyon and ahead, and the balanced rocks on the skyline to the right-front. A hike to the top of the mesa there provides access to these and other balanced rocks and odd erosional phenomena in the Moab-Entrada, and to breathtaking rim views into the extreme upper drainage of Burkholder Draw, part of which is known as "Valley of the Gnomes" because of numerous spires there of various heights topped with white caprocks, all within the Slickrock-Entrada.

3.1 Here, at a pullout just beyond where the trail crosses a stretch of slickrock, is a good place to park for the climb onto Balanced Rock Mesa, with its outstanding views of the spectacular surrounding terrain.

3.5 An inconspicuous spur trail to the right here goes toward the Valley of the Gnomes with its numerous tall towers and white-capped gnome-like spires, the tallest of which is called "King of the Gnomes." Hike beyond the end of the vehicle trail to get into the main part of the valley.

3.7 An arch is visible from here, to the right-rear near the top of the cone-shaped dome of Slickrock-Entrada. The trail here is traveling through the upper levels of Navajo Sandstone.

4.1 At a cattleguard, the trail enters the Manti-La Sal National Forest, traveling through dense pinyon-juniper forest. The few exposures of rock on the right are Entrada Sandstone.

4.4 A spur trail left here, not described in this log, goes to Porcupine Rim and a spectacular view of Castle Valley, with Adobe and Fisher mesas beyond. For the next short stretch of trail, the red sandstone exposed to the left-front and right-rear is Slickrock-Entrada.

4.7 Park here, just beyond the summit of a grade, for a hike west to the spectacular rimlands of Valley of the Gnomes. Beyond here, for a mile or more, the trail crosses patches of Entrada Sandstone and travels through a higher level of the pinyon-juniper plant community that includes tall Ponderosa pines.

5.0 The sandstone on both sides of the trail here is Moab-Entrada. Note the private dwelling on the southern horizon. The canyon between it and the trail is one drainage of upper Burkholder Draw. The Abajo Mountains are visible in the distance to the south, with the La Sal Mountains ahead, framed in tall, graceful Ponderosa pines. Here, and for the next mile or so, note the extremely heavy coatings of colorful lichens and moss on the Entrada Sandstone, due to the increased moisture at this elevation. See the sidebar, LICHENS. Near the trail along this stretch in the spring, a species of succulent plant burgeons and blooms, with tiny stalks of yellow flowers.

King of the Gnomes

5.4 Note the first exposure of aspen trees across the canyon at the base of the low, rounded cliff of Moab-Entrada.

5.5 As the trail curves right here, note the camping and picnic area among the tall pines and ornately eroded Moab-Entrada sandstone to the left. In the spring, there are usually many wildflowers in the drainage line just beyond this, until cattle are brought in and the flowers and drainage are trampled into mud. Just beyond this curve in the trail, the spur trail that goes left is the Kokopelli's Trail bike route. As an ORV trail it can be traveled to the paved La Sal Mountain Loop Road, but is rough in places and navigation is confused toward its upper end. The spur affords good views of Pinhook Valley and Upper Castle Valley. Note the small arch in the rock left of the main trail just beyond the spur trail. The drainage developing to the right enters Burkholder Draw.

5.9 A Forest Service trail goes right here to the private dwelling mentioned earlier. For a short stretch beyond here, the trail travels through a beautifully eroded exposure of Moab-Entrada set among giant Ponderosa pines, with good hiking on both sides of the trail. The upper rim of the slope to the right is especially rich in desert varnish, lichens and moss. See the sidebar, DESERT VARNISH.

6.4 Here, the trail begins to skirt around the upper end of another tributary of Burkholder Draw. The eroded sandstone in the draw is Moab-Entrada, but a few thin remnants of Morrison Formation deposits appear to the left of the trail just ahead. Note the historic log structures across the draw.

6.5 The trail crosses this tributary of Burkholder Draw. Beyond here, the trail climbs onto and travels Wilson Mesa for about a mile.

6.6 Note the small natural arch beside the trail.

7.0 Here, the trail swings left onto Wilson Mesa. Note the private summer cabins on the Entrada Sandstone rim across the canyon. The hill behind the cabins is decomposing Morrison Formation deposits. Most of Wilson Mesa is private land developed for grazing. The soils on the mesa are erosional remnants of Morrison Formation deposits held in place by native and exotic vegetation. Beyond here, the trail climbs into the Salt Wash Member of the Morrison Formation. The few eroding spur trails in the slopes to the left of the main trail were made decades ago in a search for uranium ore in the Morrison. Little was found in this vicinity, but some was produced from mines elsewhere around the mountain slopes in similar locations.

7.2 Note the historic log structures to the right of the trail just ahead. In this vicinity, large herds of elk have been observed in the meadows of Wilson Mesa in late fall and early winter, after the hunting season is over.

9.0 From a narrow pullout here, the western panorama, starting at the left, begins with the plateau of South Mesa in the near background, with the Abajo Mountains beyond to the south. The geology of South Mesa is similar to that of Wilson Mesa. Both are rimmed with Moab-Entrada, with Slickrock-Entrada slopes below that. The peaks in the distance are the Henry Mountains, about 80 miles from this viewpoint. Moab Valley is visible far below with its jagged rimland. See the logs for U.S. 191 North and U.S. 191 South, elsewhere in this book, for detailed descriptions of Moab Valley geology. The lines of vertical cliffs beyond and to the north of Moab Valley are Wingate Sandstone above Chinle, Moenkopi, and Cutler slopes that are largely covered with rock debris. Note the sloping strata beyond Moab Valley to the west.

To the northwest, this side of Moab Valley, the rounded sandstone domes are the Navajo Sandstone of the Sand Flats area. To the north, in the distance, the terraced bluffs of the Book Cliffs extend across the horizon. The red sandstone domes silhouetted against the Book Cliffs beyond Wilson Mesa are Entrada Sandstone in Arches National Park. The Arches Road is described elsewhere in this book. From here to the end of this trail, where it meets the La Sal Mountain Loop Road, the trail travels through decomposing Brushy Basin Member of the Morrison Formation. Some of the rock rubble near the trail is glacial debris left by the first three of nine local glacial episodes. See the sidebar, GLACIERS OF THE LA SAL MOUNTAINS.

9.8 Stop wherever convenient here for another view of South Mesa, which remains in sight from here to the Loop Road. The low hills above the mesa are remnants of Salt Wash Morrison.

11.1 The Sand Flats Trail ends here by joining the La Sal Mountain Loop Road. See ROUTE, earlier in this trail log, for optional routes back to Moab.

View from nearby Porcupine Rim

PETRIFIED PLAYAS

"Playa" is a Spanish word for beach or shore. Geologists use the word to mean much the same thing, but have also assigned another meaning -- a desert drylake, or the mineralized deposits left by such an ephemeral lake.

Such playas are common in the arid lands of the world, including those of the intermountain region, the immense expanse of desertland that lies between the Rocky and Sierra mountain ranges in the American West. Many of the highways that cross these vast, mountainous deserts skirt around such broad drylakes, which sometimes have a few inches of water in them following significant rainfall.

Most of the time, however, such ephemeral lakes live up to their names and are bone dry and blindingly white in the bright desert sun from their high content of soluble minerals.

When such ephemeral lakes do receive water, that water usually carries with it fine sediments and dissolved minerals from everywhere within the immense area that drains into the basin. This means that such playas accumulate a combination of materials that is ideal for making very hard rock, given the time and conditions it takes to lithify these materials.

One geologic deposit in canyon country that accumulated during an extremely long period of aridity, provided ideal conditions for the formation and preservation of desert drylake playas -- Navajo Sandstone. See the sidebar, GREAT DESERT PERIOD. As described elsewhere in this book, Navajo Sandstone is desert dune-sand, deposited when the canyon country region was a vast Sahara-like desert, replete with its own vast expanse of rolling sand dunes, numerous drylake playas, and occasional verdant oases. See the sidebar, PETRIFIED OASES.

This great, long-lasting desert also had another strange content -- its sands were rich in the kinds of minerals that originate from the sea, minerals that accumulate by evaporation in tidal lagoons and seashore salt flats. Geologists know that most of the great masses of sand represented by Navajo Sandstone came from elsewhere, carried into the canyon country region over a long period of time by strong prevailing winds, but they do not agree on the direction of those winds and hence the source of the sand. They also do not address the question of where the massive quantities of evaporites in all that dune-sand came from.

Since most Navajo Sandstone is relatively free of iron, in contrast to many of the other canyon country sandstone deposits, it is likely that the sand came from long-duration marine beachlands somewhere, and that at least some of the evaporite mineral content came from marine tidelands, although where these were remains a mystery.

Nonetheless, the combination of a vast desert of dune-sand rich in soluble minerals was ideal for the formation of desert playas. The scant rainfall, largely from howling equatorial storms, according to paleoclimatologists, collected in every shallow basin, bringing with it the usual burden of finer silts and sediments, plus all of the soluble evaporite minerals the water could dissolve from the dune-sand along the way.

Many such playas were small, but a few were immense, many miles long. Some were short-lived and thin, covered over by drifting dune sand before they had a chance to grow, or perhaps doomed because their drainage areas were too small. Others endured for years, perhaps decades, and grew quite thick, but sooner or later they were all buried deeper and deeper under wind-borne sand and more sand, until the end of the Navajo deposition period. Then they were buried still more deeply under succeeding deposits, none of which accumulated under conditions favorable for the formation of similar desert drylakes -- playas.

As each Navajo Sandstone playa was buried under sand, it began to lithify -- its silts, sands and evaporite minerals began the slow process of turning into a form of rock that was much harder than the surrounding mineralized sandstone. The playas petrified. Some geologists call the hard rock in such playas "limestone" because of its hardness and mineral content, even though it differs from the kind of limestone formed from the skeletal remains of tiny marine creatures, as with the limestone in the famous "White Cliffs of Dover," or that forms the foundation for the entire Yucatan Peninsula.

Today, as Navajo Sandstone is exposed and erodes away, these harder petrified playa layers resist erosion better than the sandstone above them, and hence form relatively thin, gray-hued caprock layers that protect the underlying sandstone and slow its erosion, too. Such truncated domes and terraces of ancient drylake deposits and Navajo Sandstone are common in the area covered by this book, and are noted in several of its road and trail logs.

There are still other mysteries, however, about the numerous Navajo Sandstone playas that appear as relatively thin "lenses" of mineral limestone within the cliffs and slopes of this deposit. Most such petrified playas are completely surrounded, above and below, with ordinary aeolian Navajo Sandstone, and the lenses always "pinch out" laterally and seamlessly into the same wind-borne dune-sand deposits.

One mystery is -- how could such mini-drylakes form on what was then a surface of pure dune-sand?

One well-known Utah geologist claims that each such petrified playa lens represents a place where the water table was just below the desert surface, and that the resulting playa was merely a surface aspect of the subsurface water table. This seems doubtful, however, because none of the countless limestone lenses that show in cross-

continued ----

section in canyon country cliffs and slopes of Navajo Sandstone exhibit any trace of a water-affected line continuing beyond the ends of the lenses, nor does the sandstone below each lens show any signs of having been saturated by a water table and hence better cemented and harder, as it would be. Further, with the water table that close to the surface, the desert would hardly have remained a desert. Plant life would have thrived there -- which it did not, for millions of years.

Another hypothesis advanced to explain these limestone playa lenses suspended in uniform dune-sand is more complex but also more probable, because it incorporates all the observable facts. This concept holds that no matter how deep the dune-sand, when the right amount of rain wets the sandy surface, two things happen -- minute amounts of soluble minerals dissolve and then form web-like crystalline growths that bind the sand granules together more tightly, and microscopic life forms thrive briefly in the thin layer of moist sand, binding its granules together with a web of minute fibers, much like certain microplants bind arid sandy soils. See the sidebar, CRYPTOGAMIC SOILS.

Slowly, with still further moisture, such thin and fragile crusts of minerals and microplants would make that crust less permeable to water, holding more moisture after each rainfall and gradually building a catch-basin capable of collecting still more water and entrained sediments and minerals, until a durable limestone lens had formed.

In most of the large desert basins in the intermountain west, the barren rocky hills and the valleys between them hold very little material in their lowlands, and each drylake drains an enormous area, so perhaps in some cases these desert valleys do accumulate enough water to make shallow water tables that surface as playas. But these conditions did not exist during the deposit of Navajo Sandstone. There were simply thousands of square miles of sand dunes. Thus, there has to be some explanation, other than surfacing water tables, for the myriad petrified playa lenses in this enormously thick dune-sand deposit.

There is still another mystery about these ephemeral desert drylakes in the Navajo. The petrified foot tracks of a wide variety of creatures, both vertebrate and invertebrate, land-dwellers and water-dwellers, have been found in such petrified playa deposits. These include the three-toed tracks of large predator dinosaurs, small mammal-like reptiles, small crocodillians, and at least two unidentified species of invertebrates.

While it is easy to presume that smaller animals would seek out a temporarily wet drylake, for its water to drink and its hatching pothole life to eat, and that larger predators would be there to hunt the smaller animals, what about invertebrates that left foot tracks two to three inches long? How did creatures this large live in an ephemeral lake? How did they survive after the water evaporated? What kind of life cycle permitted such large, slow-growing water-loving animals to thrive under such adverse desert conditions?

Dinosaur tracks are popular, but these invertebrate tracks in Navajo Sandstone petrified playas are far more intriguing! The petrified playas in this dominant canyon country sandstone are strange, and the more they are observed the stranger they become!

*One of several four-toed **Otozoum** prosauropod tracks from a playa deposit above Moab Valley.*

Measuring a theropod track discovered by the author in a playa deposit near Cane Creek Road.

TRAIL NAME: **SPRING CANYON POINT TRAIL**

SUMMARY: This off-road vehicle trail travels through spectacular high-desert panoramas of colorful geologic strata exposed in miles of soaring cliffs and buttes, plus rolling high-desert meadows set with domes and walls of ornately-eroded white sandstone. The color-banded Rainbow Rocks ridge dominates the first part of the trail, with the upper levels of deep Spring Canyon and its rugged tributaries visible in the foreground. For the last several miles, the awesomely deep Green River gorge, which defines Spring Canyon Point, is visible in the near-distance on each side of the trail. At one point, a short hike leads to a breathtaking view of the neck of Bowknot Bend, where two great meanders of the river closely approach. Several miles of the Green River gorge can be viewed from the lofty rimlands of Spring Canyon Point, where the trail ends. Altogether, this easy off-road vehicle trail offers an outstanding view of certain types of canyon country stratigraphy and erosional landforms.

GEOLOGIC STRATA VISIBLE FROM THIS TRAIL

Formation	Member	Page
Morrison Formation	Brushy Basin	Page 18
	Salt Wash	
	Tidwell	
Entrada Sandstone	Moab	Page 16
	Slickrock	
	Dewey Bridge	
Navajo Sandstone		Page 13
Kayenta Sandstone		Page 13
Wingate Sandstone		Page 12
Chinle Formation		Page 12
Moenkopi Formation		Page 12

TRAIL SURFACE: packed sediments, sand, broken rock and slickrock, with the trail eroded and rough in many places.

VEHICLE: off-road vehicles only.

MAP: ***Canyon Country* OFF-ROAD VEHICLE TRAIL MAP - Island Area.**

GUIDEBOOK: ***Canyon Country* OFF-ROAD VEHICLE TRAILS - Island Area.**

MILEAGES:

TO START: 27.0 miles, from Main and Center, Moab.
TRAIL: 14.0 miles, one-way to end.
RETURN: 41.0 miles, from end of described trail.
TOTAL: 82.0 miles, from Main and Center, round trip.

TIME: allow for a short day to explore this ORV trail and Dubinky Well Road, which provides access. Allow a full day if any spurring trails are explored.

ROUTE: Main Street (U.S. 191) north, Utah 313 west, Dubinky Well Road north, then this trail west, returning via the same route.

NAVIGATION: Drive north from Moab on Main Street (U.S. 191) for 11.3 miles, west on Utah 313 for 8.7 miles, north on Dubinky Well Road for 6.8 miles to historic Dubinky Well, then return 0.2 miles to where the Spring Canyon Point Trail spurs west. Stay on this trail to its end, then return to Moab via the same route.

GEOLOGY TRAIL LOG:

0.0 Turn west on this trail from Dubinky Well Road. Just ahead from this junction, the color-banded sandstone of Rainbow Rocks stretches across the western horizon, with the Henry Mountains to the southwest. Some of the upper rimlands of Spring Canyon and other tributaries of the Green River gorge are visible in the mid-distance. All of the white rock in that direction is Navajo Sandstone. The lighter color-banding in Rainbow Rocks represents a period during which soluble minerals, and probably decomposing plantlife, were deposited in an area that was normally dominated by coastal-plains dune sands. This may have been a local short-lived tidal lagoon. The dirt-like deposit on the summit of the color-banded pyramid, and the similar layer in the more distant Entrada Sandstone cliff, is not described in geologic books and papers. It may represent interlayering between the western extension of the Moab Member of Entrada Sandstone and the first deposits of the Tidwell Member of the Morrison Formation. The white layer of rock above that anomalous layer in the more distant cliffline ahead is Moab-Entrada.

The rolling meadows on each side of the trail for the next mile or so are largely dune sand held in place by cryptogams and remnants of native vegetation, with sand and some alluvium accumulated in drainage lines. As the trail crosses one of these drainage lines between here and Rainbow Rocks, note how accelerated erosion is cutting into these deposits. Note also, as the trail approaches Rainbow Rocks, the large living sand dunes at the base of the bluff that vegetation is beginning to hold in place.

0.4 Note to the left of the trail the bare Navajo Sandstone exposed by erosion.

2.0 As this trail rounds the end of Rainbow Rocks, the spur trail to the left leads to a breathtaking pouroff into a tributary of Spring Canyon. Just ahead, beyond the cattleguard, watch for a tall, isolated sandstone spire in the cliffline to the right-front. This spire led early settlers to name this rock prominence "The Needles," a misnomer that has survived on U.S. Geologic Survey maps even though it is inappropriate and a duplicate of nomenclature now applied to the southern part of Canyonlands National Park.

2.2 The Rainbow Rocks ORV trail angles right at a junction here. That trail is not described in this book but is worth exploring on the return trip from Spring Canyon Point, if time permits.

2.4 Stop somewhere convenient along this stretch of trail for another look at Rainbow Rocks and the isolated, sheer-walled prominence to the right-front. The upper rock formations are as described earlier, with the addition of dark red deposits of the Dewey Bridge Member of Entrada at the base of the cliffs. The large drainage line between these two sets of cliffs is white Navajo Sandstone that has eroded into a maze of small canyons. The Rainbow Rocks ORV trail noted earlier travels along the base of the southern cliffs, offering a better view of this curiously eroded area. Note that the general level of the terrain on this side of Rainbow Rocks is much lower than on its southern side, largely because prevailing winds have kept sand piled against the southern face, while this side has been water-eroded.

3.2 Somewhere along this stretch of ridgeline trail, stop and study the scene to the north and east. In the foreground, brushy meadows phase into exposed Navajo Sandstone, with one area of this white rock covered by the flat, limestone deposits of a desert playa. See the sidebar, PETRIFIED PLAYAS. The red-hued cliffs are the lower two members of Entrada Sandstone. Some of these cliffs are topped with white Moab-Entrada and rubbly slopes of Morrison Formation deposits. In the distance, to the north, the Book Cliffs dominate the entire skyline. To the west, the jagged cliffs of the San Rafael Reef, the eastern monocline of the San Rafael Swell, stretch across the horizon.

In the lowlands to the north, the main drainage is Tenmile Canyon, which is marked by a line of green cottonwood trees below one stretch of vertical cliffline. This drainage cuts deeply into the high-desert terrain farther west to form a long, deep and twisting canyon that eventually joins the Green River gorge. One tree-bordered stretch of the Green River is visible in the low country to the northwest. The Henry Mountains are visible against the southwestern skyline, although they may be obscured by coal-burning power plant smog during parts of the year. Farther to the south on the most distant skyline, the angular spire of Cleopatras Chair, west of the Maze District of Canyonlands National Park, protrudes upward above the surrounding terrain. The white rock in the near foreground to the south is all Navajo Sandstone.

3.7 The Freckle Flat ORV trail goes right here toward Dripping Spring, a verdant seep-cave in one wall of upper Tenmile Canyon. The Spring Canyon Point Trail continues generally southwestward toward the southern end of the San Rafael Reef. This trail has been two-wheel drive up to this point, but should be traveled beyond here only by off-road vehicles.

4.5 This stretch of trail travels through barren exposures of Navajo Sandstone, with the trail constructed on occasional patches of sand and soil. While traveling, note the intricate crossbedding of this aeolian sandstone caused by shifting winds in the ancient desert. Note ahead the upper rimlands of one branch of Spring Canyon.

4.7 From here, a short loop-spur angles left toward the Secret Spire, the top of which is visible in the distance to the left-front, silhouetted against the more distant upper cliffline of Spring Canyon. The Secret Spire loop is not described in this log but can be explored on the return trip without confusing mileage figures. The other end of this loop is noted, where it rejoins the main trail a mile ahead. Along the next few yards of trail, Juniper Arch tops a low mound of Navajo Sandstone straight ahead, just a few yards from the trail.

5.5 From a fork here, the Spring Canyon Point Trail goes straight ahead, while the bladed trail swings to the right. Note the rock cairn on top of the low hill ahead. This cairn was constructed from slabs of desert playa limestone found at that level. Note along this stretch of trail and on ahead, how the rolling terrain is largely dune sand barely held in place by vegetation.

5.7 The trail that goes left here is the western end of the Secret Spire loop. Beginning here, notice the upper rimlands of Spring Canyon in the lower terrain to the south. Cleopatras Chair is still visible on the distant skyline to the southwest. Walk up the short spur trail ahead to the cairn for an inspiring panoramic view from the summit of that hill. The cairn there makes a good landmark for finding this western end of the Secret Spire loop trail upon returning from Spring Canyon Point. Beyond this trail junction, the trail continues to travel through the upper levels of Navajo Sandstone.

6.9 Along this stretch of trail, note how Spring

Juniper Arch

Canyon continues to develop in the lower terrain to the left. The La Sal Mountains stand out against the southeastern skyline, with the dark red buttes of Rainbow Rocks farther east. All of the white sandstone domes and ridges visible from here are Navajo Sandstone, but the layered rimlands of Spring Canyon are Kayenta Sandstone.

7.3 Just beyond a fenceline here, the Spring Canyon Point Trail angles right, while a spur trail goes left.

9.6 From here, the upper cliffs of the Green River gorge come into view ahead and to the south. Spring Canyon Point is the high, narrow peninsula formed by a giant meander in the Green River. Excellent views down into the rivergorge are just a few yards from the trail near its end. Note the three separate island groups of the Henry Mountains ahead. From here, the trail descends into terrain in which Navajo and Kayenta sandstones are not clearly differentiated, with Navajo-like desert aeolian deposits interlayered with Kayenta-like water deposits. On clear days, from here, Thousand Lake Mountain and the Aquarius Plateau are visible in the extreme distance to the west. To the south, Cleopatras Chair and other prominences to the west of the Maze District of Canyonlands National Park are visibly silhouetted against the skyline.

10.0 A short spur left ends at a viewpoint looking south.

10.8 The trail spurring right here is the Hey Joe Canyon Rim Trail. The upper rimlands of that canyon are visible in the lowlands to the right. The trail eventually skirts around both sides of deep Hey Joe Canyon, which is a short tributary of the Green River gorge. A downcanyon spur from that trail, plus a hike along the rim of the river gorge, leads to Cliffhanger Bridge, a huge natural span that was first reported in the book ***Canyon Country* ARCHES & BRIDGES.**

Secret Spire

Cliffhanger Bridge

From this trail junction, all of the white rock visible in the panorama to the north is Navajo Sandstone above layered Kayenta Sandstone, although some reddish Entrada Sandstone can be seen beyond the Green River gorge. Parts of the more distant San Rafael Reef are also visible from here. From here to its end, the trail travels through clearly differentiated Kayenta Sandstone, although this deposit is sometimes hidden beneath soils and vegetation.

11.2 A short spur trail angles left here toward the Green River gorge, whose upper cliffline is very prominent along this stretch of the main trail.

12.6 A trail spurs left here, back toward the higher ridge of white sandstone to the left and a primitive camping area. This ridge affords a good view of the saddle between the two loops in the Green River that define Bowknot Bend. The river is visible on both sides of that gap in the cliffline. The river-level off-road vehicle trails viewed from this rim on both loops of the river are segments of the Spring Canyon Trail. From this viewpoint, the upper cliffs are light-hued Navajo Sandstone above layered deposits of Kayenta Sandstone, with sheer Wingate Sandstone cliffs continuing downward. Below these, rubble-covered Chinle and Moenkopi formation sediments slope to river level.

A short walk to the right in this vicinity, beyond the low ridge of white sandstone, leads to another rim of the

Green River gorge and an excellent view upriver. From this rim, the off-road vehicle trail visible beside the river is the upriver segment of the Spring Canyon Trail. Just beyond this stop, where the trail is crossing open slick-rock, a number of large, shallow potholes fill with water following rain.

14.0 The trail ends here at the tip of Spring Canyon Point, just beyond an upright metal pipe that marks an old drill site. Park here and walk around the canyon rim for outstanding views of the Green River gorge. Note in this vicinity the large quantities of rounded river gravel left here long ago, when the Green River was flowing at this level. Note also that where it was viewed from mile 3.2, the Green River was flowing at about the Entrada-Navajo interface. Here, the river is flowing near the base of the much older Chinle Formation. This is because all the geologic strata in this area are tilted upward toward the south. Note that from higher levels of the Point, the upper tips of the La Sal Mountains are visible to the southeast.

From anywhere around the rim of Spring Canyon Point, the geologic strata on both sides of the river are the same. The uppermost lighter sandstone is Navajo, with part of it undercut and collapsed to form vertical cliffs. Below the Navajo, the terraced Kayenta Sandstone continues downward to sheer walls of Wingate Sandstone. Terraces and slopes of Chinle Formation continue down to the water level, with the riverbanks stabilized by vegetation. Note the several immense seep-caves in the Navajo Sandstone and the large sheer-walled amphitheaters cut into the Wingate cliffs. There is a large seep-cave in the Wingate to the north of the point. On Spring Canyon Point itself, especially along its western rim, the Kayenta Sandstone has terraced down almost to the upper surface of the Wingate. For a description of how such rivers as the Green have created gigantic looping meanders, see the sidebar, ENTRENCHED MEANDERS.

The water of the Green River is generally gray-colored from silt. This burden of fine sediments is especially apparent at The Confluence, where the Green and Colorado rivers join their waters. The silt is acquired by the Green as it flows through hundreds of miles of younger sedimentary strata in southern Wyoming and northern Utah. Following rain on the more colorful deposits in this vicinity, the river takes on a red hue. The off-road vehicle trail parallel to the river that is visible below Spring Canyon Point is the upriver leg of the Spring Canyon ORV trail that ends at the old mining settlement in Hey Joe Canyon.

TIME TRAVEL TOPOGRAPHY

Historically, humanity has long assumed that the earth on which it lives is relatively unchanging, other than for a few short-term events like tides, weather variations, an earthquake or landslide now and then, or a rare burst of volcanic activity. Other than these transient occurrences, most of humanity has assumed throughout its recorded history that what it could see of its world had always existed throughout the past, and would continue to exist into the distant future.

Given humanity's short perspective in time -- just a few thousand years of recorded history -- this was not an unreasonable conclusion. Throughout most of its brief history, humanity has observed little about the environment in which it lives that was not either cyclic, like the seasons or weather, or rare and brief, like the natural disasters that occasionally occur. The lives of individuals are too short, and humanity has not yet existed long enough, to provide a more realistic perspective on the very slow pace of global geologic activities. Only knowledge gathered by scientists within the last century has permitted a broader, long-range understanding of planetary processes, and that knowledge is still not commonly available or accepted.

Strangely, there are still professional geologists who remain skeptical of the well-proven concepts of continental drift and plate tectonics, and there are still professional paleontologists who doubt that the immense meteors that occasionally impact this planet have had any serious effect on its life, despite almost overwhelming evidence that such impacts not only bring massive extinctions of life but also initiate drastic and long-lasting changes in global climate and weather, which in turn have immense impact on evolutionary trends, erosion patterns and rates, geologic deposition, tectonic activities, and myriad other natural processes.

With this much confusion even among scientists, how can others be expected to gain any real understanding of what they are seeing when they first view the spectacular geology and topography of the canyon country of southeastern Utah?

Other sidebars in this book attempt to provide some perspective to several aspects of this region's geology and its interface with the life on this planet. The purpose of this sidebar is to offer an overview of this region's stratigraphy and topography by using human imagination and ability to visualize a described scene.

The average visitor to canyon country -- who observes its unique, many-layered cliffs, bluffs, mesas and plateaus, its petrified rainbow of colorful rock formations, its wide spectrum of exposed deposits that range from solid sandstone hundreds of feet thick, to terraced, multiple-layered rock, to steep slopes of color-banded shales -- will be casually curious about these varied rock strata, about how each distinct layer was deposited and how it acquired its color.

More thoughtful observers, however, might ask more complex questions, some that remain unanswered and controversial mysteries even after decades of study by professional geologists:

- *Earlier, was this region a part of the supercontinent, Pangea, part of another land-mass, or part of the present North American continent?*
- *How was the ongoing process of continental drift, with its complex tectonic side-effects, affecting the region?*
- *What was the region like as each of the presently-exposed rock formations was being deposited?*
- *Where were the nearest mountains and other highlands?*
- *Where was the nearest ocean or inland sea?*
- *What was the region's elevation above sea level?*
- *How far was it from the equator?*
- *What was its climate like, its weather?*
- *How big an area did each deposit cover?*
- *How long did each deposit take to accumulate?*
- *What was the origin of the material being deposited, and how was it transported?*
- *What kind of life was around during each period of deposition?*
- *What happened to make the often drastic changes between one deposit and the next?*
- *What went on during those millions of years represented by the "unconformities" between some geologic strata?*
- *What forces ultimately shaped the land as it is today, and when did that shaping take place?*

The list of such fascinating questions about the environment that dominated as each canyon country geologic layer was being deposited is almost endless.

Some of these questions have been answered in this book's section on stratigraphy and in its various sidebars. Others remain a mystery, or are still highly controversial even among scientists. The region has been studied exhaustively over the past several decades in support of "economic geology," that is, in pursuit of finding its mineral wealth -- its gold, copper, radium, uranium, vanadium, gas and oil -- but all too little study has been made for the sake of simply understanding its complexity, for acquiring scientific knowledge unrelated to commercial benefit.

This sidebar is a step in that direction, oriented primarily toward one aspect of canyon country geology, that is, answering one other question that may have occurred to many of the region's visitors, a question rarely voiced or addressed in geology books, whether popular, scientific or academic --

> *"If the topography of canyon country is this spectacular now, what was it like in the past?"*

One way to answer that question is to take two imaginary flights, one to gain an overall view of what the terrain is like now, then another into the distant past to see what it was like over the period of time represented by the geologic strata presently exposed in the region and described in this book.

For a good overview of present topography, let's take an imaginary flight in a helicopter, with the chopper flying to an elevation of 11,000 feet above sea level, about 5,000 feet above Grand View Point on Island-in-the-Sky, then hovering there while we scan the entire panoramic view, from directly below to the most distant horizons.

Beneath us are the vertical cliffs of the sprawling, many-armed peninsula of Island-in-the-Sky, with deep, sheer-walled canyons radiating in every direction from The Knoll, a high point a few miles north. These canyons join the equally deep and spectacular gorges of the Green and Colorado, as these ancient rivers approach from the north and northeast, with the Green emerging from a distant gap in the soaring ramparts of the Book Cliffs that dominate the entire northern horizon, and with the Colorado cutting through the canyon-slashed terrain northwest of the nearby La Sal Mountains.

Where the rivers join, deep within their colorful, terraced gorges to the south, the surrounding land is a tortuous maze of rocky slopes, twisting canyons and sheer cliffs of color-banded sandstones like nowhere else in the world. This dramatically varied land ascends in great red and white terraces to Island-in-the-Sky below, toward the elevated peninsulas of Hatch Point and Harts Point to the east and southeast, toward the maze of sandstone walls and needles on the massive lower slopes of the Monument Uplift and the Abajo Mountains to the south, and to the sheer-walled ramparts of the Orange Cliffs to the west.

Still farther west, the soaring peaks of the Henry Mountains loom to the left of the long, jagged eastern monocline of the San Rafael Swell. The faint, more distant outlines of the Aquarius Plateau, Thousand Lake Mountain and other distant highlands terminate the view in that direction.

To the northeast, the complex and colorful rock spires, ridges and fins of Arches National Park outline Salt Valley beyond where the erosion-scarred sandstone slopes of Poison Spider Mesa, Gold Bar Rim and Arths Rim angle northwestward. Still farther away, in the same direction, the receding Book Cliffs appear to phase into the wooded heights of the massive Uncompahgre Uplift and the sharp, steep slopes of the La Sal Mountains to the east.

Left of these glacier-carved ramparts, the deep canyons of Castle Valley, Professor Valley and Fisher Valley slash the terrain between the mountains and the Colorado River gorge, widening it for many miles to form colorful Richardson Amphitheater. The slender peninsula of Fisher Mesa separates Fisher and Professor valleys, and Adobe Mesa phases into the soaring spire of Castle Tower and isolated Parriott Mesa. A hint of the Dolores River gorge is visible in the distance, beneath the massive spine of the Uncompahgre.

West of the La Sal Mountains, still other canyons angle downward toward the Colorado River, Moab Valley, and the immense drainages of Cane Creek Canyon and Hatch Wash Canyon, and each of these has still other deep and sheer-walled tributaries. Just east of the Colorado River gorge are the northern tip of Hatch Point, the lofty walls of Cane Creek Canyon, the maze of sandstone fins and domes of Behind-the-Rocks and the Hunter Canyon drainage, with Moab Valley just beyond these. Still farther away, the rounded domes of the Sand Flats area and the elevated, red-walled terraces of Wilson Mesa and South Mesa phase into the western slopes of the La Sal Mountains.

To the east and southeast, beyond the canyon-slashed peninsulas of Hatch and Harts points, the broad sage flats of Dry Valley, with their isolated ridges and mesas of bare rock, recede toward a still higher complex of canyons, slopes, ridges and domes of red and white sandstones that are tilted upward by salt anticlines, with hints of the distant Rocky Mountains, San Juan Range and Sleeping Ute Mountain in the extreme distance.

With all of this magnificent and varied geography and geology in full 360-degree panorama, the question again occurs:

> *"If the topography of canyon country is this spectacular now, what was it like in the past?"*

And tentative answers are bound to come to mind:

> *"It must have been this way forever."*

Or perhaps,

> *"It was probably even more spectacular in the past."*

To find out what the breathtaking land now visible was *really* like in the past, it is necessary to stretch still further the human ability to visualize scenes described in words. It is necessary to convert our imaginary helicopter into a time machine that can hover in similar manner, with its elevation above the surface below automatically controlled by radar.

Why? Because over past geologic ages, the visible surface of the region that is now canyon country has varied drastically from sea level to thousands of feet above sea level, and it would not do for our time machine to become trapped underground during some phase of its coming excursion into the geologic past.

To begin this trip, let's take our hovering time machine straight back for 320 million years, but hold its position directly above where Grand View Point will eventually be. Let's return to just before any of the geologic strata presently exposed in the general Moab area were deposited. Then let's pause there for a few moments, while we study the scene below and to the far horizon in all directions.

320 MILLION YEARS AGO (MYA)

What do we see? Directly below, and for as far as we can see to the north, west and south, we observe nothing but open sea -- water to the far horizons -- unrelieved except to the east, where there are hints of land just barely visible as a dark, irregular band shimmering just above the sea, like a mirage in the desert. The first of the Ancestral Rocky Mountains are thrusting upward -- not the Rockies we know, but a predecessor range that formed along the same gigantic faults in the continent's basement rock.

What continent? North America? No -- another mass of continental crust, shaped nothing like the present North America, despite what is shown in popular articles about continental drift. This land mass is moving very slowly -- an inch or so a year -- northward across the planet's single globe-girdling ocean toward its coming collision with other unrecognizable masses of land, and carrying with it the Colorado Plateau, a still more ancient remnant of continental crust. The Colorado Plateau, now completely inundated by the sea visible below us, sits directly on the equator, and the climate is tropical, with hot and humid weather. See the sidebar, COLORADO PLATEAU.

The shallow sea is rich in marine life, but the distant hills of the Ancestral Rockies host little recognizable land life, and the air above both land and sea is entirely empty of flying creatures. These, whether insects, pterosaurs or the birds that came later, have yet to evolve, although artists who attempt to depict such ancient seascapes are sometimes tempted to add creatures in flight to give their scenes a touch of life.

Were we to visit the low hills visible in the distance to the east, we would see very little familiar life. There would only be primordial plants, wingless insects and very primitive reptiles and amphibians. Along the shores, equally primordial fish species would inhabit the shallows, but some of the other marine creatures and plants living in the tidelands might be recognizable.

After a brief look at the featureless seascape of the Colorado Plateau and canyon-country-to-come 320 million years ago, at the shallow seas that have already dominated the region for tens of millions of years, it is time to set out on our return journey through time. As we time-travel, at a rate of one million years per minute, we will watch everything below and all around to see how the topography changes as the megayears speed by. Even at this rapid rate, the trip back to our present will take 5 hours and 20 minutes, plus any pauses along the way for longer looks at various scenes or developing features.

THE FIRST 13 MINUTES

As our time machine hovers over the same spot and travels forward at 1 million years per minute, we watch below and all around for signs of change. For the first 13 minutes nothing appears to change, but beneath the surface of the sea the terrain is slowly altering. An immense area of seafloor, centered just a few miles to the east of our vantage point -- about where the La Sal Mountains will eventually appear -- is slowly sinking, and a ridge is rising around the sinking area, creating a landlocked basin cut off from the surrounding ocean.

This low, encircling ridge consists of one range of the Ancestral Rockies, which are growing into respectable mountains by now, plus other uplifts whose ancient remnants will eventually be named Uinta, Emery, Kaibab, Zuñi and Kaibab. Most of these young uplifts, however, are too distant to be visible, even from our point of view 5,000 feet above the surface of the sea.

Even so, they initiate a geologic process that will have profound effect on the future topography of the Colorado Plateau. With the huge basin of sea water cut off from the rest of the ocean, with only occasional flooding from fluctuating sea levels, the entire basin becomes a gigantic evaporation pond, with cycle after cycle of flooding, evaporation to salt marsh, total aridity, then more flooding, for an eventual total of as many as 34 complete evaporation cycles. See the sidebar, SALT ANTICLINES.

STOP

We stop our plunge forward through time to take a better look at the terrain below, toward the end of one evaporation cycle. Below, saline marshland dominates the scene, almost as far as the eye can see. Strange plants growing in clusters lend a touch of green to shallow brine that is mottled with patches of bright color that represent vast colonies of microscopic life. Altogether, the great salt marsh does not resemble any we have ever seen, because the plantlife we are familiar with has yet to evolve.

As evaporation continues, the marsh ecosystem we see will die, leaving behind decay products that will eventually become convenient marker beds for counting cycles of evaporation.

GO

As we continue our time travel through the first 13 megayears of our trip, life evolves in the swampy lower terrain of the distant lands and in the salt marshland that occasionally appears below. Reptiles and amphibians adapt better to life on dry land, developing into more sophisticated species. Plants progress into a wider range of forms, and certain insect species evolve wings, to become the first flying creatures on the planet. Most marine creatures also develop, but more slowly.

AFTER 17 MINUTES, 303 MYA

Then, about 17 minutes from the start of our return trip, the same deep continental forces that have thrust up the ridges that encircle the immense evaporation pond below produce another ridge, this one near enough to see very clearly. The first uprising of the Uncompahgre Uplift cuts diagonally across the seascape to the east of our viewpoint. It extends out of sight to the northwest and southeast, dividing the great dead sea into two basins, later to be named Paradox and Maroon by geologists. See the sidebar, UNCOMPAHGRE UPLIFT.

As this first low rise of the Uncompahgre appears, the floor of the new Paradox Basin below us slowly sinks even more, and the evaporites produced during the intervals between new flooding become tinged with the scant sediments from the low uplift. Not for long, however, because the ridge soon erodes away and disappears beneath the periodically flooding sea, leaving us with nothing visible nearby but still more flooding and evaporation. At our rate of travel through time, these would appear as nothing but rapid cycles of reflective sea below, with the water quickly shrinking to a smaller and smaller pond fringed with green and surrounded by blinding white salts, then a final central oval of green that would in turn fade and again be flooded by the sea.

AFTER 19 MINUTES, 301 MYA

This repeated cycle of Paradox salts deposition finally begins to phase out about 19 minutes into our return trip, as the Uncompahgre rises again above the sea, this time to soar to great height above the immense solar evaporation basin, permanently separating the Paradox Basin from the Maroon Basin. As we watch, several more evaporation cycles take place during the next two minutes, but the period of Paradox salt deposition slowly ends as land sediments from the Uncompahgre intertongue with marine deposits in the basin, creating upper Hermosa strata. Land deposits from the Uncompahgre eventually dominate.

Then, a climate change, probably triggered by some continental drift event that affected the pattern of the world's major ocean currents, brings several million years of slow erosion to the low tideland and tidal plains below. For these several minutes, we watch as this blurs the land we can see, and the lofty ridge of the Uncompahgre to the east erodes away visibly.

As the climate once again changes and the deposition of sediments from the Uncompahgre resumes, this time producing the varying strata of the Cutler Group, the vast tidelands and coastal plains below our viewpoint are alternately dominated by land sediments and shallow marine deposits. From our time-traveling viewpoint, the surface below shimmers quickly from wet to dry and back to wet again, with the nearby green-tinted Uncompahgre appearing to melt slowly away.

Although most Cutler deposits eventually become dark red in color, geologists disagree about whether they were red when being deposited. See the sidebar, ROCK COLORS. The sediments from the uplift were certainly rich in iron minerals, but these may or may not have been mineral forms that were red, so the land below our time machine may be red, or it may be light-hued sand, but it is not green with vegetation. The low-lying coastal plains that appear at intervals are still very near the equator and are hot, arid, saline and inhospitable to land life.

During this period, faults deep below the flat, visible terrain are starting another process that will ultimately help shape canyon country. Crustal compression, doubtless from lateral pressures caused by continental drift, is further fracturing and faulting the area just west of the Uncompahgre. These parallel deep faults, in concert with the immense weight of younger sediments accumulated by now, are creating a series of upwelling salt anticlines from the deep beds of Paradox evaporites.

These anticlines may be barely and briefly visible from our time machine as low, parallel ridges in the surface below that quickly fade away when erosion and invading tidelands alternately smooth the otherwise featureless surface. For the next 150 minutes -- 2-1/2 hours -- of our journey forward through time, various of these same anticlines will again become active for brief intervals, visible as low ridges that are quickly eroded away or buried beneath new sediments.

AFTER 40 MINUTES, 280 MYA

About 40 minutes into our trip back through time, during the middle of the Cutler deposition period, the Uncompahgre rises further, a third uplift. Toward the end of Cutler deposition, a tongue of sea intrudes from the west, stopping several miles east of our vantage point, then retreats, leaving the beach sands of White Rim Sandstone. Then, another climate change triggers a long period of erosion, during which the barren coastal plains visible below, including some of the White Rim deposits, slowly erode away. This period of erosion continues until 1 hour and 15 minutes into our trip, with little visible to us but the Uncompahgre to the east and flat, featureless plains below.

AFTER 75 MINUTES, 245 MYA

Then, something drastic happens, probably triggered by a meteor impact somewhere on the planet. As we watch, the air darkens very briefly, then all below appears to continue as before, but elsewhere on the globe stupendous changes are taking place. Approximately 95% of the species of life on the planet are extinguished by the darkening that to us, traveling through time at 1 million years per minute, was brief, but that in the real world lasted for centuries, perhaps millennia. See the sidebar, GREAT EXTINCTIONS. The same catastrophe has initiated a global ice age and a series of tectonic events, including a fourth rise in the Uncompahgre Uplift.

Erosion continues to score the flat terrain below our time-travel viewpoint for about 3 more minutes, then phases into about 3 minutes of alluviation, slowly building the Moenkopi Formation deposits, again perhaps their ultimate reddish color, or maybe just light-hued sands mixed with tiny dark granules of some iron-bearing mineral that will later oxidize to a red-hued iron compound. Still, all we can see below and to the most distant horizon is the Uncompahgre and low, arid coastal plains almost entirely bereft of animal life and with very little vegetation.

AFTER 93 MINUTES, 226 MYA

By 18 or 19 minutes after the major extinction and climate change -- and after as many millions of years have passed in real time -- the climate has improved to the point that the terrain below is rich with life, which we note in passing as a blur of green and the shifting sheen of long-lasting lakes and rivers. This lasts for 1 minute, then is interrupted briefly by a minor extinction that we probably see only as a very brief flicker of darkness, probably caused by the dusts thrown up by another meteor impact, or perhaps by a period of intense volcanic activity somewhere -- or both. Whatever the cause, it triggers another small uplift of the Uncompahgre, a fifth.

Below, what we see continues to be the flicker of lakes and rivers, with a tinge of green from rich communities of vegetation. Animals also inhabit the terrain, with therapsids, the reptilian predecessors of the mammals, dominating the ecosystem and with the dinosaur progenitors still small and inconspicuous. Chinle Formation deposits are accumulating. In the distance, beyond the eroding Uncompahgre, the Ancestral Rocky Mountains are also wearing away.

For the next 17 minutes of our time journey, little seems to change. The land slowly accumulates sediments from the Uncompahgre and that once-soaring uplift gradually melts away, finally to disappear entirely. The several salt anticlines between our viewpoint and the Uncompahgre become inactive. Low-lying, monotonously flat terrain dominates in all directions, with the only exception being the eroding Ancestral Rockies on the far eastern horizon. Almost all change in the visible world seems to have halted, although occasional subliminal flickers of darkness tell us that something is happening somewhere.

STOP

We pause for a few minutes in our headlong flight forward through time to take a better look at the scene below as the Chinle Formation is being deposited, and to see what those brief instants of darkness mean.

From 5,000 feet above the land, it is a patchwork of vegetation, streams, swamps and wide, shallow lakes, as far as the eye can see in any direction. Even the now-familiar ridge of the Uncompahgre Uplift is gone, buried beneath Chinle sediments. As we watch, strange creatures move slowly across the land and in the shallow waters. Both herbivores and carnivores appear, in many unfamiliar shapes and sizes. They are vaguely reptilian and as big as the dinosaurs we know about, but are not dinosaurs. They are therapsids, the mammal-like reptiles that eventually evolve into our mammal lineage. The reptilian dinosaur predecessors are in the scene below, but are small and inconspicuous in order to survive in a world dominated by large, carnivorous therapsids.

About 25 miles to the southeast of our viewpoint, too far away to be seen, several mid-sized therapsids are walking through a patch of mud, leaving foot tracks that will be exposed millions of years later on the floor of a canyon called Shay, within sight of a road labeled Utah 211.

About 35 miles to the northeast, we barely glimpse where a river joins a large, shallow lake, leaving a broad fan-shaped delta of silts and gravel deposits that in the distant future will intrigue geologists and be visible as an anomaly in the Chinle Formation that can be viewed from a road labeled Utah 128.

And while we study the scene below, the sky slowly darkens as a vast cloud of ash from a distant volcanic eruption reaches the area. This explains the strange, flickering instants of darkness we had seen while traveling through time. Fallout from these frequent eruptions adds minerals to the land below, and rain runoff from large areas of the land concentrates these minerals in the lakes we see. The river and lake in the distance are both gray with fine, suspended volcanic ash.

GO

We resume our trip through time, as Chinle deposits continue to accumulate in the level, blurred landscape below. The therapsids continue to evolve and dominate the animal world below, which is still near-equatorial in latitude, with a hot and humid climate. Flickers of darkness brought on by volcanism continue.

AFTER 112 MINUTES, 208 MYA

Now our view darkens again, for a second or so, a time period representing many years in real time. Once again, a meteor from outer space has impacted Earth, this one in what will eventually become Quebec, Canada, leaving an immense crater many miles wide. Fallout from the resulting stupendous explosion darkens the atmosphere for a long time, killing vegetation, and starving herbivores and the carnivores who prey upon them.

The resulting chaos of destruction to the planet's entire ecosystem extinguishes some 60% of the animal species, including all of the larger, more specialized therapsids, leaving only the smaller species of therapsids and reptiles. Beginning from this catastrophe, the reptiles start their long evolution along dinosaurian lines, taking over the dominating position formerly held by therapsids. Many of the smaller therapsids survive, but remain small and inconspicuous as the evolving dinosaurs fill the ecological niches vacated by the now-extinct large therapsids.

The tables have been turned, by a meteor impact. Animal life on the planet subsequently develops in new directions for more than 140 million years.

Although we are traveling too fast through time to see it directly, the meteor impact also drastically changed the global climate. What we do see, however, is that the land below becomes barren of the tinges of green that mean vegetation, and deposits accumulated over millions of years are melting away. A long period of erosion has begun that will later appear as an unconformity in the stratigraphic record.

AFTER 117 MINUTES, 203 MYA

Some 5 million years after the meteor impact -- after 5 more minutes of our trip through time -- the land below changes. Erosion of the dreary land that extends as far as we can see appears to stop, to be replaced by a blur of ceaselessly-shifting sand dunes, again as far as we can see in any direction, except for the distant, dwindling range of the Ancestral Rockies.

Desert dune-sand is slowly building the Wingate Sandstone of the future, the deep deposit that in our present erodes to form hundreds of miles of sheer cliffs. Sand dune deserts will dominate the land we see below for most of 40 million years, beginning now. See the sidebar, GREAT DESERT PERIOD. Toward the end of the Wingate Sandstone deposition period, a few small three-toed dinosaurs venture into the region below during the brief times when it is slightly wetter, leaving their foot tracks in moist silts. Some of these will be found in our present, on terraces high above a Wingate-walled gorge called Little Canyon.

Meantime, far to the west, the sea lies quiescent, waiting to reclaim the Colorado Plateau.

AFTER 124 MINUTES, 196 MYA

About 7 minutes after the first dunes of the Great Desert Period became apparent, a climate change floods the terrain below with sediments borne by a network of freshwater rivers and streams, probably originating from the distant remnants of the Ancestral Rockies, adding several hundred feet of Kayenta Sandstone strata.

We can tell by the tinges of green below that at least some vegetation is now thriving. The petrified remains of trees found in Kayenta deposits will verify this in the distant future, and numerous dinosaur tracks in Kayenta Sandstone will tell that these creatures also found the wetter environment hospitable.

AFTER 129 MINUTES, 191 MYA

After another 5 minutes, the land below again becomes desert -- a somewhat different desert, however -- and the Great Desert Period resumes. The dune sands now accumulating to great depths are being transported into the region from somewhere else, and will ultimately lithify into rock later called Navajo Sandstone. If we could see beyond the endless expanse of dunes below, we would probably spot tongues of the sea to the north, south and west, with the sands and marine tideland minerals of those distant beaches being carried into our region by strong near-equatorial trade winds and tropical storms.

STOP

We halt our time machine at some random time during the deposition period of Navajo Sandstone, to take a better look at its desert ecosystem. Below our vantage point, great sand dunes stretch to the far horizon. Arid desert drylakes -- playas -- of several sizes are visible as flat patches of blindingly bright minerals, accumulated there by runoff from the occasional violent tropical storms that originate over the seas to the west, then cross the low-lying desert region. Scattered patches of green trees and shrubs set around small ponds of spring water appear here and there as desert oases.

Amazingly, we spot several small animals in the

muddy shallows of one playa, and a few others near an oasis. Those around the playa appear to be catching something in its rapidly evaporating waters, probably types of crustaceans and insects that briefly swarm in such ephemeral lakes when they are wet, then burrow into the mud during the long periods of heat and aridity to come, to survive in either egg, larval or adult form. In the distant future, these strange, tiny creatures will continue to thrive in the drylakes and sandstone potholes in other deserts, and will be called "cryptobiotic life" because of their puzzling and unique abilities to withstand extreme heat and dehydration.

And there, striding purposefully along the playa shoreline, is a bipedal theropod, a big carnivorous dinosaur who is apparently intent on dining on some of the smaller animals -- some of them reptiles, some of them small therapsids -- who are scrambling for the swarming drylake creatures.

Some of the tracks in the playa mud being made by the big carnivore and his intended prey will be filled by blowing sand, lithified, then revealed again in the distant future in full and intricate detail, on big slabs of playa deposits that will be found near Moab. See the sidebar, PREHISTORIC TRACKS. Further, dozens of the lithified remains of the desert oases visible at any time within the Navajo Sandstone period will also be reported, many associated with the fossilized remains of the trees we now see growing around them. See the sidebars, PETRIFIED PLAYAS and PETRIFIED OASES.

GO

As we resume our trip through time, the Great Desert Period also continues. Navajo Sandstone deposits accumulate to great depths but, slowly, changes are taking place far beyond our range of vision. As far as we can see from our vantage point, desert dunes dominate clear to the lower slopes of the dwindling Ancestral Rockies in the distance, with one exception.

To the east, the dunes terminate at the base of a long, rounded ridge that is the above-ground expression of the immense Paradox Valley salt anticline. If the other salt anticlines in the area are also active, they do not show above the surface of the deep desert sands.

Also unseen from our vantage point 5,000 feet above the desert dunes, something else is happening that is affecting the area we are observing, and will affect it still more in the future. The seashores that are the source of the Navajo sands and minerals are slowly being drowned by the rising sea level, thus inhibiting any further accumulation of sand below us. It is not clear whether this is initiating a long period of erosion, or one in which the existing sands will just be moved around by shifting winds, with no net gain or loss in deposition. Either alternative could create the major unconformity that will separate Navajo Sandstone deposits from those that follow.

AFTER 140 MINUTES, 180 MYA

About now, major tectonic events far to the west are also changing the ancient desert scene we view as we continue toward the still-distant time dominated by humanity. Continental drift events are triggering mountain-building that will create new lands and highlands, thus cutting off the Colorado Plateau from the vast western sea. This will affect the region's climate, making the land still more arid and barren. Other continental drift activities will continue raising the world sea level, eventually allowing ocean water to encroach once again into the flat, sprawling desert below.

AFTER 170 MINUTES, 150 MYA

After about 170 minutes of travel through time, we see a rapid change below, as a brief invasion of the sea inundates the desert dunes, then retreats, leaving behind the tideland mud deposits of the Dewey Bridge Member of Entrada Sandstone and its equivalent Carmel Formation farther west.

Then, almost immediately, the Great Desert Period resumes, with the land below again dominated by broad expanses of coastal dunes that end at the edge of the intruding sea, just out of sight to the west. There, the sand dunes that will become the Slickrock Member of Entrada Sandstone merge into tidal flats and muds that lie approximately where the San Rafael Swell will later rise.

Some of the sand in the dunes we now watch accumulating to great depths are the tideland muds of the Dewey Bridge reworked by the continual winds. The rest of it is brought into the region by water and winds from elsewhere. The visible landscape below continues to be dreary and unchanging, just a vast blur of dune-sand desert, relieved only for split seconds by wet sheens that represent immense tidal surges created by occasional great tropical storms.

Then, as our time-journey continues, the climate changes once again, probably because of the rising mountain ranges to the west, but the scene visible from our time machine remains virtually static for long minutes -- with one exception.

As we watch, a tongue of the great inland sea that now dominates the continent to the east and north briefly intrudes into the area we can see to the northeast of our vantage point. As it retreats, it leaves the white beach sands of the Moab Member of Entrada Sandstone interlayered with the tidal flats muds of the Summerville Formation invading from the west. Dinosaurs of all sizes walk these Moab-Entrada beaches, leaving thousands of tracks to be studied by scientists in later ages, as they are uncovered along both sloping flanks of the Salt Valley anticline.

As we watch, the Great Desert Period ends, as Summerville tidal-flat and coastal-plain sediments build to considerable depth over everything in sight. Then another period of erosion arrives -- a stratigraphic unconformity -- and the Summerville deposits disappear from the visible area, down almost to the upper surfaces of the two now-lithified Entrada Sandstone members.

AFTER 163 MINUTES, 157 MYA

Again unseen from our time machine, global tectonic events are triggering volcanism to the west, and causing the sea level to change. This tectonic activity is probably a byproduct of the great dance of continental blocks now taking place as Pangea, the supercontinent of which the Colorado Plateau has been part for many millions of years, slowly shifts and breaks up into a new set of continents that will eventually evolve into familiar shapes -- in another 100 million years.

This tectonic activity brings another brief invasion of the sea into the terrain below our time machine. For a few moments, the arid, eroding terrain is flooded, leaving the thin muds of the Tidwell Member of the Morrison Formation covering everything in sight except, perhaps, the summit of the distant Paradox Valley salt anticline. Tidwell silts fill the exposed dinosaur tracks on the upper surface of the Moab-Entrada.

The retreat of the brief Tidwell sea leaves the entire region ready for the next major phase of deposition, one that will eventually cover immense areas, including all of the region's salt anticlines, the already buried Uncompahgre Uplift, and most of the low remnants of the Ancestral Rockies.

This next phase of deposition is triggered by rising land to the south that cuts off all further marine invasions for a long time, and creates a great inland basin. A wetter climate introduces more precipitation to the region which, in turn, brings massive runoff into the immense basin, mostly from the north and west, slowly depositing the Salt Wash Member of the Morrison Formation. Volcanic activity to the west and northwest sheds massive fallout on the region, which accumulates in the sandy deposits. Some of this fallout contains compounds of heavy metals, such as uranium, that will later play a large part in man's economic and social history. See the sidebar, URANIUM.

All of these major changes, however, are taking place somewhere else, even though they will greatly affect the region we are watching. From our vantage point, the flat, arid, eroding land below simply phases into flat, alluviating land that glistens with constantly shifting streams and shallow lakes, with an overall hint of green from vegetation.

This continues for almost an hour -- 50 million years of real time -- changing only in one way, and that not visible from our time machine. When the immense basin first starts filling, the gradients of the streams and rivers that feed into it are fairly steep. This allows the rushing water to carry coarser particles -- sands and fine gravels -- together with lighter silts. As the immense basin slowly fills, the stream gradients reduce, and the slower-moving water can transport only volcanic ash, dissolved minerals and finer silts, thus depositing the Brushy Basin Member of the Morrison Formation.

AFTER 211 MINUTES, 109 MYA

This pattern of deposition, which is virtually invisible to time travelers, is broken briefly by something -- possibly just one gigantic storm -- that brought great quantities of water-borne sand into the region we can see, depositing the material that becomes the sandstone ledge that marks the end of Morrison deposits and the beginning of the overlying Cedar Mountain - Burro Canyon formations, which are essentially the same deposits with two names. At least to our eyes, as we journey on up through time, the land below continues to look much the same as far as we can see in any direction -- flat, open terrain and shallow, highly-mineralized marshes and lakes fed by slow-moving streams laden with fine silts and volcanic ash.

By now, however, the immense basin that the Morrison, Cedar Mountain and Burro Canyon sediments are filling has reached its capacity and is overflowing intermittently somewhere. As we watch, for several minutes there is little net gain of deposits in the land below. Filling and erosion take turns, but not visibly at our rate of time travel.

Then, the long period of placid alluviation, unbroken by changes in the topography within sight, ends with a groan. The westward moving North American continental plate has encountered resistance, and that triggers two sequential periods of tectonic activity in the continent's western region.

AFTER 220 MINUTES, 100 MYA

After some 220 minutes of travel through time, the Sevier Orogeny begins, followed almost immediately by the Laramide Orogeny. Actually, both are part of a continuous pattern of tectonic activity in the western part of the continent. The Sevier is caused by stupendous eastward crustal pressures, and affects the region immediately west of our area of interest. As these pressures slowly travel on eastward, they initiate tectonic effects in what later becomes eastern Utah and western Colorado. See the sidebar, LARAMIDE OROGENY.

Evidently the Sevier Orogeny is part of a pattern of global tectonic activity because, at about that time, erosion increases in our area of interest, and massive seafloor spreading brings on a long period of rising sea level. This produces an effect visible below us, as the sea once again invades, inundating the land as far as we can see in any direction.

As the sea encroaches, its shoreline deposits will later become Dakota Sandstone, and the marine sediments that bury these will later be called Mancos Shale. Once again, we can see nothing below but inland sea in all directions, with the surface of the water unbroken by any kind of land. The last marine invasion of canyon-country-to-be is here. Only in the extreme distance to the west are there faint traces of land on the horizon.

AFTER 238 MINUTES, 82 MYA

After watching this featureless sea for about ten minutes, as we zoom upward through time, freshwater sediments from higher terrain out of sight to the north start intruding. These Mesaverde Group deposits eventually reach far enough south that we can see the distant shoreline they are creating. Then the next phase of tectonic activity begins -- the Laramide Orogeny.

AFTER 240 MINUTES, 80 MYA

The Laramide Orogeny produces enormous, highly visible changes in our region of interest, some during the remaining minutes of Mesaverde deposition, some during the following period of erosion. Volcanism triggered by the earlier Sevier Orogeny continues fitfully to darken our view of the terrain below, and sediments from rising highlands in all directions meet the Mesaverde deposits from the north, covering the last deposits of the Mancos Shale sea.

As we watch, fascinated, our rate of travel through time allows us actually to see the enormous Monument Uplift rise above the general level of the flat terrain to the south. To the east, the Uncompahgre Uplift thrusts upward for the sixth time. Farther east, a new range of Rocky Mountains is forming, rising through the sediments that long ago buried the Ancestral Rockies.

A change of direction in the movement of the North American continental plate is resisted briefly by the ancient Colorado Plateau and, far out of sight to the north, the east-west range of the Uinta Mountains climbs toward the sky.

Despite all of this nearby tectonic activity, or perhaps because of it, the land below our time machine is undergoing a period of erosion, creating the unconformity between the Mesaverde and coming Wasatch Formation deposits. As highlands rise all around us, the land below is eroding away. Then, as though nature had a grudge against life on this planet, another catastrophe occurs.

AFTER 254 MINUTES, 66 MYA

As our time machine approaches the 66 mega-year point, we slow its rate of travel, watching for an event we expect. At the first flicker of darkness we stop, retreat slowly through time, then --

STOP

-- where the scene below is darkest, as dark as a moonless night. Another great meteor has just smashed into Earth, in the Gulf of Mexico to the south. This has produced planet-wide devastation. Skies everywhere darken for decades, perhaps centuries. Blocked from life-giving sunlight by dust and the soot from continent-wide forest fires, unburned plants die, as do the animals who eat the vegetation and the predators who eat the herbivores. Another great extinction is taking place.

This one will extinguish perhaps 75% of the species of life on the planet, among them all the giant sea reptiles and the entire dinosaurian population. Small, relatively unspecialized reptiles -- land and sea snakes and turtles, some crocodilians and a few odd lizard-like creatures that will eventually evolve into birds -- will survive, as will the variety of small mammals that have evolved from the therapsids that survived the previous big extinction.

The spectacular pterosaurs, and all dinosaurs large and small that have dominated since that time, will die, leaving room for mammals to diversify and grow larger, to fill the vacant ecological niches. One of these developing lines of mammals will eventually become humans. Once again the balance of nature has tipped.

About 142 million years earlier, therapsids -- the mammal predecessors -- dominated, then were almost wiped out by a meteor, giving rise to the dinosaurs. Now, another meteor has eliminated the dinosaurs, giving the mammals another chance. Doubtless, the balance will shift yet again. The only questions are -- when, and what kinds of creatures will take over next? It's the reptiles turn again, but perhaps another genetic group will inherit the planet next time.

With this sobering thought, we --

GO

-- to see a brilliant flash of light almost immediately. Had we not been looking southwestward, we would have been blinded momentarily. And had we not been moving through time, we would have been vaporized instantly, as a relatively small meteor created a nuclear-bomb-sized explosion by impacting just a few miles to the north and a little west of our viewpoint. We back up for a few seconds in time, then --

STOP AGAIN

-- to take a first look at the wide, shallow, rubble-filled crater that will eventually become Upheaval Dome -- a misnomer if there ever was one -- after hundreds of feet of sediments and fractured rock have eroded away to expose deformed strata now still deeply buried, strata thrust back upward in the center by "isostatic rebound," and followed by still deeper Paradox salts to form the low dome that some geologists will mistakenly think *caused* the upthrust, rather than simply being another effect of the meteor impact. We then --

GO

-- on into the future, to find that erosion of the land below is continuing, as is the Laramide Orogeny, probably enhanced by the tectonic aftereffects of the two meteor impacts. In addition to the new highlands visible to the east, we watch as the San Rafael Swell rises from the flat terrain to the west, like a gigantic surfacing whale.

AFTER 263 MINUTES, 57 MYA

Then we note that erosion has halted and deposits are once again accumulating, flowing into the region from the new highlands that Laramide tectonics have thrust upward. Some of these sediments to the north of us will later be called the Wasatch Formation, others the Green River Formation, but those below and on to the south will never be named because they will erode completely away later in our trip through time.

The topography nearby is still flat and relatively featureless, but higher terrain is visible in the distance, in all directions. We see occasional tinges of green on the land, indicating that land life is recovering from its latest setback.

AFTER 272 MINUTES, 48 MYA

As we watch, the slow but continuing Laramide Orogeny produces another highland, this one to the southwest. A continental drift event, doubtless related to the current orogeny, is affecting the entire western half of the North American continent, creating widespread volcanism. The thick, tough basement rock of the Colorado Plateau resists several upwelling columns of lava, which do not quite reach the surface of the land. Instead, they spread laterally in great vertical faults and joints to form dikes, and between layered strata to form sills and laccoliths. The distant Henry Mountains range is born.

AFTER 282 MINUTES, 38 MYA

After about ten more minutes of our voyage through time, with the terrain below continuing to be boringly featureless, the San Rafael Swell to the west thrusts still higher. To the east, the Uncompahgre shudders upward again, for the seventh time.

AFTER 292 MINUTES, 28 MYA

In another ten minutes, more columns of molten rock try to reach the surface, but are once again frustrated by the tough Colorado Plateau rock. We watch the laccolithic Abajo Mountains range rise to the south, through the eastern slopes of the immense Monument Uplift but reaching even higher. See the sidebar, LACCOLITHIC MOUNTAINS.

Then, about four minutes later the land immediately to the east of our viewpoint swells upward too, powered by the same deep igneous forces. The La Sal Mountains are created as we watch, although at this stage they more closely resemble hills because the surrounding terrain is much higher than it will be later.

For the next fourteen minutes the terrain below changes little -- for that many millions of years. Although the region is now surrounded by highlands, runoff from these is minimal. The three great ranges of mountains to the west -- the Wasatch Mountains and their Wasatch Plateau southerly extension, the Sierra Nevadas and the Coast Range -- have severely reduced precipitation in the low-lying terrain between the various uplifts of the Colorado Plateau. This has left the region a low-lying desert, with few distinct topographic features. Its elevation to high-desert status is yet to come.

The canyon country we know is still in the future. Were we to stop here for a look around, we would see the La Sals, Abajos and Henrys as clusters of low, rounded hills, with the gentle contours of the Uncompahgre Uplift farther to the east. The equally inconspicuous hump of the San Rafael Swell would lie to the west, with nothing but more flat terrain to the north.

We would see no great rivers crossing the arid land -- just a few desert drywashes. We would see no traces of the salt anticline valleys yet to come on each side of the La Sals except, perhaps, a long, low mound where Paradox Valley will later form.

AFTER 310 MINUTES, 10 MYA

Only now, 310 million years after our return trip through time began, does the unique topography of the canyon country we know have its beginning. That start is slow and inconspicuous, however.

The first thing we notice is that the automatic, radar-controlled mechanism for keeping our altitude at 5,000 feet above the terrain below is active. We hold our distance above the land below, but our elevation above sea level is changing. We are rising, as the land below is being raised by some mysterious plate tectonic activity. Some evidence suggests that this continental lifting might be caused by a chunk of overridden plate passing beneath the western regions of the North American plate, but the raising might have some other cause. Geophysicists are not yet sure.

But the general rise of several thousand feet of much of western North America, including the Colorado Plateau, does happen as we watch from the vantage point of our trip through time, and this dramatically changes the future of the arid, drab land below. Among other things, the general rise is accompanied by another shrug upward of the Uncompahgre Uplift, its eighth.

The climate becomes a little wetter. Increased rain and snow on the higher uplifts and mountains within and surrounding the Plateau begins to have an effect, not rapidly in normal time, but quite noticeably from our rate of travel through time. Limited erosion of the highlands and arid terrain below our vantage point begins.

Small drainages from the higher slopes join to form wide, shallow drywashes that are wet only following seasonal storms, or the occasional rapid melting of snow in the spring. A few such washes have trickles year around, and in some areas several of these join to form small rivers. As we watch for several more minutes -- several millions of years of real time -- three of these rivers begin to penetrate the still-arid high-desert lowlands below us, before disappearing into the equally arid desert to the southwest.

The Dolores and Colorado rivers are born, and are joined by the Green River sometime during this long, slow preamble to the drama yet to come. Their joined waters, far beyond our range of vision to the southwest, at first flow into a great inland basin out of sight in what will eventually become the state of Arizona.

Then, as the land in that region is uplifted by another powerful plate tectonic event, the river is diverted into a more westerly course, to start sculpting the Grand Canyon, and the lake it had reached earlier drains backward, emptying much of its contents back into the diverted Colorado and creating the north-flowing Little Colorado River.

As our flight through time continues, we decide to --

STOP

-- for a better look at the land below and all around, about 318 million years along our return journey, still some 2 million years from journey's end.

Over the last 8 million years, the lowlands have already lost several thousand feet of younger deposits through very slow erosion. The unnamed sediments have disappeared, and the Green River, Wasatch and Mesaverde formations are visible only in the distance to the north, where the first signs of the Book Cliffs are beginning to appear as a low ridge. The deep deposits of the underlying Mancos Shale are exposed some places, almost gone in others. Out of sight to the east, the Gunnison River is still flowing through the deep notch it has cut through the Uncompahgre Uplift -- Unaweep Canyon -- and still joins the Dolores River northeast of the La Sal Mountains.

The distant Uncompahgre and San Rafael uplifts are visible as gently-rounded highlands. The nearby La Sal and Abajo mountains, and the Henry Mountains in the distance to the southwest, do not have their familiar sharp peaks and ridges, although their central stocks and a few dikes, sills and laccoliths have been exposed here and there by erosion.

To the north, we see where the Green River emerges from the low bluffs that will later become the Book Cliffs, then continues south through a shallow canyon toward its confluence with the Colorado. This river, coming from the northeast and joined by the out-of-sight Dolores River north of the La Sals, has already cut a significant gorge, as it wends its twisting course across the broad, open high-desert terrain below. So have the Dolores and Green, above their confluences with the Colorado. See the sidebar, ENTRENCHED MEANDERS. Although we cannot see it from our vantage point, the Colorado River gorge is already some 800 feet deep where it cuts through the gently-sloping alluvial plain north of the La Sals.

Even though the river gorges are there, the surrounding landscape still does not even faintly resemble the spectacular system of canyons with which we are familiar. Beyond the river gorge rims, the land is relatively flat, uncut by equally deep tributary canyons because most of the water in the rivers comes from various distant mountain ranges, not from local precipitation.

A wide, shallow depression is all that appears where the colorful sandstone fins and ridges of Arches National Park will eventually appear. The maze of red and white terraces and canyons that typify the Canyonlands National Park we know do not exist. Nor have the great, sheer-walled salt anticline valleys that will later surround the La Sals formed yet, although there are indications in the land between our viewpoint and the Uncompahgre of where the larger anticlines and synclines will later appear.

The stage is set, the preamble is over. It is time for the main actors to appear. The drama has already started. A new planetary ice age is building stupendous masses of glacial ice in both polar regions, and these continent-sized masses of ice are growing and flowing, spreading toward the equator. The cooler, wetter climate that is creating this ice age will also bring glaciers to all western mountains and to the higher areas of the Colorado Plateau -- including its three island ranges, the La Sals, Henrys and Abajos. These glacial episodes -- a total of nine in the La Sals, fewer in the lower Abajos and Henrys -- are the main actors in the shaping of canyon country. For more details than can be related here, see the sidebar, GLACIERS OF THE LA SAL MOUNTAINS.

What brought on the ice age? Perhaps a major shift in ocean currents, or the aerial debris from massive volcanic activity, or dusts and vapors from another meteor strike. Scientists are not sure. What triggered the various glacial episodes within the ice age? Perhaps either volcanic activity or meteor impacts. There are known meteor craters that date to 300 thousand years ago (in Australia), 52 thousand years ago (in India) and 50 thousand years ago (in Arizona), that might have been responsible for three of the episodes.

GO

Again, we start forward through time, but now at a much slower rate, so we can watch the land below closely. Rather than finish the last 2 million years of our long trip through time in two minutes, let's take twenty minutes. Even then, we will have to observe carefully in all directions to see everything that happens.

As we watch the La Sals, where most of the action will take place, the first glaciers flow down the gently sloping valleys that radiate from the various rounded summits, gouging out deep, narrow gorges and fanning out across the open lowlands. One glacier reaches toward our viewpoint as far as where Black Ridge will someday be, but the canyons that will later form that ridge do not yet exist.

As the decades pass by quickly, the land beyond the reach of the glaciers flickers from white to green as the annual seasons change. During the green seasons, we can see that the timberline in the ice-shrouded mountains has lowered to very little above the level of the surrounding terrain.

As the climate warms and the first glaciers recede, the timberline climbs and the melting ice begins the process of cutting into the open slopes below the mountains, creating drainage lines that cut diagonally across places where canyons will later appear. The massive erosion that will ultimately shape canyon country has just barely begun to configure the open, gently-sloping land that flanks the La Sals and the Abajos on all sides, higher than the future upper surfaces of Wilson and South mesas and Black Ridge. As the new streams reach the nearby river gorges, they plunge steeply down to the rivers.

Shortly, the glaciers return, this time covering even greater areas. The next interglacial period is also short, then glaciers flow down the deepening mountain valleys for a third time. Each pulse of glaciers leaves behind its debris as it melts, then some of this debris is spread broadly across the lower slopes by the copious water flowing from the melting ice.

During the longer interval that follows the first three glaciations, trees and other vegetation completely cover the mountain peaks, and while the mountain tops still flicker from green to white with each season, the climate is warmer and the geologic deposits in the lowlands are melting away as we watch, working toward the extreme topography that we know is coming. The river gorges cut deeper still. See the sidebar, ROCK EROSION.

Two more glacial cycles, each milder than the first three, bring still more erosion during their warmer intervals. The following longer interglacial period again allows trees to conquer the highest La Sal peaks.

Somewhere about now, the Uncompahgre lifts again, for the ninth time. Its rise blocks the flow of the Gunnison River through Unaweep Canyon, forcing it to divert north, to join the Colorado River about where it does in our present.

The sixth and seventh glacial episodes repeat the same cycle of events, but the glaciers are smaller yet. The land surrounding the La Sals and Abajos continues to erode away. Canyons now radiate in every direction from the river gorges, cut by the greatly increased local precipitation. To the south of us, the vast maze of colorful terracing on all sides of the Green-Colorado confluence is appearing. The heart of what will become Canyonlands National Park is taking shape. The salt anticlines that surround the La Sals are becoming obvious, as ground water from the melting snows and greater rainfall dissolves the underground salt anticline ridges, allowing their surfaces to subside.

After another longer interglacial period, two final sets of glaciers form and melt in the La Sals, but not in the much lower Abajos. As the last of these small ice masses finally melts away, the severe topography of the land we are familiar with appears in all of its spectacular, chromatic beauty! And, if we were a bit farther east, we would see the first white settlement in Moab Valley, in 1855. A few seconds later we reach --

THE PRESENT

-- and are now ready to answer the question we set out to resolve:

> *"If the topography of canyon country is this spectacular now, what was it like in the past?"*

We now know that for almost all of the past 320 million years, and even earlier than that, the landscape that would ultimately become canyon country was almost totally flat -- either unrelieved seascapes, endless flat tidelands or coastal plains, or sediments slowly filling almost equally flat and endless basins, with all this monotony only briefly relieved by occasional rounded upthrusts of highlands that soon either eroded away, or were covered by leveling deposits.

Canyon country did not offer much interesting topography until the last two million years, when the ice age that is still slowly receding initiated the current unconformity-in-the-making, and the water that the wetter climate brought to the region carved the present system of canyons from the ancient sediments that we watched being deposited in such neat, level layers.

THE FUTURE

And if we could take our imagined time machine on into the future, would we see still more glaciers, still more mass extinctions of life? Would humanity itself still be there for more than a few seconds, at our normal rate of time travel? Would any human works survive for long if another meteor struck?

These are fascinating questions that as yet have no answers.

The pouroff into Moonflower Canyon, Mile 3.1 in the Cane Creek Road geology log.

COMBINATION GEOLOGY TOURS

Following are suggestions for some practical combination geology tours of roads and off-road vehicle trails. They are grouped according to the kinds of vehicles needed to complete the entire suggested tours. For further details about the roads and trails listed, refer to the summaries at the head of each separate log. For detailed navigation, refer to the logs for the noted roads and trails, beginning with the first ones listed.

– ANY VEHICLE –

U.S. 191 SOUTH & UTAH 211

This combination geology tour passes through most of the beautiful and diverse geology that lies between the southeastern Utah towns of Moab and Monticello, then heads west through additional varied geology to end in the heart of the unique erosional phenomena that dominate the Needles District of Canyonlands National Park. This trip requires a full day.

NEEDLES OVERLOOK ROAD & ANTICLINE OVERLOOK ROAD

This trip provides an overview of the geology of the Hatch Point District of Canyon Rims Recreation Area, plus visits to two of the most spectacular panoramic overlooks in the canyon country region. For those prepared to camp out, there are two developed campgrounds along the route. Otherwise, the trip takes a full day.

BLACK RIDGE ROAD & UTAH 46 - COLORADO 90

This tour offers a panoramic overview of the unique geology of Moab Valley and vicinity from a lofty vantage point, then skirts around the southern flanks of the La Sal Mountains before dropping into a colorful sandstone canyon. It ends, as described, in an immense salt anticline valley in western Colorado. Part of the log for U.S. 191 South can be covered on this route. This trip takes a full day.

UTAH 128, CASTLE VALLEY ROAD & LA SAL MOUNTAIN LOOP ROAD

This loop route travels up the scenic Colorado River gorge, then ascends spectacular Castle Valley with its lofty sheer cliffs and soaring sandstone towers, before climbing still further to travel the elevated western slopes of the La Sal Mountains. This combination trip requires a full day. It can be shortened by excluding the portion of road that goes to the end of Utah 128 description before traveling the Castle Valley Road.

UTAH 313 & ISLAND ROAD

This popular combination trip travels up scenic Sevenmile Canyon, crosses panoramic highlands, then goes to breathtaking plateau-rim viewpoints in Dead Horse Point State Park and the Island-in-the-Sky District of Canyonlands National Park. It also provides a close look at the unique geology of an immense meteorite crater. The elevated overlooks provide outstanding views of the highly diverse geology of an immense area of canyon country. This trip requires a long day.

UTAH 279, POTASH - SHAFER TRAIL & ISLAND ROAD

This combination trip goes down the Colorado River gorge, continues through colorful and highly eroded redrock terrain, climbs steeply through a variety of geologic strata, then travels lofty Island-in-the-Sky to various elevated overlooks in Canyonlands National Park and visits the Upheaval Dome meteorite crater. It returns via Utah 313 and U.S. 191 North. This outstanding tour takes a long day.

UTAH 279, LONG CANYON TRAIL & ISLAND ROAD

This combination is similar to the one above except that it ascends a deep and narrow canyon to reach the high terrain of Island-in-the-Sky. This much shorter route still requires most of a day. As an option, return to Moab from the Island Road via the Potash-Shafer Trail, although traveling that route in this direction would require following its geology log in reverse.

Utah 279

DUBINKY WELL ROAD & SPRING CANYON POINT TRAIL

This combination of graded dirt road and fairly long ORV trail provides a good look at some of the colorful and geologically interesting terrain north of Canyonlands National Park, as well as breathtaking views of the layered geologic strata of the deep Green River gorge. The trip requires a full day.

UTAH 128 & THE ONION CREEK - THOMPSON CANYON - POLAR MESA TRAIL

This sequence of roads and ORV trails travels through a wide variety of geologic formations and erosional landforms, including deep gorges, high rimlands and a complex maze eroded through ancient upthrust marine salt deposits and dark red sandstone. The trip requires a long day.

CANE CREEK ROAD & HURRAH PASS - CHICKEN CORNERS TRAIL

This route travels the very popular route into magnificent Cane Creek Canyon, then goes over spectacular Hurrah Pass and along the rimlands of the Colorado River gorge, offering outstanding geologic highlights and colorful scenic views along the entire way. This trip requires a long day.

ARTHS PASTURE ROAD - GEMINI BRIDGES TRAIL & BULL CANYON TRAIL

The trip to Gemini Bridges is very popular with users of all kinds of off-road vehicles. This route goes first to the tops of the big natural bridges, then travels the beautifully eroded sandstone canyons below them, offering views of a wide variety of erosional forms, including several other natural rock openings, plus cliff-wall exposures of unique Roberts Rift, a miles-long joint filled with rock from far below. This combined trip takes a long day.

SAND FLATS ROAD & SAND FLATS TRAIL

The Sand Flats Road is popular because it offers access to a large recreation area near Moab. The ORV trail extension of this road climbs steeply through colorful sandstone strata into higher terrain, eventually to connect with the equally popular La Sal Mountain Loop Road. The return trip via this road allows at least part of its geology log to be added to this combination tour. The geology along this route is stratigraphically varied, with a wide variety of tectonic features and erosional landforms. The trip requires a long day.

MONITOR & MERRIMAC TRAIL & HIDDEN CANYON RIM TRAIL

The Monitor & Merrimac Trail, with its spectacular monoliths and other erosional landforms of Navajo and Entrada sandstones is popular with both mountain bikers and four-wheelers. Still further geologic variety is offered by adding the Hidden Canyon Rim Trail as a side-trip, either before or after exploring the better known trail. This combination takes a long day.

U.S.191 SOUTH, UTAH 211 & LAVENDER CANYON TRAIL

This long combination route provides an excellent overview of canyon country geology, plus an intimate look at one of its most spectacular and beautiful canyons, with a wide variety of geologic strata and erosional landforms visible along the way. Lavender Canyon offers the chance to view a number of unusual natural arches and to hike colorful and beautifully eroded Cedar Mesa Sandstone. The trip requires either a very long day, or two days with primitive overnight camping in or near Lavender Canyon. Three days and two nights will provide a better chance to study the geologic variety this outstanding combination tour offers, and to more fully appreciate its unique natural beauty.

INDEX OF ROADS AND TRAILS

PAVED ROADS

PARTLY PAVED ROADS

GRAVELED ROADS

GRADED DIRT ROADS

OFF-ROAD VEHICLE TRAILS

INDEX OF SIDEBARS

Reservoir beside Utah 211, Mile 19.9.

OTHER CANYON COUNTRY PUBLICATIONS

Those who wish to know more about the unique and fascinating canyon country of southeastern Utah will find other books and maps in the ***Canyon Country*** series both useful and informative. They are stocked by many visitor centers and retail outlets in the region.

The listed books are profusely illustrated with photographs, charts, graphs, maps and original artwork. The maps are also illustrated with representative photographs.

GENERAL INFORMATION

Canyon Country **HIGHWAY TOURING** by F. A. Barnes. A guide to the highways and roads in the region that can safely be traveled by highway vehicles, plus descriptions of all the national and state parks and monuments and other special areas in the region.

Canyon Country **EXPLORING** by F. A. Barnes. A brief history of early explorations, plus details concerning the administration of this vast area of public land and exploring the region today by land, air and water.

Canyon Country **CAMPING** by F. A. Barnes. A complete guide to all kinds of camping in the region, including highway pull-offs, developed public and commercial campgrounds, and backcountry camping from vehicles and backpacks.

Canyon Country **GEOLOGY** by F. A. Barnes. A summary of the unique geologic history of the region for the general reader, with a list of its unusual land-forms and a section on rock collecting.

Canyon Country **PREHISTORIC INDIANS** by Barnes & Pendleton. A detailed description of the region's two major prehistoric Indian cultures, with sections telling where to view their ruins, rock art and artifacts.

Canyon Country **PREHISTORIC ROCK ART** by F. A. Barnes. A comprehensive study of the mysterious prehistoric rock art found throughout the region, with a section listing places where it can be viewed.

Canyon Country **ARCHES & BRIDGES** by F. A. Barnes. A complete description of the unique natural arches, bridges and windows found throughout the region, with hundreds depicted.

CANYONLANDS NATIONAL PARK - ***Early History & First Descriptions*** by F. A. Barnes. A summary of the early history of this uniquely spectacular national park, including quotes from the journals of the first explorers to see and describe it.

Canyon Country's **CANYON RIMS RECREATION AREA** by F. A. and M. M. Barnes. A description of the natural and human history and outstanding scenic beauty in the area to the south and east of Canyonlands National Park, plus a summary of its outstanding recreational opportunities.

BACKCOUNTRY GUIDE BOOKS AND MAPS

***Canyon Country* HIKING & Natural History** by F. A. Barnes. A summary of the unusual natural history of the region, plus descriptions of most of the established hiking trails and a sampler of trail-less routes.

***Canyon Country* OFF-ROAD VEHICLE TRAILS - Arches & La Sals Areas** by F. A. Barnes. Descriptions of most of the backcountry roads and off-road vehicle trails in these two canyon country areas.

***Canyon Country* OFF-ROAD VEHICLE TRAILS - Canyon Rims & Needles Areas** by F. A. Barnes. Descriptions of most of the backcountry roads and off-road vehicle trails in these canyon country areas.

***Canyon Country* OFF-ROAD VEHICLE TRAILS - Island Area** by F. A. Barnes. Descriptions of most of the backcountry roads and off-road vehicle trails in this canyon country area.

***Canyon Country* OFF-ROAD VEHICLE TRAILS - Maze Area** by Jack Bickers. Descriptions of most of the backcountry roads and off-road vehicle trails in this canyon country area.

***Canyon Country* OFF-ROAD VEHICLE TRAILS - Canyon Rims Recreation Area** by F. A. Barnes. Descriptions of the backcountry roads and selected off-road vehicle trails in Canyon Rims Recreation Area.

***Canyon Country* OFF-ROAD VEHICLE TRAIL MAP - Arches & La Sals Areas** by F. A. Barnes. Special topographic maps showing the named roads and off-road vehicle trails in these two areas.

***Canyon Country* OFF-ROAD VEHICLE TRAIL MAP - Canyon Rims & Needles Areas** by F. A. Barnes. Special topographic map showing the named roads and off-road vehicle trails in these two areas.

***Canyon Country* OFF-ROAD VEHICLE TRAIL MAP - Island Area** by F. A. Barnes. Special topographic map showing the named roads and off-road vehicle trails in this area.

***Canyon Country* OFF-ROAD VEHICLE TRAIL MAP - Maze Area** by F. A. Barnes. Special topographic map showing the named roads and off-road vehicle trails in this area.

***Canyon Country* OFF-ROAD VEHICLE TRAIL MAP - Canyon Rims Recreation Area** by F. A. Barnes. A special topographic map depicting the roads, ORV trails and other features of the magnificent Canyon Rims Recreation Area.

***Canyon Country* OFF-ROAD VEHICLE TRAIL MAP - Moab Area** by F. A. Barnes. A map of the roads and popular off-road vehicle trails in the immediate vicinity of Moab, Utah.

***Canyon Country* MOUNTAIN BIKING** by F. A. Barnes & Tom Kuehne. Background information useful to mountain bikers, detailed descriptions of 23 good trails for biking, and basic information about the other major trails in the canyon country region.

UTAH-COLORADO MOUNTAIN BIKE TRAIL SYSTEM - Route I - Moab to Loma - *"Kokopelli's Trail"* by Peggy Utesch. A complete mile-by-mile guide to this 138-mile mountain bike trail, with elevation charts and additional useful information.

MOUNTAIN BIKING in Canyon Rims Recreation Area by Peggy & Bob Utesch. A sampler of the numerous mountain biking routes in Canyon Rims Recreation Area, which surrounds Canyonlands National Park on two sides, plus a description of the recreation area's 235-mile *"Mountain Bike Challenge Route."*

***Canyon Country* SLICKROCK HIKING & BIKING** by F. A. Barnes. Introduces an approach to hiking and mountain biking that is unique to the canyon country of southeastern Utah, then describes and gives detailed directions for reaching many places where these new forms of recreational activity can be sampled.

HIKING THE HISTORIC ROUTE of the 1859 MACOMB EXPEDITION by F. A. Barnes. A detailed guide to hiking the route of the first American expedition to enter southeastern Utah and describe the spectacular land now within Canyonlands National Park.

CAMEO CLIFFS Biking - Hiking - Four-Wheeling by F. A. & M. M. Barnes. A detailed guide to the beautiful Cameo Cliffs public land recreation area adjacent to Canyon Rims Recreation Area, with practical guidance for mountain bikers, hikers and four-wheelers.

Map of CAMEO CLIFFS Biking - Hiking - Four-Wheeling by F. A. Barnes. An illustrated map of the CAMEO CLIFFS recreation area, with roads, ORV trails and scenic highlights indicated, plus text describing the area's geography and geology.

North & South Windows, Arches Road.

Lower Castle Valley

Jughandle Arch, Utah 279.

The Titan - Fisher Towers, near Utah 128.